Microwave & Wireless Communications Technology

Joseph J. Carr

Newnes

Boston   Oxford   Johannesburg   Melbourne   New Delhi   Singapore

Newnes is an imprint of Butterworth-Heinemann.

Copyright © 1997 by Butterworth-Heinemann

ℛ A member of the Reed Elsevier group

∞ Recognizing the importance of preserving what has been written, Butterworth–Heinemann prints its book on acid-free paper whenever possible.

Library of Congress Cataloging-in-Publication Data
Carr, Joseph J.
   Microwave & wireless communications technology / Joseph J. Carr
      p.   cm.
   ISBN 0-7506-9707-5 (pkb.)
   1. Microwaves.   2. Microwave communication systems.   3. Wireless communication systems.   I. Title.
   TK7876.C385   1996
   621.381'3—dc20   96-14591
                                                       CIP

British Library Cataloguing-in-Publication Data
A catalogue record for this book is available from the British Library

The publisher offers discounts on bulk orders of this book.
For information, please write:

Manager of Special Sales
Butterworth-Heinemann
313 Washington Street
Newton, MA 02158-1626
Tel: 617-928-2500
Fax: 617-928-2620

For information on all Newnes electronics publications available, contact our World Wide Web home page at: http://www.bh.com

10 9 8 7 6 5 4 3 2 1

Printed in the United States of America

CONTENTS

PREFACE

Electronic communications is exploding today, even though it was only a few years ago that doomsayers were warning that we were coming to the end of spectrum usage, and hence of communications possibilities. But as soon as one service became saturated, another technology popped up, increasing the number of people who could use electronic communications of one sort or another.

An entirely new galaxy of products are on the market today that were either nonexistent or saw very limited use a few years ago. Many high-school kids and parents carry beepers (little radio receivers) so that they can keep in touch with each other. Twenty years ago beeper service was limited, and only a few critical people (for example, physicians) carried them.

And look what's happened in voice communications! Handheld telephones, cellular telephones, cordless telephones, mobile telephones, and a host of other products allow you to project your voice anywhere in the world for pennies. It's not unreasonable for you to hold a telephone conversation while walking to a luncheon, and be talking to someone in London or Tokyo.

Data communications exploded in the early 1980s. In the previous decade there were only a few modems (called "data sets" then), and most of them were supplied by the local telephone company. Today, there are scores upon scores of modems available for a wide variety of digital applications. You can now connect a computer to a telephone line or radio transceiver with ease. And how many of us have not sent or received a fax? There was a time when fax machines were large, noisy, and smelled faintly of ozone. Those machines were used mostly by newspapers.

Another historical perspective is that it was only the 1930s when transatlantic telephone became available, and it was carried on shortwave radio bands. Because of the frequencies used, these telephone circuits were unreliable when sun acted up. A sudden ionospheric disturbance (SID) could wipe out transcontinental telephone for days at a time. It was 1955 when the first transatlantic telephone cable was layed across the Atlantic. Yet Telstar, the first telecommunications satellite, was only a few years behind the undersea wire.

If you are just beginning a career in communications you will find that much of it is done on ultra-high-frequency (UHF) and microwave bands. Electronics technology in the microwave bands is a bit different from other radio frequencies. The reasons are that the wavelengths approximate the physical size of ordinary components, and the values of inductance and capacitance needed for resonance, filters, and other purposes are smaller than the stray values found in ordinary electronic circuits. These facts make microwave electronics very different from lower-frequency electronics.

This book takes two approaches. The first fourteen chapters deal with some of the basics of microwave technology. You will learn about microwave signal propagation, transmission lines, waveguides, and a host of solid-state microwave devices. Later in this book you will learn about microwave systems such as transmitters, receivers, radars and wireless communications devices.

Joseph J. Carr, MSEE
P.O. Box 1099
Falls Church, VA 22041

CHAPTER 1

Introduction to Microwaves

OBJECTIVES

1. Learn the definition of microwaves.
2. Understand why microwave technology differs from ordinary high-frequency technology.
3. Learn the designations for various microwave bands.
4. Understand why ordinary lumped constant components don't work properly at microwave frequencies.

1-1 PREQUIZ

These questions test your prior knowledge of the material in this chapter. Try answering them before you read the chapter. Look for the answers (especially those you answered incorrectly) as you read the text. After you have finished studying the chapter, try answering these questions again, and those at the end of the chapter.

1 _____ _____ time is a serious limitation of vacuum tubes that prevents operation at microwave frequencies.

2. The microwave spectrum is generally recognized as those frequencies from about _____ MHz to _____ GHz.

3. The _____ _____ refers to the fact that alternating currents flow only near the surface of conductors, especially at VHF and above.

4. Calculate the wavelength of a 11.7-GHz microwave signal in Teflon®, which has a dielectric constant of 9.6.

1-2   WHAT ARE MICROWAVES?

Micro means very small, and *waves* in this case refers to electromagnetic waves. Microwaves are, therefore, very small (or very short) electromagnetic waves, that is, electromagnetic waves of very short wavelength.

The electromagnetic spectrum (Fig. 1-1) is broken into bands for the sake of convenience and identification. The microwave region officially begins above the UHF region, at 900 or 1000 megahertz (MHz) depending on the source authority. For the purposes of this book, however, we will recognize the high end of the UHF region as microwave because it shares much in common with the actual microwave region. In addition, certain applications that are usually microwave are sometimes also used in the UHF region (an example is long-range radar). For our present purposes, therefore, we alter the usual definition of microwaves to include all frequencies above 400 MHz and below infrared (IR) light.

The student may well ask how microwaves differ from other electromagnetic waves. Microwaves become a separate topic of study because at these frequencies the wavelength approximates the physical size of ordinary electronic components. Thus, components behave differently at microwave frequencies than they do at lower frequencies. At microwave frequencies, a half-watt metal film resistor, for example, looks like a complex *RLC* network with distributed *L* and *C* values and a surprisingly different *R* value. These distributed components have immense significance at microwave frequencies, even though they can be ignored as negligible at lower frequencies.

Before examining microwave theory, let's first review some background and fundamentals.

1-3   DEVELOPMENT OF MICROWAVE TECHNOLOGY

The microwave spectrum was relatively undeveloped until World War II, when radar and other electronics technologies needed for the war effort caused the U.S. government (and its allies) to pour a large amount of money into research. Vac-

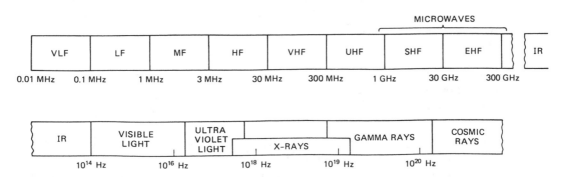

Figure 1-1   Electromagnetic spectrum from VLF to cosmic rays.

uum tubes, especially those used in the 1920s and 1930s, did not operate in the UHF or microwave regions. The two main problems that prevented operation at those frequencies were *interelectrode capacitance* and *electron transit time* between the cathode and anode. Attempts to reduce electrode size and spacing resulted in other problems that proved unacceptable in practical circuits.

The interelectrode capacitance of vacuum tubes was reduced by clever geometries, as well as closer interelectrode spacing (a factor in transit time), so that by World War II frequencies up to 200 MHz were possible, with 500 to 800 MHz being achieved by the end of the war and 800 MHz by the early 1950s. In 1920, the team of Barkhausen and Kurz (Germany) achieved oscillation at 700 MHz by working with, rather than against, the electron transit time (see Chapter 10). Although the Barkhausen–Kurz oscillator (BKO) achieved high-frequency oscillations, it was very inefficient. The grid element of a tube used in the BKO mode glowed white hot, with resultant low reliability.

A solution to the heating caused by poor efficiency was proposed in 1921 by A.W. Hull. According to the system used by Hull, the grid in the vacuum tube was replaced with a magnetic field. Hull's device came down to us in modified form as the *magnetron* (Chapter 11). The magnetron was the first practical high-power microwave generator and is still used today in some radars and in microwave ovens. Unfortunately, the magnetron is a narrow-band device, so one often has to make a trade-off between output power and frequency agility.

The power versus frequency dilemma seemed unsolvable for about a decade, but in the mid-1930s several researchers simultaneously reached similar solutions. W.W. Hansen (Stanford University) and A. Heil and O. Heil began to turn the transit time dilemma into an advantage through the use of *velocity modulation* of the electron beam. Russell Varian and Sigurd Varian extended Hansen's work in 1937 to produce the first *klystron* device, a vacuum tube that amplified microwaves (Chapter 11).

In the 1950s, vacuum tubes began to be replaced with solid-state devices such as bipolar transistors. Although *PN* junction semiconductor diodes were made to work at low microwaves during World War II (for example, the 1N23 microwave detector diode), other solid-state devices resisted the upper end of the frequency spectrum. Semiconductor material exhibits a phenomenon similar to the vacuum tube transit time limitation: *electron saturation velocity*. In 1963, however, IBM scientist John Gunn was experimenting with *N*-type gallium arsenide (GaAs) material when he noted a negative resistance characteristic. From that initial observation came the Gunn diode and other transferred electron devices (TEDs). Other solid-state microwave devices soon followed. It is now possible to purchase for about 1$ a dc to 2-GHz monolithic microwave integrated circuit (MMIC) device (see Chapter 13) that produces 13 to 20 decibels (dB) of gain (depending on the model). Other solid-state microwave devices operate to 20 GHz and more.

Modern electronics designers have a wide variety of solid-state devices to select from, and microwaves are no longer the stepchild of electronics technology. Radar, wireless communications, electronic navigation, and satellite TV links all operate in the microwave region. In addition, modern medical diathermy (tissue heating) equipment long ago moved out of the 11-meter Citizen's Band up to microwave frequencies.

TABLE 1-1   METRIC PREFIXES

Metric Prefix	Multiplying Factor	Symbol
tera	10^{12}	T
giga	10^{9}	G
mega	10^{6}	M
kilo	10^{3}	K
hecto	10^{2}	h
deka	10	da
deci	10^{-1}	d
centi	10^{-2}	c
milli	10^{-3}	m
micro	10^{-6}	μ
nano	10^{-9}	n
pico	10^{-12}	p
femto	10^{-15}	f
atto	10^{-18}	a

1-4   UNITS AND PHYSICAL CONSTANTS

In accordance with standard engineering and scientific practice, all units in this text will be in either the CGS (centimeter–gram–second) or MKS (meter–kilogram–second) systems unless otherwise specified. Because the metric (CGS and MKS) systems depend on using multiplying prefixes on the basic units, we include a table of common metric prefixes (Table 1-1). Other tables are as follows: Table 1-2 gives the standard physical units; Table 1-3 shows physical constants of interest, including those used in the problems in this and other chapters; and Table 1-4 lists some common conversion factors.

TABLE 1-2   UNITS

Quantity	Unit	Symbol
Capacitance	farad	F
Electric charge	coulomb	Q
Conductance	mhos (siemens)	Ω
Conductivity	mhos/meter	Ω/m
Current	ampere	A
Energy	joule (watt-sec)	J
Field	volts/meter	E
Flux linkage	weber (volt-second)	ψ
Frequency	hertz	Hz
Inductance	Henry	H
Length	meter	m
Mass	gram	g
Power	watt	W
Resistance	ohm	Ω
Time	second	s
Velocity	meter/second	m/s
Electric potential	volt	V

TABLE 1-3 PHYSICAL CONSTANTS

Constant	Value	Symbol
Boltzmann's constant	1.38×10^{-23} J/K	K
Electric charge (e^-)	1.6×10^{-19} C	q
Electron (volt)	1.6×10^{-19} J	eV
Electron (mass)	9.12×10^{-31} kg	m
Permeability of free space	$4\pi \times 10^{-7}$ H/m	U_o
Permittivity of free space	8.85×10^{-12} F/m	ϵ_o
Planck's constant	6.626×10^{-34} J · s	h
Velocity of electromagnetic waves	3×10^8 m/s	c
Pi (π)	3.1416	π

TABLE 1-4 CONVERSION FACTORS

1 inch	= 2.54 cm
1 inch	= 25.4 mm
1 foot	= 0.305 m
1 mile	= 1.61 km
1 nautical mile	= 6080 ft
1 statute mile	= 5280 ft
1 mil	= 2.54×10^{-5} m
1 kg	= 2.2 lb
1 neper	= 8.686 dB
1 gauss	= 10,000 teslas

1-5 WAVELENGTH AND FREQUENCY

For all forms of wave, the velocity, wavelength, and frequency are related such that the product of frequency and wavelength is equal to the velocity. For microwaves, this relationship can be expressed in the form

$$\lambda F\sqrt{\varepsilon} = c \qquad (1\text{-}1)$$

where

λ = wavelength in meters (m)
F = frequency in hertz (Hz)
ε = dielectric constant of the propagation medium
c = velocity of light (300,000,000 m/s)

The dielectric constant (ε) is a property of the medium in which the wave propagates. The value of ε is defined as 1.000 for a perfect vacuum, and very nearly 1.0 for dry air (typically 1.006). In most practical applications, the value of ε in dry air is taken to be 1.000. For mediums other than air or vacuum, however, the velocity of propagation is slower, and the value of ε relative to a vacuum is higher. Teflon®, for example, can be made with dielectric constant values (ε) from about 2 to 11.

Equation (1-1) is more commonly expressed in the forms of Eqs. (1-2) and (1-3):

$$\lambda = \frac{c}{F\sqrt{\varepsilon}} \qquad\qquad (1\text{-}2)$$

and

$$F = \frac{c}{\lambda\sqrt{\varepsilon}} \qquad\qquad (1\text{-}3)$$

All terms are as defined for Eq. (1-1).

A design standard requires that a parabolic-dish microwave antenna used in communications or radar be built with a surface smoothness of 1/12 wavelength. For a 2.2-GHz signal, therefore, the surface should be built to within 13.6 cm/12, or 1.13 cm of the mathematically exact curve. Some of these antennas have diameters on the order of 15 meters (1,500 cm), so you can see the precision to which the parabolic shape must be maintained: this is one reason why microwave design can be difficult.

The wavelengths of microwave signals in air vary from 75 cm at 400 MHz to 1 mm at 300 GHz. Microwave frequencies above 20 GHz are sometimes referred to as *millimeter waves* because of their extremely short wavelengths.

In media other than dry air or a vacuum, the microwave velocity decreases. In order to retain the constant velocity c required by Eq.(1-1), the wavelength must decrease if frequency is held constant.

The wavelength of an 18-GHz microwave signal in air is 0.17 m, and in Teflon® it is about 0.0053 m. Changing the medium reduced the wavelength by a considerable amount.

During World War II, the U.S. military began using microwaves in radar and other applications. For security reasons, alphabetic letter designations were adopted for each band in the microwave region. Because the letter designations became ingrained, they are still used throughout industry and the defense establishment. Unfortunately, some confusion exists because at least three systems are currently in use: (1) pre-1970 military (Table 1-5), (2) post-1970 military (Table 1-6),

TABLE 1-5   OLD U.S. MILITARY MICROWAVE FREQUENCY BANDS

Band Designation	Frequency Range
P	225–390 MHz
L	390–1550 MHz
S	1550–3900 MHz
C	3900–6200 MHz
X	6.2–10.9 GHz
K	10.9–36 GHz
Q	36–46 GHz
V	46–56 GHz
W	56–100 GHz

TABLE 1-6 NEW U.S. MILITARY MICROWAVE
FREQUENCY BANDS

Band Designation	Frequency Range
A	100–250 MHz
B	250–500 MHz
C	500–1000 MHz
D	1000–2000 MHz
E	2000–3000 MHz
F	3000–4000 MHz
G	4000–6000 MHz
H	6000–8000 MHz
I	8000–10,000 MHz
J	10–20 GHz
K	20–40 GHz
L	40–60 GHz
M	60–100 GHz

and (3) an IEEE-industry standard (Table 1-7). Additional confusion is created because the military and defense industry use both pre- and post-1970 designations simultaneously, and industry often uses military rather than IEEE designations. The "old military" designations (Table 1-5) persists as a matter of habit.

1-6 SKIN EFFECT

There are three reasons why ordinary lumped constant electronic components do not work well at microwave frequencies. The first, mentioned earlier in this chapter, is that component size and lead lengths approximate microwave wavelengths. The second is that distributed values of inductance and capacitance become significant at these frequencies. The third is a phenomenon called *skin effect*.

Skin effect refers to the fact that alternating currents tend to flow on the surface of a conductor. While dc currents flow in the entire cross section of the conductor, ac flows in a narrow band near the surface. Current density falls off

TABLE 1-7 IEEE/INDUSTRY STANDARD
MICROWAVE FREQUENCY BANDS

Band Designation	Frequency Range
HF	3–30 MHz
VHF	30–300 MHz
UHF	300–1000 MHz
L	1000–2000 MHz
S	2000–4000 MHz
C	4000–8000 MHz
X	8000–12,000 MHz
Ku	12–18 GHz
K	18–27 GHz
Ka	27–40 GHz
Millimeter	40–300 GHz
Submillimeter	Above 300 GHz

exponentially from the surface of the conductor toward the center (Fig. 1-2). At the *critical depth* (δ), also called *depth of penetration,* the current density is 1/e, or 1/2.718 = 0.368, of the surface current density. The value of (δ) is a function of operating frequency, the permeability (μ) of the conductor, and the conductivity (σ). Equation (1-4) gives the relationship.

$$\delta = \sqrt{\frac{1}{2\pi F \sigma \mu}} \qquad\qquad (1\text{-}4)$$

where

      δ = critical depth
      F = frequency in hertz (Hz)
      μ = permeability in Henrys per meter (H/m)
      σ = conductivity in mhos per meter

1-7   ELECTRONIC COMPONENTS

Because current only flows in a small part of the conductor (instead of the entire conductor, or nearly so as at lower frequencies), the ac resistance of a conductor at microwave is higher than the dc resistance by a considerable amount. Because skin effect is a function of frequency, we find that conductors will perform as they do at dc for very low-frequency ac, but may be useless at microwave frequencies.

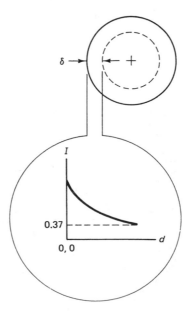

Figure 1-2
Skin effect depth falls off exponentially from the surface.

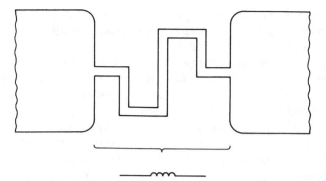

Figure 1-3
Example of a microwave inductor.

For these reasons, microwave circuits must consider the distributed constants of *R*, *L*, and *C* for electronic components. In addition, specially designed lumped constant components, such as thin-film inductors, strip-line inductors (Fig. 1-3), chip resistors, and capacitors, must be used. A combination of these will be found in many microwave circuits. We will discuss these components in a later chapter.

1-8   MICROWAVE SAFETY CONSIDERATIONS

Microwave signals are a form of electromagnetic energy. For several reasons, that energy constitutes a biohazard to humans and animals. Part of the danger results from the fact that microwave wavelengths approximate biostructure sizes, so resonances can occur. In other cases, the hazard arises from more subtle physiological interactions. These effects are so powerful that industry is able to offer homeowners microwave ovens to cook food. If you doubt the potency of microwave energy, then think about how little time it takes to cook a meal in the microwave oven compared to the time required in an ordinary heat oven.

High-powered microwave sources can cook you. Several years ago, the author interviewed an engineer who worked at the Earth station terminal for a major television network. The network owned a power klystron-based microwave transmitter that fed a high-gain 15-m diameter dish antenna. The engineer had to replace a woodpecker-damaged focal-point feed horn, so he rented a cherry picker to raise himself up to the height required for doing the work. After the repair was completed, he sent his technician into the station and started the engine to lower the cherry picker to the ground. But the gasoline engine failed, and he was stranded just outside the main lobe of the antenna radiation pattern . . . but well within the first major sidelobe. Thinking that the engineer was safely on the ground, the transmitter technician "lit-off" the transmitter and fed a high-power microwave signal to the antenna. After about three minutes the error was realized, and the engineer was rescued.

Initially there did not appear to be any ill effects. But by the following morning the engineer was passing blood and solid matter in his urine, so he went to the doctor. The physician found blood in his stool, and massive bowel adhesions.

These serious medical problems continued for at least several years . . . from only a few minutes' exposure to microwave energy. The doctors told him that he would be lucky if he were ever able to have children.

Even low levels of microwave energy are hazardous. Damage to eyes and other sensitive organs can occur with either short, high-intensity doses or lower, long-term doses. Some authorities recommend an occupational dosage limitation of 10 mW/cm². Other experts insist on lower levels.

A common myth among male workers in microwave communications and radar is that a few minutes' exposure in appropriate locations of the body will render the man sterile for a day or so. This is a myth, and the power levels at which the desired effect occurs are sufficient to cause serious, permanent damage to the man, and possibly damage to unborn progeny in the event that the temporary sterilization doesn't work.

Although specific rules will vary with type of equipment, some general guidelines are valid. For example, *never* stand in front of a radiating microwave antenna. Remember that antennas have *sidelobes*, so regions that you might believe are safe may be hazardous. Window frames and glass windows reflect microwave signals. Unless the window is made of special materials that are translucent to microwaves, aiming the antenna outdoors through a window may raise the amount of energy in the room to a dangerous level.

Never operate a microwave signal source with protective shields removed. Be sure that the RF gasket around the edge of the shield is clean and in good repair when the unit is closed. Never attempt to defeat a safety precaution.

There is no reason to fear for your health if you work around microwave equipment. If you understand the hazards and follow the rules, then safety will follow: *work smart*.

1-9   SUMMARY

1.   Microwaves are those electromagnetic waves with frequencies of 900 MHz to 300 GHz, with wavelengths of 33 cm to 1 mm.

2.   Conventional electronic components do not work well at microwave frequencies because wavelengths approximate component sizes, distributed constants (*LC*) that are negligible at lower frequencies are critical at microwave frequencies, and the skin effect produces an ac resistance that is greater than the dc resistance.

3.   Ordinary vacuum tubes do not work at microwaves because of excessive interelectrode capacitance and electron transit times greater than the period of the microwave signal.

4.   Low-frequency solid-state devices are frequency limited because of electron saturation velocity.

5.   Electromagnetic waves propagate at a velocity of 300,000,000 meters per second (3×10^8 m/s), or about 186,200 miles per second.

6.   The skin effect is a phenomenon in which ac flows only in the region close to the surface of a conductor. Skin effect is a function of frequency.

RECAPITULATION

Now return to the objectives and prequiz questions at the beginning of the chapter and see how well you can answer them. If you cannot answer certain questions, place a check mark by each and review the appropriate parts of the text. Next, try to answer the following questions and work the problems using the same procedure.

QUESTIONS AND PROBLEMS

1. Calculate the wavelength in free space (in centimeters) of the following microwave frequencies:

    (a) 1.5 GHz, (b) 6.8 GHz, (c) 10.7 GHz, (d) 22.5 GHz.

2. Calculate the wavelength in free space (in centimeters) of the following microwave frequencies:

    (a) 900 MHz, (b) 2.25 GHz, (c) 8 GHz, (d) 15 GHz.

3. Microwave wavelengths below about 20 GHz are typically measured in units of

    _____.

4. Microwave wavelengths above 20 GHz are sometimes referred to as _____ waves.

5. Calculate the frequency of the following wavelengths; express the answer in megahertz (MHz):

    (a) 14 cm, (b) 32 mm, (c) 10 cm, (d) 175 mm.

6. Calculate the frequency (in gigahertz) of the following wavelengths:

    (a) 90 mm, (b) 3.7 cm, (c) 21 cm, (d) 0.15 m.

7. A newly developed microwave dielectric material has a dielectric constant (ε) of 8. Calculate the wavelength of a 4.7-GHz microwave signal

    (a) in free space, (b) in the new dielectric medium. Compare the two calculations.

8. A Teflon® dielectric material ($\varepsilon = 10.5$) passes a 12-GHz microwave signal. Calculate the wavelength

    (a) in the dielectric material, (b) in free space.

9. List three applications for microwaves.

10. List the six properties of sinusoidal waves.

11. Events per unit of time is _____.

12. Answer the following:

    (a) 2300 MHz = _____ GHz, (b) 12.6 MHz = _____ GHz, (c) 900 MHz = _____ kHz, (d) 30,000 MHz = _____ GHz

13. Calculate the time (in nanoseconds) required for electromagnetic waves of the following frequencies to travel a distance of one wavelength:

    (a) 950 MHz, (b) 2200 MHz, (c) 6.6 GHz, (d) 9500 MHz.

14. A 3.3-cm microwave signal in free space requires _____ nanoseconds to travel one wavelength.

15. Calculate the length of time required for a microwave signal to travel one nautical mile (1 n. mile = 6080 ft). *Note*: 1 meter = 3.28 feet.

16. A 10-GHz microwave signal travels _____ yards in 8.15 microseconds (1 yard = 0.915 meter).
17. Calculate the period in seconds of the following microwave signals:
    (a) 900 MHz, (b) 1.275 GHz, (c) 6.2 GHz, (d) 20 GHz.
18. Calculate the period in seconds of the following microwave signals:
    (a) 0.95 GHz, (b) 2300 MHz, (c) 7 GHz, (d) 15 GHz.
19. The velocity of an electromagnetic wave in free space is _____ m/s.
20. The velocity of an electromagnetic wave in air ($\varepsilon = 1.006$) is _____ m/s.
21. A _____ diode is an example of a transferred electron device (TED).
22. A _____ uses a magnetic field to control the electron flow in a special vacuum tube.
23. The klystron is a microwave amplifier that depends on _____ modulation of the electron beam.
24. Electron _____ velocity is a fundamental limitation on microwave operation of ordinary solid-state devices.
25. Calculate the *critical depth* of a 9.75-GHz microwave signal flowing in a gold conductor ($\sigma = 4 \times 10^7$ mhos/m) that is 3 mm in diameter. Assume a permeability of $4\,\pi \times 10^{-7}$ H/m.
26. Calculate the critical depth of a 21-GHz microwave signal flowing in an aluminum conductor ($\sigma = 3.81 \times 10^7$ mhos/m) as a percentage of its 1.5-mm diameter.

KEY EQUATIONS

1.
$$c = \lambda F \sqrt{\in}$$

2.
$$\lambda = \frac{c}{F\sqrt{\in}}$$

3.
$$F = \frac{c}{\lambda\sqrt{\in}}$$

4.
$$T = \frac{1}{F}$$

5.
$$\delta = \sqrt{\frac{1}{2\pi F \sigma \mu}}$$

CHAPTER 2

Microwave Propagation

OBJECTIVES

1. Learn the basic principles of electromagnetic wave behavior.
2. Understand propagation phenomena.
3. Learn the different modes of propagation applicable to microwaves.

2-1 PREQUIZ

These questions test your prior knowledge of the material in this chapter. Try answering them before you read the chapter. Look for the answers (especially those you answered incorrectly) as you read the text. After you have finished studying the chapter, try answering these questions again and those at the end of the chapter.

1. Polarization of an electromagnetic wave is, by definition, the direction of the _____ field.
2. The velocity of the microwave signal in free space is about _____ m/s.
3. List three wave behavior properties that affect propagation at microwave frequencies.
4. An aircraft flying at an altitude of 10,000 ft sees a radio horizon of _____ nautical miles (*Hint:* 1 n. mile = 6080 ft = 1.15 statute miles).

2-2 THE ELECTROMAGNETIC FIELD

Radio signals are *electromagnetic* (EM) waves exactly like light, infrared, and ultraviolet, except for frequency. The EM wave consists of two mutually perpendicular oscillating fields (see Fig. 2-1A) traveling together. One field is an *electric* field, while the other is a *magnetic* field.

MICROWAVE PROPAGATION

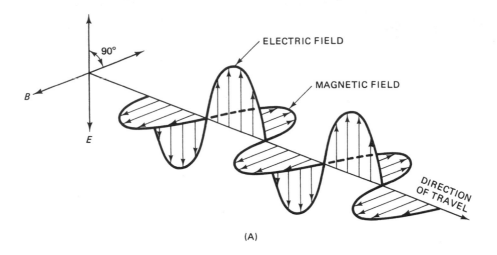

(A)

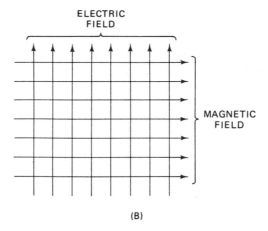

(B)

Figure 2-1   A) An electromagnetic wave consists of co-traveling electrical and magnetic fields oscillating 90 degrees out of phase with each other, and arranged orthogonally to each other; B) the front of the wave, showing vectors for electrical and magnetic fields.

In dealing with both antenna theory and radio-wave propagation, we sometimes make use of a theoretical construct called an *isotropic source* for the sake of comparison and simpler arithmetic. An isotropic source assumes that the radiator (that is, the antenna) is a very tiny spherical source that radiates equally well in all directions. The radiation pattern is thus a sphere with the isotropic antenna at the center. Because a spherical source is uniform in all directions and its geometry is easily determined mathematically, signal intensities at all points can be calculated from basic principles.

The radiated sphere gets ever larger as the wave propagates away from the isotropic source. If, at a great distance from the center, we take a look at a small slice of the advancing wavefront, we can assume that it is essentially a flat plane, as in Fig. 2-1B. This situation is analogous to the apparent flatness of the prairie, even though the surface of Earth is a near-sphere. We would be able to "see" the electric and magnetic field vectors at right angles to each other (Fig. 2-1B) in the flat plane wavefront.

The *polarization* of an EM wave is, by definition, the *direction of the electric field*. In Fig. 2-1, we see vertical polarization because the electric field is vertical with respect to Earth. If the fields were swapped, then the EM wave would be horizontally polarized.

These designations are especially convenient because they also tell us the type of antenna used: vertical antennas (common in landmobile communications) produce vertically polarized signals, while horizontal antennas produce horizontally polarized signals. Some texts erroneously state that antennas will not pick-up signals of the opposite polarity. Such is not the case, especially in the HF and lower VHF regions. At microwave frequencies, a loss of approximately 20 dB is observed due to cross-polarization.

An EM wave travels at the speed of light, designated by the letter c, which is about 300,000,000 m/s (or 186,200 miles/s if you prefer English units). To put this velocity in perspective, a radio signal originating on the Sun's surface would reach Earth in about 8 minutes. A terrestial radio signal can travel around Earth seven times in 1 second.

The velocity of the wave slows in dense media, but in air the speed is so close to the free-space value of c that the same figures are used for both air and outer space in practical problems. In pure water, which is much denser than air, the speed of radio signals is about one-ninth the free-space speed. This same phenomenon shows up in practical work in the form of the velocity factor (V) of transmission lines. In foam dielectric coaxial cable, for example, the value of V is 0.80, which means that the signal propagates along the line at a speed of $0.80c$, or 80% of the speed of light.

In our discussions of radio propagation, we will consider the EM wave as a very narrow "ray" or "pencil beam" that does not diverge as it travels. That is, the ray remains the same width all along its path. This convention makes it easy to use ray-tracing diagrams in our discussions. Keep in mind, however, that the real situation, even when narrow-beamwidth microwave signals are used, is much more complicated. Real signals, after all, are sloppier than textbook examples: they are neither infinitesimally thin nor nondivergent.

2-3 PROPAGATION PHENOMENA

Because EM waves are waves, they behave in a wavelike manner. Figure 2-2 illustrates some of the wave behavior phenomena associated with light and radio waves: *reflection*, *refraction*, and *diffraction*. All three play roles in radio propagation. In fact, many propagation cases involve all three in varying combinations.

Reflection and refraction are shown in Fig. 2-2A. Reflection occurs when a wave strikes a denser reflective medium, as when a light wave strikes a glass mirror. The incident wave (shown as a single ray) strikes the interface between less

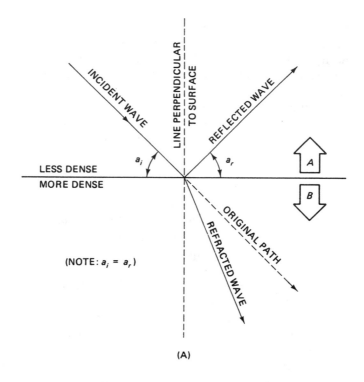

(A)

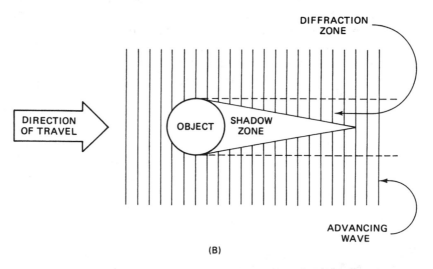

(B)

Figure 2-2 At the boundary of two mediums of differing density, some of the incident wave is reflected from the boundary, and some is refracted as it penetrates the second medium.

dense and more dense mediums at a certain angle of incidence (a_i), and is reflected at exactly the same angle, now called the angle of reflection a_r. Because these angles are equal, we can often trace a reflected microwave signal back to its origin.

Refraction occurs when the incident wave enters the region of different density and thereby undergoes both a velocity change and a directional change. The amount and direction of the change are determined by the ratio of the densities between the two mediums. If zone B is much different from zone A, then bending is great. In radio systems, the two mediums might be different layers of air with different densities. It is possible for both reflection and refraction to occur in the same system.

Diffraction is shown in Fig. 2-2B. In this case, an advancing wavefront encounters an opaque object (for example, a steel building). The shadow zone behind the building is not simply perpendicular to the wave, but takes on a cone shape as waves bend around the object. The *umbra region* (or diffraction zone) between the shadow zone (*cone of silence*) and the direct propagation zone is a region of weak (but not zero) signal strength. In practical situations, signal strength in the cone of silence never reaches zero. A certain amount of reflected signals scattered from other sources will fill in the shadow a little bit.

2-4   PROPAGATION PATHS

There are four major propagation paths: *surface wave, space wave, tropospheric,* and *ionospheric* (some textbooks break the categories up a little differently). The ionospheric path is not important to microwave propagation, so it will not be discussed in detail here. We will, however, discuss ionospheric phenomena that are not strictly microwave in nature, but are included for perspective. The space wave and surface wave are both *ground waves,* but they behave differently enough to warrant separate consideration. The surface wave travels in direct contact with Earth's surface and it suffers a frequency-dependent attenuation due to absorption into the ground. Because the absorption increases with frequency, we observe much greater surface wave distances in the AM broadcast band (540 to 1,700 kHz) than in the Citizen's Band (27 MHz), or VHF/UHF/microwave bands. For this reason, broadcasters prefer frequency assignments on the lower end of the AM band.

The space wave is also a ground wave phenomenon, but is radiated from an antenna many wavelengths above the surface. No part of the space wave normally travels in contact with the surface; VHF, UHF, and microwave signals are usually space waves. There are, however, two components of the space wave in many cases: *direct* and *reflected* (see Fig. 2-3).

The tropospheric wave is lumped with the direct space wave in some textbooks, but has properties that actually make it different in practical situations. The troposphere is the region of Earth's atmosphere between the surface and the stratosphere, or about 4 to 7 miles above the surface. Thus, all forms of ground wave propagate in the troposphere. But because certain propagation phenomena caused mostly by weather conditions only occur at higher altitudes, we need to consider tropospheric propagation as different from other forms of ground wave.

The ionosphere is the region of Earth's atmosphere that is above the stratosphere and is located 30 to 300 miles above the surface. The peculiar feature of the ionosphere is that molecules of air gas (O_2 and N) can be ionized by stripping away electrons under the influence of solar radiation and certain other sources of energy. The electrons are called negative ions, while the formerly neutral atoms they were removed from are now positive ions. In the ionosphere the air density is

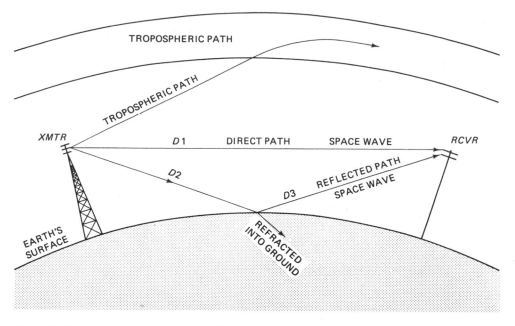

Figure 2-3   Propagation paths: tropospheric, direct, and reflected.

so low that ions can travel relatively long distances before recombining with oppositely charged ions to form electrically neutral atoms. As a result, the ionosphere remains ionized for long periods of the day, even after sunset. At lower altitudes, however, air density is greater and recombination thus occurs rapidly. At those altitudes solar ionization diminishes to nearly zero immediately after sunset, or never achieves any significant levels, even at local noon. Some authorities believe that ionization and recombination phenomena in the ionosphere add to the noise level experienced at VHF, UHF, and microwave frequencies.

2-5   GROUND WAVE PROPAGATION

The ground wave, naturally enough, travels along the ground (or at least in close proximity to the surface). There are two forms of ground wave: *space* and *surface*. The space wave does not actually touch the ground. As a result, space wave attenuation with distance in clear weather is about the same as in free space. Above the VHF region, weather conditions add attenuation not found in outer space. The surface wave is subject to the same attenuation factors as the space wave, but in addition it also suffers ground losses. These losses are due to ohmic resistive losses in the conductive earth. In other words, the signal heats up the ground!

Surface wave attenuation is a function of frequency and increases rapidly as frequency increases. In the AM broadcast band, for example, the surface wave operates out to about 50 to 100 miles. At Citizen's Band frequencies (27 MHz, which is in the upper HF region), on the other hand, attenuation is so great that communication is often limited to less than 20 miles. It is common in the upper HF region to be able to hear stations across the continent, or internationally, while local stations only 25 to 30 miles distant remain unheard or are very weak. For both forms of ground wave, communication is affected by these factors: *wavelength, height of both receive and transmit antennas, distance between antennas, terrain,* and *weather along the transmission path.*

Ground wave communication also suffers another difficulty, especially at VHF, UHF, and microwave frequencies. The space wave is made up of two components (Fig. 2-3): *direct* and *reflected waves.* If both of these components arrive at the receive antenna, they will add algebraically to either increase or decrease signal strength. There is always a phase shift between the two components because the two signal paths have different lengths (that is, D1 is less than D2 + D3). In addition, there may be a 180° phase reversal at the point of reflection (especially if the incident signal is horizontally polarized), as in Fig. 2-4. The following general rules apply in these situations:

1. A phase shift of an odd number of half-wavelengths causes the components to add, increasing signal strength (constructive interference).
2. A phase shift of an even number of half-wavelengths causes the components to subtract (see Fig. 2-4), reducing signal strength (destructive interference).
3. Phase shifts other than half-wavelength add or subtract according to relative polarity and amplitude.

A category of reception problems called *multipath* phenomena exists because of interference between the direct and reflected components of the space wave. The form of multipath phenomenon that is, perhaps, most familiar to most readers is *ghosting* in television reception. Some multipath events are transitory in nature (as when an aircraft flies through the transmission path), while others are permanent (as when a large building or hill reflects the signal). In mobile communications, multipath phenomena are responsible for dead zones and "picket fencing." A dead zone exists when destructive interference between direct and reflected (or multiple reflected) waves drastically reduces signal strengths. This problem is most often noticed when the vehicle is stopped, and the solution is to move the antenna one half-wavelength (a few cm!). Picket fencing occurs as a mobile unit moves through successive dead zones and signal-enhancement (or normal) zones, and sounds like a series of short noise bursts.

At VHF and above, the space wave is limited to so-called line-of-sight distances. The horizon is theoretically the limit of communications distance, but the

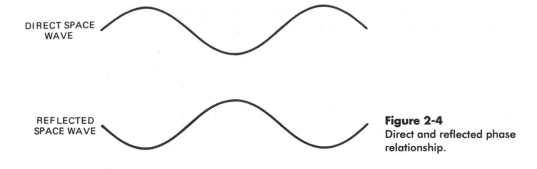

DIRECT SPACE
WAVE

REFLECTED
SPACE WAVE

Figure 2-4
Direct and reflected phase relationship.

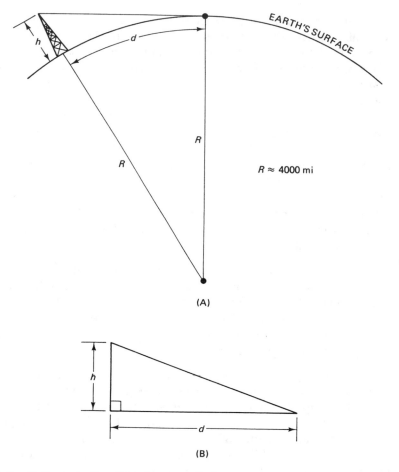

Figure 2-5   A) Geometry for calculating the radio horizon; B) reduced to a simple triangle.

radio horizon is actually about 60% farther than the optical horizon. This phenomenon is due to bending in the atmosphere. Although more sophisticated models are available today, we can still make use of the traditional model of the radio horizon shown in Fig. 2-5. The actual situation is shown in Fig. 2-5A. Distance d is a curved path along the surface of Earth. But because Earth's radius R is about 4000 miles and is thus much larger than practical antenna height h, we can simplify the model to that shown in Fig. 2-5B. The underlying assumption is that Earth has a radio radius equal to about four-thirds its actual physical radius.

The value of distance d is found from the expression

$$d = \sqrt{2Rh}$$

where

        d = distance to the radio horizon in statute miles
        R = normalized radius of Earth
        h = antenna height in feet

When all constant factors are accounted for, the expression reduces to

$$d = 1.42\sqrt{h}$$

All factors are as defined previously.

EXAMPLE 2-1

A radio tower has a UHF radio antenna mounted 150 ft above the surface of Earth. Calculate the radio horizon (in miles) for this system.

Solution

$d = 1.42(h)^{1/2}$
$= (1.42)(150 \text{ ft})^{1/2}$
$= (1.42)(12.25)$
$= 17.4 \text{ miles}$

2-6   TROPOSPHERIC PROPAGATION

The troposphere is the portion of the atmosphere between the surface of Earth and the stratosphere, or about 4 to 7 miles above the surface. Some older texts group tropospheric propagation with ground wave propagation, but modern practice requires separate treatment. The older grouping overlooks certain common propagation phenomena that do not happen with space or surface waves.

Refraction is the mechanism for most tropospheric propagation phenomena. Recall that refraction occurs in both light and radio wave systems when the wave passes between mediums of differing density. In that situation, the wave path will bend an amount proportional to the difference in density.

Two general situations are typically found, especially at VHF and above. First, because air density normally decreases with altitude, the top of a beam of radio waves typically travels slightly faster than the lower portion of the beam. As a result, those signals refract a small amount. Such propagation provides slightly longer surface distances than are normally expected from calculating the distance to the radio horizon. This phenomenon is called *simple refraction*.

A special case of refraction called *superrefraction* occurs in areas of the world where warmed land air goes out over a cooler sea. Examples are areas that have deserts adjacent to a large body of water; the Gulf of Aden, the southern Mediterranean, and the Pacific Ocean off the coast of Baja, California are examples. VHF/UHF/microwave communications to 200 miles are reported in such areas.

The second form of refraction is weather related. Called *ducting*, this form of propagation (Fig. 2-6) is actually a special case of superrefraction. Evaporation of seawater causes *temperature inversion regions* to form in the atmosphere, that is, layered air masses in which air temperature is greater than in the layers below. (Air temperature normally decreases with altitude, but at the boundary with an inversion region it increases.) The inversion layer forms a duct that acts similarly to microwave waveguides. In Fig. 2-6, the distance D1 is the normal radio horizon distance, while D2 is the distance over which duct communications can occur.

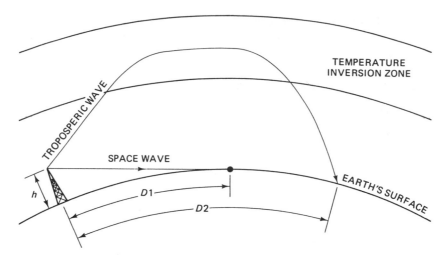

Figure 2-6 Tropospheric duct propagation.

Ducting allows long-distance communication at VHF through microwave frequencies, with 50 MHz being a lower practical limit and 10 GHz being an ill-defined upper limit. Airborne operators of radar and other microwave equipment can sometimes note ducting at even higher microwave frequencies, but it is uncommon.

Antenna placement is critical for ducting propagation. Both receive and transmit antennas must be either (1) inside the duct physically (as in airborne cases) or (2) able to propagate at such an angle that the signal gets trapped inside the duct. The latter is a function of antenna radiation angle. Distances up to 2500 miles or so are possible through ducting. Certain paths where ducting occurs frequently have been identified: the Great Lakes to the Atlantic seaboard; Newfoundland to the Canary Islands; across the Gulf of Mexico from Florida to Texas; Newfoundland to the Carolinas; California to Hawaii; and Ascension Island to Brazil.

Another condition is noted in the polar regions where colder air from the landmass flows out over warmer seas. Called *subrefraction*, this phenomenon bends EM waves away from Earth's surface, thereby reducing the radio horizon by about 30% to 40%.

All tropospheric propagation that depends on air mass temperatures and humidity shows diurnal (that is, over the course of the day) variation due to the local rising and setting of the sun. Distant signals may vary 20 dB in strength over a 24-hour period.

These tropospheric phenomena explain how TV, FM broadcast, and other VHF signals can propagate at great distances, especially along seacoast paths, at some times while being weak or nonexistent at others.

2-7   UNUSUAL PROPAGATION MODES

Several modes of propagation exist that are opportunistic in nature, so they only exist when certain events occur. *Ionospheric scatter* occurs when waves that would normally pass through the ionosphere into outer space are scattered (a random reflection event) by clouds of randomly ionized particles. Ionization in the upper ionosphere occurs from several sources; passing meteorites, dust, and artificial satellites are examples. Although opportunistic in nature, meteor scatter is routinely used for communications in some systems. During a 2-year period, records of false-alarm complaints on a 150-MHz VHF radio paging system showed that distinct peaks of two to three times normalized monthly averages occurred in August and December, corresponding to the annual Perseids and Geminides meteor showers, respectively.

The auroral lights (northern lights and southern lights) occur when a *proton event* on the sun's surface showers the upper atmosphere with charged particles. The lights are generated because these particles cause luminescence of air particles in lower levels of the atmosphere. They also act as a radio mirror to scatter VHF, UHF, and microwave signals. Communication over great distances is possible along the front created by the auroral lights because of the "angle of reflection equals angle of incidence" rule.

Backscatter and *sidescatter* occur in all forms of propagation. These modes result from energy being reflected at random angles from terrain and atmospheric irregularities. These forms of propagation are too numerous and too random to discuss in explicit detail.

2-8   IONOSPHERIC EFFECTS

The upper portion of the atmosphere is called the ionosphere because it tends to be easily ionized by solar and cosmic phenomena. The reason for the ease with which this region (30 to 300 miles above the surface) ionizes is that the air density is very low. Energy from the sun strips away electrons from the outer shells of oxygen and nitrogen molecules. The electrons become negative ions, while the remaining portion of the atom forms positive ions. Because the air is so rarified at those altitudes, these ions can travel great distances before recombining to form electrically neutral atoms again.

Several sources of energy will cause ionization of the upper atmosphere. Cosmic (or *galactic* as it is sometimes called) radiation from outer space causes some degree of ionization, but the vast majority of ionization is caused by solar energy. The role of cosmic radiation was first noticed during World War II when British radar operators discovered that the distance at which their equipment could detect German aircraft depended on whether or not the Milky Way was above the horizon. Intergalactic radiation raised the background microwave noise level, thereby adversely affecting the signal-to-noise ratio.

Two principal forms of solar energy affect communications: electromagnetic radiation and charged solar particles. Most of the radiation is above the visible spectrum in the ultraviolet and X-ray/gamma-ray regions of the spectrum. Because electromagnetic radiation travels at the speed of light, solar events that release radiation cause changes to the ionosphere about 8 minutes later. Charged particles, on the other hand, have finite mass and so travel at a considerably slower velocity. They require 2 or 3 days to reach Earth.

Various sources of both radiation and particles exist on the sun. Solar flares may release huge amounts of both radiation and particles. These events are unpredictable and sporadic. Solar radiation also varies on an approximate 27-day period, which is the rotational period of the sun. The same source of radiation will face Earth once every 27 days, so events tend to be repetitive.

Solar and galactic noise affects the reception of weak signals, while solar noise will also either affect radio propagation or act as a harbinger of changes in propagation patterns. Solar noise can be demonstrated by using an ordinary radio receiver and a directional antenna, preferably operating in the VHF/UHF/microwave regions of the spectrum. Aim the antenna at the sun on the horizon at either sunset or sunrise. A dramatic change in background noise will be noted as the sun slides across the horizon. Galactic noise is similar in nature to solar noise.

Sunspots.   A principal source of solar radiation, especially the periodic forms, is sunspots. Sunspots can be 70,000 to 80,000 miles in diameter, and generally occur in clusters. The number of sunspots varies over a period of approximately 11 years, although the actual periods since 1750 (when records were first kept) have varied from 9 to 14 years. The sunspot number is reported daily as the *Zurich smoothed sunspot number*, or *Wolf number*. The number of sunspots greatly affects radio propagation via the ionosphere, although at microwave frequencies propagation is not generally affected. The low sunspot number was about 60 (in 1907), while the high was about 200 (1958).

Another indicator of ionospheric propagation potential is the *Solar flux index* (SFI). This measure is taken in the microwave region (wavelength of 10.2 cm, or 2.8 GHz), at 1700 UT at Ottawa, Canada. The SFI is reported by the National Bureau of Standards (NBS) radio stations WWV (Fort Collins, Colorado) and WWVH (Maui, Hawaii).

2-9   SUMMARY

1.   Microwave signals are electromagnetic (EM) waves that propagate at a velocity of 3×10^8 m/s in free space. Polarization of an EM wave is defined as the direction of the electric field vector. The magnetic field vector is always orthogonal to (90° from) the electric field vector.

2.   EM waves exhibit four major propagation paths: *surface wave*, *space wave*, *tropospheric wave*, and *ionospheric wave*. The space wave has two components: direct and reflected.

3.   Ground wave communications are affected by these factors: *wavelength*, *height of both antennas, distance between antennas, terrain*, and *weather along the transmission path*.

4.   When both direct and reflected space wave signals reach the receiver antenna, either constructive or destructive interference will be exhibited.

5.   Tropospheric propagation occurs mostly by refraction and takes place 4 to 7 miles above the surface.

6.   Various tropospheric propagation modalities are recognized: simple refraction, superrefraction, ducting, and subrefraction.

7.   Because of temperature and humidity dependencies, tropospheric propagation phenomena show diurnal variation of up to 20 dB or so.

8.   Ionospheric scatter occurs when regions of the ionosphere become heavily ionized locally and so reflect signals that would otherwise propagate to outer space.

9.   Radio signal reflection can occur from ionized air in the lower atmosphere during periods of northern or southern lights displays.

10.   Ionization caused by solar radiation and charged particles can increase the microwave background noise level.

11.   Galactic noise increases when the Milky Way is above the horizon.

2-10 RECAPITULATION

Now return to the objectives and prequiz questions at the beginning of the chapter and see how well you can answer them. If you cannot answer certain questions, place a check mark by each and review the appropriate parts of the text. Next, try to answer the following questions and work the problems using the same procedure.

QUESTIONS AND PROBLEMS

1.   Name the two oscillating fields present simultaneously in an electromagnetic wave.
2.   The two fields in an EM wave are mutually _____ to each other.
3.   Radio propagation and antenna theory use an _____ radiator (or point source) for purposes of simpler calculation and systems comparisons.
4.   A microwave signal radiated into free space will travel _____ miles in 46 microseconds.
5.   How many seconds are required for a microwave signal to travel 56 km?
6.   Find the velocity of a microwave signal in a transmission line that has a velocity factor of 0.75.
7.   A cross-polarization loss of about _____ dB can be expected in microwave systems.
8.   In reflection phenomena, the angle of reflection equals the angle of _____.
9.   List three wave behavior phenomena found in microwave propagation.
10.   The principal factor in _____ phenomena is differing densities between two propagation mediums, or between regions of the same medium.
11.   _____ behavior results from the fact that microwave signals bend around radio-opaque barriers.
12.   The cone of silence is also called a _____ zone.
13.   List four major propagation paths.
14.   Which of the paths in question 13 is not normally important to microwave propagation?

15. A _____ wave signal suffers a frequency-dependent attenuation due to ground absorption of the radiated signal.

16. The _____ consists of two components: direct and reflected.

17. The _____ wave is found in two forms: space and surface.

18. List the six principal factors affecting ground wave communications.

19. During _____ reception, signal anomalies occur due to interference between direct and reflected signals.

20. Direct and reflected space wave components arrive exactly six half-wavelengths apart. The resultant interference is (constructive/destructive), and results in (increased/decreased) signal levels.

21. The microwave radio horizon is about _____ % farther than the optical horizon.

22. A microwave radio data link antenna is mounted on a 200-ft tower that is perched on top of a 5800-ft mountain peak. Assuming that there are no higher mountains for hundreds of miles around, what is the distance to the radio horizon?

23. An airplane with a microwave data link onboard flies at an altitude of 37,000 ft. What is the distance to the radio horizon for this system?

24. _____ is a propagation phenomenon caused by the fact that lower air density at high altitude causes the top of a signal to travel at a slightly greater velocity than the bottom.

25. _____ is caused by warm air from a desert region landmass blowing out over a cooler sea.

26. The formation of _____ _____ layers in the atmosphere causes the phenomenon called *ducting*.

27. A 1-GHz airborne signal generated in the sky above Newfoundland is detected in the Canary Islands. Operators suspect that _____ propagation took place.

28. In ducting propagation, signals in the troposphere behave as if trapped in a microwave _____.

29. Solar and galactic events can cause increased _____ levels in the microwave region.

30. List two principal sources of solar energy that affect communications.

KEY EQUATIONS

1. Velocity of electromagnetic waves:

$$c = 3 \times 10^8$$

2. Distance to the radio horizon:

$$d = \sqrt{2Rh}$$

and

$$d = 1.42\sqrt{h} \quad \text{(miles)}$$

CHAPTER 3

Physical Foundations of Microwave Technology

OBJECTIVES

1. Review the elements of electron dynamics in a vacuum.
2. Review the quantum mechanical model of electrons and other particles.
3. Learn the principles by which lasers and masers operate.

3-1 PREQUIZ

These questions test your prior knowledge of the material in this chapter. Try answering them before you read the chapter. Look for the answers (especially those you answered incorrectly) as you read the text. After you have finished studying the chapter, try answering these questions again and those at the end of the chapter.

1. An electron moving at a velocity $0.45c$ in an electrical field has a mass of _____.
2. The final velocity of an electron accelerated from rest in a 1200-volt E field is _____.
3. Nonrelativistic electrons are those with velocities that approach/are much less than (select one) c.
4. Calculate the de Broglie wavelength for a 1-kilogram object.

3-2 INTRODUCTION

All electronic technology is based on the principles of physics. But in microwave electronics, certain aspects of physics stand out more clearly and therefore need to be studied in greater depth. Consider the electron, for example. In low-frequency electronics, we may safely model the electron as a tiny point-source mass similar to a billiard ball with a negative electrical charge. Some experiments seem to confirm that view. At microwave frequencies, however, devices and their operating principles often must appeal to relativistic and quantum physics in addition to classical "billiard ball" physics for a proper and complete explanation.

In this chapter we will examine both particle and wave mechanical phenomena in order to provide you with a foundation for studying microwave theory. For some students, this treatment will be a review, and for them the material may seem hopelessly simple. We beg their indulgence and recommend skipping ahead to the next assigned chapter. For other students, this material will open up new ways of looking at all electronic theory. For all students this material forms a foundation for the study of microwaves.

3-3   "BILLIARD BALL" ELECTRON DYNAMICS

The classical view of the electron that is normally taught in elementary electronics technology textbooks models this subatomic particle as a point-source mass (9.11×10^{-28} gram) with a unit negative electrical charge (1.6×10^{-19} coulomb). The other common subatomic particle in electronics is the proton, and it has a unit positive charge that exactly balances the electronic charge to produce electroneutrality (although the proton mass is 1835 times greater than the electron mass). Let's consider the action of electron particles under various conditions.

3-3.1 Motion in Linear Electrostatic Fields

Assume that an electric field E is impressed across a space, s, between two conductive plates. Assume that the space is a perfect vacuum so that no inadvertent collisions take place between the electron and gas molecules. The strength of the electric field is E/s (V/cm). An electron injected into the space interacts with the field in a specific manner. The electron carries a negative charge, so it is repelled from the negative plate, while being attracted to the positive plate.

The electrostatic field exerts a force on the electron, as is proved by the fact that the electron moves. The mechanical work (W) done on the electron is

$$W_m = F \times s \text{ ergs} \tag{3-1}$$

and the electrical work is

$$W_e = E \times e \text{ joules (J)} \tag{3-2}$$

The value of the force (F) is

$$F = \frac{10^7 Ee}{s} \text{ dynes} \qquad (3\text{-}1)$$

When the electron is in an electrostatic field, the force is constant if (1) space s is constant and (2) the electric field potential E is held constant. A particle experiencing a constant force will exhibit a constant acceleration of

$$F = \frac{10^7 Ee}{sm} \qquad (3\text{-}4)$$

In a perfect vacuum, the electron velocity is a function of the applied field strength. Because the acceleration is constant, the average velocity V_{av} will be one-half the difference between final velocity (V_f) and initial velocity (V_i):

$$V_{av} = \frac{V_f - V_i}{2} \qquad (3\text{-}5)$$

or, if for simplicity's sake we assume that initial velocity $V_i = 0$, Eq. (3-5) reduces to

$$V_{av} = \frac{V_f}{2} \qquad (3\text{-}6)$$

The time required to traverse the field is the distance divided by average velocity:

$$t = \frac{s}{A_{av}} \qquad (3\text{-}7)$$

Substituting Eq. (3-6) into Eq. (3-7),

$$t = \frac{s}{\left(\dfrac{V_f}{2}\right)} \qquad (3\text{-}8)$$

$$= \frac{2s}{V_f} \qquad (3\text{-}9)$$

Final velocity V_f is the product of acceleration and time:

$$V_f = at \tag{3-10}$$

Or, by substituting Eqs. (3-4) and (3-9) into Eq. (3-10),

$$V_f = at \tag{3-11}$$

$$= \frac{10^7 Ee2}{sm} \times \frac{2s}{V_f} \tag{3-12}$$

and, by combining terms,

$$V_f^2 = \frac{10^7 Ee2}{m} \tag{3-13}$$

or

$$V_f = \frac{\sqrt{10^7 Ee2}}{m} \tag{3-14}$$

Note that the terms in Eq. (3-14) are all constants except for field strength E. This equation therefore reduces to

$$V_f = \frac{\sqrt{10^7 Ee2}}{m}$$

$$= \sqrt{\left(\frac{(10^7)(1.6 \times 10^{-9} c)(2)}{9.11 \times 10^{-28} g}\right)} \times \sqrt{E} \tag{3-15}$$

$$= 5.93 \times 10^7 \ \sqrt{E} \ \ cm/s$$

EXAMPLE 3-1

An electron is injected into an electric field of 1200 V. Calculate the final velocity, assuming that the initial velocity is zero.

Solution

$$V_f = 5.93 \times 10^7 \times \sqrt{E} \text{ cm/s}$$
$$= 5.93 \times 107 \times \sqrt{(1,200 \text{ V})} \text{ cm/s}$$
$$= 5.93 \times 10^7 \times 34.6 \text{ cm/s}$$
$$= 2.05 \times 10^9 \text{ cm/s}$$

A moving electron of mass m has a kinetic energy (U_k) of

$$U_k = \frac{1}{2} mV^2 \qquad (3\text{-}16)$$

or

$$U_k = 10^7 Ee \qquad (3\text{-}17)$$

In a later section we will use these relationships to describe other physical phenomena.

In the preceding material, we assumed that the electron travels in a perfect vacuum. Such an electron can theoretically accelerate until its velocity reaches the speed of light. While this assumption is reasonable for analysis purposes, it fails somewhat in real devices. Because of residual internal gas pressure, all real vacuum electronic devices contain atoms or molecules that can collide with individual electrons. Thus, in a stream of electrons flowing in a vacuum device, the average velocities reach an upper limit; this limit is called the *saturation velocity* (V_{sat}). Saturation velocity is also found in conductors because of electron interaction with the atoms of the conductor or semiconductor material.

Saturation velocity becomes important in microwave devices because it limits the speed of travel across devices (this is called *transit time*) and therefore limits the operating frequency unless a means is found to overcome this difficulty. For example, in a vacuum tube, the electron must traverse a vacuum space between a cathode and anode. As the period of a waveform approaches the device transit time, the ability of the device to amplify breaks down. Similarly, in a semiconductor device, the saturation velocity is on the order of 10^7 m/s in normal semiconductors, and there is an upper frequency limit for any given path length through the semiconductor material.

3-3.1.1 *Motion in an Orthogonal Electric Field*

In the previous section, we discussed the case where an electron was injected into an E field parallel to the electric lines of force (Fig. 3-1). Here we consider the case where the electron is injected into the E field orthogonal to the electrical lines of force (Fig. 3-2).

In Fig. 3-2, the electron enters the E field from the right with an initial velocity V_i. The negatively charged electron is repelled from the negative end of the field and is attracted to the positive end. It thus deflects from its original path to a new

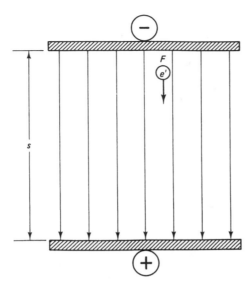

Figure 3-1 Experiment to show the movement of an electron in an electrical field.

curved path in the direction of the positive end of the field. The degree of deflection, measured by angle θ, is a function of the strength of the E field and the length of the interaction region 1.

During its excursion through the E field, the electron retains its initial velocity, but also picks up a *translational velocity* (V_t) in the direction of the positive end

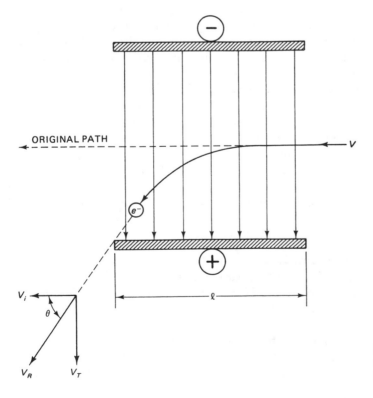

Figure 3-2 Path of an electron curve as it enters an electrical field.

of the field. The resultant velocity (V_r) along the line of travel is a vector sum of the two components, or in trigonometric terms,

$$\Theta = \tan^{-1} \frac{V_t}{V_i} \qquad (3\text{-}18)$$

and

$$V_r = V_i \cos \Theta \qquad (3\text{-}19)$$

3-3.2 Relativistic Effects

The electronic mass quoted in most textbooks (9.11×10^{-28} g) is the *rest mass* of the particle, and its value is usually symbolized by m_0. For velocities $V << c$, where c is the velocity of light and other electromagnetic waves (3×10^{10} cm/s), the rest mass is a good value to use. As velocity increases, however, the mass changes, and parameters calculated from mass begin to show errors. As the velocity increases above about $0.1c$, the electron becomes heavy enough to begin to show errors in calculations. The new mass, according to the theory of relativity, is

$$m = \frac{m_0}{\sqrt{1 - \dfrac{V^2}{c^2}}} \qquad (3\text{-}20)$$

where

m = relativistic mass at velocity V
m_0 = rest mass (9.11×10^{-28} g)
V = electron velocity (cm/s)
c = speed of light (3×10^{10} cm/s)

At potentials used in some high-power microwave devices, the final velocity of electrons will reach significant relativistic levels. The potentials at which these effects become important are surprisingly low, as you will find if you carry out the exercises at the end of the chapter. The general rule is that relativistic effects begin to be important at potentials above 5,000 V. Therefore, designers sometimes use relativistic mass in their deliberations.

3-3.3 Electron Motion in Magnetic Fields

An electron in motion constitutes an *electrical current,* so we may reasonably conclude that it will react when it encounters a magnetic field. Consider Fig. 3-3,

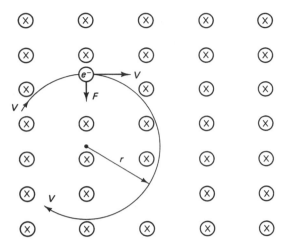

$(\times)$ = H-FIELD AWAY FROM
OBSERVER AND IS UNIFORM

Figure 3-3 Operation of a magnetic field on an electron.

where an electron is injected into a magnetic field of H oersteds. In this figure, the arrows pointing into the page indicate that the magnetic lines of force are *away* from the observer. The electron is deflected into a circular path of radius r by the magnetic field. The electron does not alter tangential velocity, kinetic energy, or magnitude of momentum.

The deflecting force (F) is derived from the force exerted on a current-carrying conductor by a magnetic field:

$$F = \frac{HVe}{10} \quad \text{dynes} \qquad (3\text{-}21)$$

The *angular velocity* (ω) of the electron is given by V/r, while its acceleration (a) is $V \times \omega$. The acceleration is directed inward toward the center of the circle along the same vector line as force F. We know from Newton's laws that $F = ma$, so we may conclude that $F = ma = mV\omega$. Therefore,

$$mV\omega = \frac{HVe}{10} \qquad (3\text{-}22)$$

or

$$\omega = \frac{He}{10m} \qquad (3\text{-}23)$$

But, because e and m are constants,

$$\omega = \frac{H(1.6 \times 10^{-19}\,C)}{10(9.11 \times 10^{-28}\,g)} \tag{3-24}$$

$$= 1.76 \times 10^7\,H \text{ rad/s} \tag{3-25}$$

An implication of Eq. (3-25) is that angular velocity ω is invariant under changes in velocity V. This fact becomes important when we study certain vacuum tube microwave power generators.

The path radius (r) is given by $r = V/\omega$. Thus, by substituting the expressions for V and ω in Eqs. (3-15) and (3-25), respectively, we arrive at

$$r = \frac{V}{\omega} \tag{3-26}$$

$$= \frac{5.93 \times 10^7 \sqrt{E}}{1.76 \times 10^7 H} \text{ cm} \tag{3-27}$$

$$= \frac{3.37\sqrt{E}}{H} \text{ cm} \tag{3-28}$$

3-3.4 Electron Motion in Simultaneous *E* and *H* Fields

Thus far we have studied the effects of both electric and magnetic fields on the trajectory of a moving electron. Many electronic devices can be understood with only this information. But some devices apply both electric and magnetic fields simultaneously. We must, therefore, examine the path of an electron in simultaneous E/H fields. Two cases are considered: *parallel fields* and *crossed fields*. It is not surprising that two different classes of microwave vacuum tubes depend on these cases.

3-3.4.1 Motion in Parallel Fields

Consider the case where two fields (E and H) are parallel to each other in the same space, as in Fig. 3-4A. The trajectory of the electron in this field depends on the velocity vector at the site of injection. If this vector is parallel to the E and H fields (for example, V1 in Fig. 3-4A), then the electron possesses no tangential velocity and

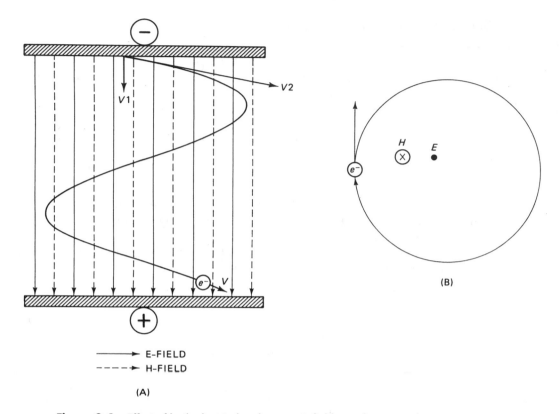

E-FIELD
H-FIELD

(A)

(B)

Figure 3-4   Effect of both electrical and magnetic fields on electron path.

the magnetic field therefore cannot act on it. But if the injection velocity vector is not parallel to the fields (crossed fields; for example, V2), then there is a tangential component to velocity and the electron interacts with the H field. The path will be helical, as shown in side aspect in Fig. 3-4A and end aspect in Fig. 3-4B.

3-4   QUANTUM MECHANICAL DESCRIPTION OF ELECTRONS

The "billiard ball" (or naive) description of electrons works well in some areas of electronics technology, but falls down utterly in other areas. For example, the naive description works well in describing vacuum tube devices at low frequencies and also in some transistor devices. However, no reasonable explanation of the tunneling phenomenon found in the Esaki (tunnel) diode can be derived from the naive description. For some devices and phenomena, therefore, we must appeal to the *quantum mechanical description* of matter (which includes electrons).

The quantum mechanics (QM) revolution, which brought us the many benefits of electronics technology (especially in the semiconductor world), began on December 14, 1900, at a meeting of the German Physical Society in Berlin. Max Planck presented a paper that night that revolutionized the world of physics. For decades, scientists had been working on the problem of black-body radiation, that is, the ra-

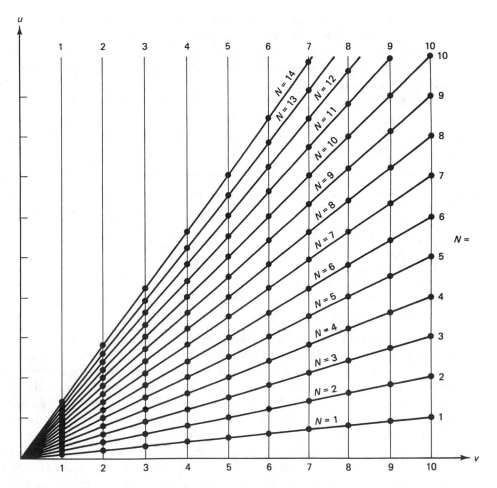

Figure 3-5 Permitted values of electron energy.

diation given off by a black body (such as a lump of iron) when it is heated. The theories that scientists believed at that time did not fit the experimental data. But Planck demonstrated that the equations worked well if he allowed only certain discrete amounts of energy. These discrete energy levels are defined by

$$U = nh \qquad (3\text{-}29)$$

where

U = energy in ergs
h = Planck's constant (6.63×10^{-27} erg-s)
n = an integer (1, 2, 3 . . .)

In 1905, Albert Einstein demonstrated that Planck's theory applied to the photoelectric effect. In this phenomenon, the energy of electrons emitted when special photosensitive metallic plates are exposed to light is not a function of light intensity (as expected), but rather of light frequency. This discovery opened a disquieting question for scientists who had (since Maxwell) viewed light as a wave: the photoelectric phenomenon made light look like a particle. It is now believed that light has a dual nature: it behaves as a particle in some experiments and as a wave in others. The two properties are complementary, not contradictory. In the 1920s, Danish physicist Neils Bohr formalized this proposition in his *complementarity principle*. The relationship between the energy of the light photon and the frequency (color) of the light is

$$U = nh\nu \qquad\qquad (3\text{-}30)$$

where

U = energy in ergs
h = Planck's constant (6.63×10^{-27} erg-s)
ν = frequency of the light in hertz (Hz)
n = an integer ($1, 2, 3 \ldots$)

It had been noted in the nineteenth century that certain materials would emit light when excited above the ground state. Applying a certain amount of energy to a system would cause the emission of specific colors of light as the system returned to equilibrium. Several sets of light spectra were noted, most notably the Rydberg series, the Lyman series, the Paschen series, and the Balmer series. Bohr explained the several series by applying Planck's theory of quantized energy to the electrons orbiting the nucleus in an atom. When an electron at rest receives energy from an external source, its energy level (hence orbital radius, r) increases only in certain allowable discrete levels defined by Planck's constant ($U2$, $U3$, and so on, in Fig. 3-6). The energetic state is unstable, and the electron soon returns to ground state. Because of conservation of energy, however, that energy must go somewhere, so it is emitted as a photon of light with a frequency according to Eq. (3-30).

The intuitive expectation was that an electron would fall from the excited state back to its initial state in one movement. This motion would cause a single color of light to be emitted equal to $U_{6,3}$ (in the case of Fig. 3-6), that is, the difference of energy levels U_6 and U_3. But in the experiments there was more than one color in the spectra, indicating that a different scenario was taking place. Bohr explained the existence of multiple colors by postulating that the electrons dropped back to the original state in more than one step. In Fig. 3-6, there are two steps ($U_{6,4}$ and $U_{4,3}$), that is, from a conservation point of view, functionally equivalent to $U_{6,3}$, which allowed for the existence of the multiple color emissions.

The phenomenon explained by Bohr is the basis of a number of devices today, including maser (microwave amplification by stimulated emission of radiation) and laser devices. It is also the operant phenomenon in such mundane devices as neon glow lamps, fluorescent lamps, and certain semiconductor devices that we will explore further in later chapters. In the laser the simultaneous emission of energy from an extremely large number of electrons is coordinated by an external energy source (such as a xenon flash tube) so that the emissions occur in

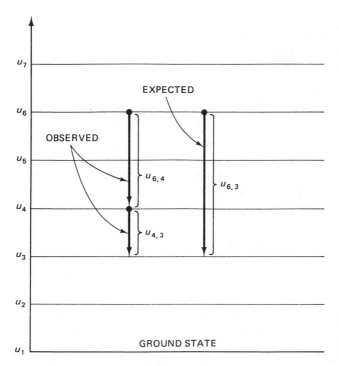

Figure 3-6 Observed versus expected electron energy levels in excited electrons.

phase with each other, leading to the phenomenon of *coherent light*, for which the laser is famous.

The Bohr model of the atom assigned electrons to orbits around the nucleus at specific quantized distances prescribed by their respective energies. Bohr's solar system model of the atom is still accepted today for naive (but useful) explanations, even though modern physicists know that the real situation is somewhat more complex. During the 1920s, quantum mechanics became more formalized into the theory that is essentially used today (with modifications).

In 1924, Louis de Broglie proposed that not only light, but matter also has a dualistic, complementary wave-particle nature. The wave function of matter is usually symbolized by ψ. According to de Broglie, the wavelength of a particle such as an electron is given by

$$\lambda = \frac{h}{mv} \qquad (3\text{-}31)$$

or

$$\lambda = \frac{h}{\sqrt{2Um}} \qquad (3\text{-}32)$$

λ = wavelength
h = Planck's constant
U = energy of the particle
m = mass of the particle
v = velocity of the particle

Combining the Bohr and de Broglie theories produced a model in which the simple orbital electron actually became a standing wave in which an integer number of de Broglie waves fit into the space allocated by the Bohr theory for an electron of a given energy level (Fig. 3-7). According to de Broglie's theorem, we can deduce the following relationship in which the circumferences of the orbits are integer multiples of the de Broglie wavelength:

$$2\pi r = \frac{nh}{mv} \tag{3-33}$$

Erwin Schrödinger disputed Bohr's solar system model of the atom. Instead of a billiard ball nucleus surrounded by billiard ball electrons, Schrödinger proposed an entirely new model based on a probabilistic interpretation of the wave function. Like light waves, elementary subatomic particles sometimes behave like particles and at other times like waves. Again we have a complementary system, but in this case the waves are the "matter waves" postulated by de Broglie. According to Schrödinger's view, the atom consists of a matter wave nucleus surrounded by matter wave electrons. Schrödinger's wave equation describes the matter waves (ψ) in terms of probability (ψ^2). It is important to realize that matter wave equations do not describe a real chain of events the way water wave equations describe real movement by real water particles. The equations describe only the *probabilities* of finding a real particle at a specific place at a given time.

To these factors we must now add another facet: the *uncertainty principle*. In 1927, Werner Heisenberg proposed the uncertainty principle, which holds that certain pairs of properties of atomic particles cannot both be measured with accuracy

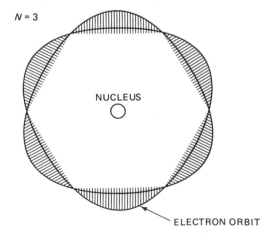

Figure 3-7 De Broglie waves in electron orbit.

simultaneously. For example, it is impossible to precisely measure both the position and the momentum of an electron. Momentum is the product of mass and velocity, so it follows that we cannot know both where the electron is located and how fast it is traveling. Stated mathematically, Heisenberg's uncertainty principle is

$$\Delta P_x \, \Delta X \geq h \qquad (3\text{-}34)$$

Do not confound the uncertainty principle with the mere inability to measure some parameters due to some kind of disturbance effect. Many physical measurements are inaccurate because the act of measurement (or the nature of the instruments) disturbs the system and thereby changes the value of the measurement enough to introduce very large errors. For example, a low-impedance voltmeter disturbs a high-impedance circuit enough to introduce serious errors. What the QM scientist is telling us, however, is that the electron *actually does not possess* both a precise location and a precise momentum. Truly astounding!

The probability interpretation of Schrodinger leads to some disquieting problems. Consider the tunnel diode, for example. The usual tunnel diode explanation goes something like this (see Fig. 3-8). In an ordinary *PN* junction, the transition region between *N*-type and *P*-type semiconductor material consists of a dipole layer created by relatively immobile, unneutralized electrons and holes. This dipole layer has an associated electric field of up to 10 kV/cm in an unbiased junction, and even more in a reverse-biased junction. Although the band gap energy is the

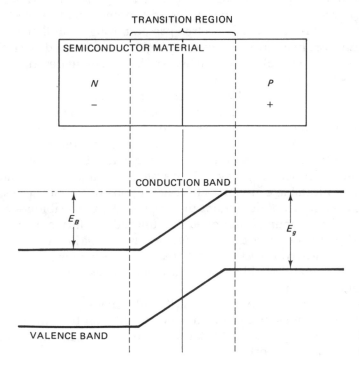

Figure 3-8 Characterization of semiconductor diode.

same on both sides of the junction, a difference in the potential energy on the two sides results in an *electrical potential barrier* (E_B) energy. The potential barrier forms a blockade, an electrical brick wall, to electrons trying to pass over the junction. Unless an electron has sufficient energy, it cannot leap over the potential barrier.

Now consider the Esaki tunnel diode. The transition region between *P*-type and *N*-type semiconductor material in a tunnel diode is on the order of 100 angstroms, rather than the 10,000 angstroms that is typical in ordinary *PN* junction diodes. In addition, very high doping levels result in much greater charge carrier concentrations. These two conditions taken together alter the situation so that certain quantum effects can take place. As an electron approaches the junction, there is a small, but finite, probability that it will somehow pass through the junction even though it lacks sufficient energy to pass over the top of the "brick wall." Furthermore, the electron has the same energy on the other side of the barrier as it had to begin with.

Let's consider a macro-world analogy to the tunnel diode to see what the scientists are actually claiming. Suppose there is a player on a racquet ball court. The particle is not an electron, but rather it is the racquet ball. The ball receives its energy when it is smacked by the player. Assume that the walls of the court are so high that the ball cannot fly high enough to get over them no matter how hard the player whacks it. According to the standard wisdom, that ball should spend eternity bouncing around the one racquet ball court.

But now suppose that the court is a special quantum court. According to QM, there is a small but finite chance that the ball will approach the wall, seem to pass right through it, and strike your neighbor on the head. Furthermore, the ball will have the same energy on both sides of the wall. This tunneling phenomenon does not really exist in the way we normally learn it. What QM is really saying is that an electron *disappears* on one side of the barrier, and another identical electron *appears* on the other side. It is not that the electron passed through, but that something like magic happened. In electronics, we tend to glibly toss around electrons as if they were micro ping-pong balls. While it "works" in our circuit descriptions, it is not what the physicists are really telling us.

During the first three decades of the twentieth century, Bohr and a relatively small band of scientists working in Copenhagen devised a system that seems to explain atomic phenomena. Physicists developed a world of weirdness where matter (including electrons), the hard stuff of reality, has a complementary wave-particle nature. Instead of a mass of tiny ping-pong balls, electrons and protons are a ghostly dance of probability waves. That system is quantum mechanics, and it displaced the comfortable cause-and-effect definitions of classical physics and put in their place a new set of definitions that are based on probabilities and "tendencies to exist." The seeming paradox is that unpredictability and uncertainty appear intrinsic to the universe at the deepest levels. Yet QM is accepted by scientists as the mathematical construct that best predicts the behavior of matter at the subatomic level. Even though problems with the theory persist, QM became for scientists the operant paradigm.

Part of the problem faced by the scientists is the utter inadequacy of human language to express the realities of the quantum world. It's not that something mystical happens, as asserted by certain popular writers on science, in phenomena such as tunneling: the equations work.

3-4.1 Selections for Further Reading in Quantum Mechanics

Gamow, George. *Thirty Years That Shook Physics: The Story of Quantum Mechanics*. New York: Dover Publications, Inc., 1966.

Gribbon, John. *In Search of Schrödinger's Cat: Quantum Physics and Reality*. New York: Bantam Books, 1984.

Herbert, Nick. *Quantum Reality: Beyond the New Physics*. Garden City, NY: Doubleday, Inc., 1987.

Pagels, Heinz. *The Cosmic Code*. New York: Simon & Schuster, 1982; Bantam Books paperback edition, 1983.

3-5 SUMMARY

1. In microwave electronics technology, both the classical "billiard ball" and quantum descriptions of electrons must be used.

2. Electron trajectory in E and H fields deflects according to the forces on the particle. The exact forces are a function of the direction and strengths of the respective fields.

3. At potentials above about 5,000 V, electron velocity becomes a larger fraction of the speed of light, so relativistic mass must be used in calculations.

4. Quantum mechanics is a branch of physics that recognizes that energy levels of entities such as photons of light and electrons must be quantized, that is, allowed only certain discrete values.

5. The complementarity principle holds that light photons, electrons, and the like have a dual wave-particle nature. Louis de Broglie postulated that electrons behave in a wavelike manner.

6. According to Schrödinger's wave equation, only a certain probability of existence can be ascribed to an electron until it is actually measured.

7. Heisenberg's uncertainty principle holds that it is not possible to simultaneously know both the position and the momentum of an electron. To the extent that one is known, the other will become unknown.

3-6 RECAPITULATION

Now return to the objectives and prequiz questions at the beginning of the chapter and see how well you can answer them. If you cannot answer certain questions, place a check mark by each and review the appropriate parts of the text. Next, try to answer the following questions and work the problems using the same procedure.

EXERCISES FOR THE STUDENT

1. Use Eq. (3-15) to calculate the velocity of an electron at potentials between 1 and 270 kV. Create a graph of these velocities using c as a hard limit above which no velocity is permitted. Use at least ten data points in the graph and calculations.
2. Calculate and graph the relativistic mass of an electron at velocities from $0.1c$ to $1.0c$. Use at least ten data points.
3. Calculate and graph the quantity m/m_o versus velocity and electrical potential (two scales are needed) for velocities of $0.1c$ to $1.0c$.

QUESTIONS AND PROBLEMS

1. An electron is injected into an electrostatic field of 230 V. If the initial velocity is zero, what is the final velocity?
2. An electron is injected into an electrostatic field of 2000 V. If the initial velocity is zero, what is the final velocity?
3. Calculate the kinetic energies of the electrons in problems 1 and 2.
4. An electron enters an orthogonal electric field with a velocity of 2.3×10^6 cm/s. Calculate the resultant velocity if it is found that the electron deflected 24°.
5. An electron accelerates to a velocity of $0.4c$. Calculate its relativistic mass.
6. What is the Planck energy (in ergs) of a microwave signal of 100 Ghz?
7. Calculate the de Broglie wavelength of an electron that has a velocity of 6.55×10^6 m/s.

KEY EQUATIONS

1. $c = \lambda F \sqrt{\varepsilon}$

2. $\lambda = \dfrac{c}{F\sqrt{\varepsilon}}$

3. $F = \dfrac{c}{\lambda\sqrt{\varepsilon}}$

4. $T = \dfrac{1}{F}$

5. $\delta = \sqrt{\dfrac{1}{2\pi F \sigma \mu}}$

CHAPTER 4

Microwave Transmission Lines

OBJECTIVES

1. Be able to define transmission line parameters.
2. Understand the response of transmission lines to step functions and ac steady-state signals.
3. Understand the operation of transmission lines.
4. Understand applications of transmission lines.

4-1   PREQUIZ

These questions test your prior knowledge of the material in this chapter. Try answering them before you read the chapter. Look for the answers (especially those you answered incorrectly) as you read the text. After you have finished studying the chapter, try answering these questions again and those at the end of the chapter.

1. Calculate the VSWR if an 80-Ω load is connected to a 50-Ω transmission line.
2. A 50-Ω transmission line as a velocity factor of 0.66; find the physical length of a section that must be $3\lambda/2$ at 1.8 GHz.
3. Calculate the VSWR if the coefficient of reflection (P) is 0.30.
4. Calculate the characteristic impedance of a transmission line if the capacitance per unit length is 12 pF and the inductance is 0.01 μH.

4-2   INTRODUCTION

Transmission lines and *waveguides* (Chapter 5) are conduits for transporting RF signals between elements of a system. For example, transmission lines are used between an exciter output and transmitter input, and between the transmitter output and the antenna. Although often erroneously characterized as a "length of shielded wire," transmission lines are actually complex networks containing the equivalent of all the three basic electrical components: resistance, capacitance and inductance. Because of this fact, transmission lines must be analyzed in terms of an *RLC* network.

In this chapter we will consider both waveguides and transmission lines, beginning with the latter. Both step function and sine-wave ac responses will be studied. Because the subject is both conceptual and analytical, we will use both analogous and mathematical approaches to the theory of transmission lines.

Figure 4-1 shows several basic types of transmission line. Perhaps the oldest and simplest form is the *parallel line* shown in Fig. 4-1A. This type of transmission line consists of two identical conductors parallel to each other and separated by a dielectric (that is, an insulator). A familiar example of a parallel transmission line from common experience is the twin lead used on many television broadcast receiver antennas. For years the microwave application of parallel lines was limited to educational laboratories, where they are well suited to performing experiments (to about 2 GHz) with simple, low-cost instruments. Today, however, printed circuit and hybrid semiconductor packaging has given parallel lines a new lease on life, if not an overwhelming market presence.

The second form of transmission line, which finds considerable application at microwave frequencies, is *coaxial cable* (Fig. 4-1B through 4-1E). This form of line consists of two cylindrical conductors sharing the same axis (hence "co-axial") and separated by a dielectric. For low frequencies (in flexible cables), the dielectric may be polyethylene or polyethylene foam, but at higher frequencies, Teflon® and other materials are used. Dry air and dry nitrogen are also used in some applications.

Several forms of coaxial line are available. Flexible coaxial cable is perhaps the most common form. The outer conductor in such cable is made of either braid or foil. Again, television broadcast receiver antennas provide an example of such cable from common experience. Another form of flexible or semiflexible coaxial line is *helical line*, in which the outer conductor is spirally wound. *Hardline* is coaxial cable that uses a thin-wall pipe as the outer conductor. Some hardline coax used at microwave frequencies uses a rigid outer conductor and a solid dielectric.

Gas-filled line is a special case of hardline that is hollow (Fig. 4-1C), the center conductor being supported by a series of thin ceramic or Teflon® insulators. The dielectric is either anhydrous (that is, dry) nitrogen or some other inert gas.

Some flexible microwave coaxial cable uses a solid air-articulated dielectric (Fig. 4-1D), in which the inner insulator is not continuous around the center conductor, but rather is ridged. Reduced dielectric losses increase the usefulness of the cable at higher frequencies. Double-shielded coaxial cable (Fig. 4-1G) provides an extra measure of protection against radiation from the line and EMI from outside sources from getting into the system.

A variant that seems to combine the advantages of both parallel and coaxial concepts is shown in Fig. 4-1C. This form of transmission line is called *shielded parallel line*. As in the parallel line, the two conductors are spaced a certain distance (*S*) apart and are parallel to each other. In the shielded variety, however, an outer conductor (a shield) is also provided.

Stripline, also called *microstripline* (Fig. 4-1D), is a form of transmission line

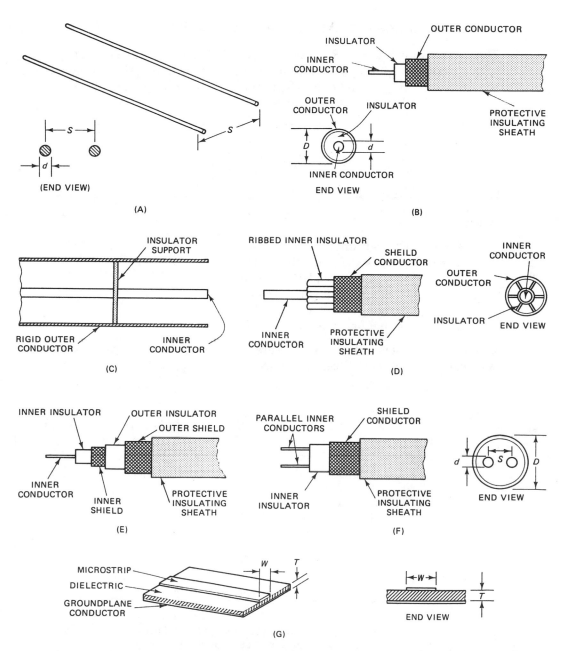

Figure 4-1   Types of transmission line: A) parallel line; B) standard coaxial cable; C) gas-filled rigid coaxial line; D) articulated coaxial line; E) double-shielded coaxial cable; F) shielded parallel conductor line; G) strip line.

used at high UHF and microwave frequencies. The stripline consists of a critically sized conductor over a ground plane conductor and separated from it by a dielectric. Some striplines are sandwiched between two groundplanes, and are separated from each by a dielectric.

4-2.1 Transmission Line Characteristic Impedance (Z_o)

The transmission line is an *RLC* network (see Fig. 4-2), so has a *characteristic impedance*, Z_o, also sometimes called *surge impedance*. Network analysis will show that Z_o is a function of the per unit of length parameters *resistance* (*R*), *conductance* (*G*), *inductance* (*L*), and *capacitance* (*C*), and is found from

$$Z_o = \sqrt{\frac{R + j\omega L}{G + j\omega C}} \qquad (4\text{-}1)$$

where

    Z_o = characteristic impedance in ohms
    R = resistance per unit length in ohms
    G = conductance per unit length in mhos
    L = inductance per unit length in henrys
    C = capacitance per unit length in farads
    ω = angular frequency in radians per second ($2\pi F$)

In microwave systems, the resistances are typically very low compared with the reactances, so Equation (4-1) can be reduced to the simplified form

$$Z_o = \sqrt{\frac{L}{C}} \qquad (4\text{-}2)$$

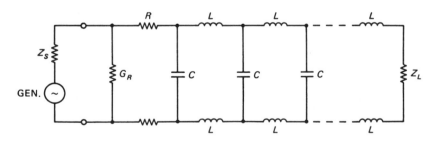

    G = CONDUCTANCE PER UNIT OF LENGTH, IN MHOS
    R = RESISTANCE PER UNIT LENGTH, IN OHMS
    L = INDUCTANCE PER UNIT LENGTH, IN HENRYS
    C = CAPACITANCE PER UNIT LENGTH, IN FARADS
    Z_L = LOAD IMPEDANCE IN OHMS
    Z_S = GENERATOR SOURCE IMPEDANCE, IN OHMS
    Z_O = TRANSMISSION LINE CHARACTERISTIC IMPEDANCE, IN OHMS

If $X \gg R$, THEN: $Z_O = \sqrt{\frac{L}{C}}$

Figure 4-2 Equivalent circuit for a transmission line.

EXAMPLE 4-1

A nearly lossless transmission line (R very small) has a unit length inductance of 3.75 nH and a unit length capacitance of 1.5 pF. Find the characteristic impedance, Z_o.

Solution

$$Z_o = \sqrt{\frac{L}{C}}$$

$$= \sqrt{\frac{3.75\text{nH} \times (1\text{H}/10^9 \text{ nH})}{1.5 \text{ pF} \times (1 \text{ F}/10^{12} \text{ pF})}}$$

$$= \sqrt{2.5 \times 10^3} = 50 \, \Omega$$

The characteristic impedance for a specific type of line is a function of the conductor size, the conductor spacing, the conductor geometry (see again Fig. 4-1), and the dielectric constant of the insulating material used between the conductors. The dielectric constant (ε) is equal to the reciprocal of the velocity (squared) of the wave when a specific medium is used:

$$\varepsilon = \frac{1}{v^2} \qquad\qquad (4\text{-}3)$$

where

  ε = dielectric constant
  v = velocity of the wave in the medium

For a perfect vacuum, $\varepsilon = 1.000$.
  Parallel Line

$$Z_o = \frac{276}{\sqrt{\varepsilon}} \log\left(\frac{2S}{d}\right) \qquad\qquad (4\text{-}4)$$

where

  Z_o = characteristic impedance in ohms
  ε = dielectric constant
  S = center-to-center spacing of the conductors
  d = diameter of the conductors

Coaxial Line

$$Z_o = \frac{138}{\sqrt{\varepsilon}} \log\left(\frac{D}{d}\right) \qquad\qquad (4\text{-}5)$$

where

D = diameter of the outer conductor
d = diameter of the inner conductor

Shielded Parallel Line

$$Z_o = \frac{276}{\sqrt{\varepsilon}} \log\left(2A\,\frac{1-B^2}{1+B^2}\right) \tag{4-6}$$

where

$A = s/d$
$B = s/D$

Stripline

$$Z_o = \frac{377}{\sqrt{\varepsilon}}\,\frac{T}{W} \tag{4-7A}$$

where

ε_t = relative dielectric constant of the printed wiring board (PWB)
T = thickness of the printed wiring board
W = width of the stripline conductor

The relative dielectric constant (ε_t) used here differs from the normal dielectric constant of the material used in the PWB. The relative and normal dielectric constants move closer together for larger values of the ratio W/T.

EXAMPLE 4-2

A stripline transmission line is built on a 4-mm-thick printed wiring board that has a relative dielectric constant of 5.5. Calculate the characteristic impedance if the width of the strip is 2 mm.

Solution

$$Z_o = \frac{377}{\sqrt{\varepsilon_t}}\,\frac{T}{W}$$

$$= \frac{377}{\sqrt{5.5}}\,\frac{4mm}{2mm}$$

$$= \frac{377}{2.35}\,(2) = 321\,\Omega$$

In practical situations, we usually do not need to calculate the characteristic impedance of a stripline, but rather design the line to fit a specific system impedance (for example, 50 ohms). We can make some choices of printed circuit material (hence dielectric constant) and thickness, but even these are usually limited in practice by the availability of standardized boards. Thus, stripline *width* is the variable parameter. Equation (4-7A) can be rearranged to the form

$$W = \frac{377T}{Z_o\sqrt{\varepsilon}} \qquad (4\text{-}7\text{B})$$

The 50-Ω impedance is accepted as standard for RF systems, except in the cable TV industry. The reason is that power-handling ability and low-loss operation do not occur at the same characteristic impedance. For coaxial cables, for example, the maximum power-handling ability occurs at 30 Ω, while the lowest loss occurs at 77 Ω; 50 Ω is therefore a reasonable trade-off between the two points. In the cable TV industry, however, the RF power levels are minuscule, but lines are long. The trade-off for TV is to use 75 Ω as the standard system impedance to take advantage of the reduced attenuation factor.

4-2.2 Velocity Factor

In the discussion preceding this section, we discovered that the velocity of the wave or signal in the transmission line is less than the free-space velocity, that is, less than the speed of light. Further, we discovered in Eq. (4-3) that the velocity is related to the dielectric constant of the insulating material that separates the conductors in the transmission line. Velocity factor (v) is usually specified as a decimal fraction of c, the speed of light (3×10^8 m/s). For example, if the velocity factor of a transmission line is rated at 0.66, then the velocity of the wave is 0.66c, or (0.66)(3 $\times 10^8$ m/s) = 1.98×10^8 m/s.

Velocity factor becomes important when designing things like transmission line transformers or any other device in which the length of the line is important. In most cases, the transmission line length is specified in terms of *electrical length*, which can be either an angular measurement (for example, 180° or π radians) or a relative measure keyed to wavelength (for example, one half-wavelength, which is the same as 180°). The *physical length* of the line is longer than the equivalent electrical length. For example, let's consider a 1-GHz half-wavelength transmission line.

A rule of thumb tells us that the length of a wave (in meters) in free space is 0.30/F, where frequency (F) is expressed in gigahertz; therefore, a half-wavelength line is 0.15/F. At 1 GHz, the line must be 0.15 m/1 GHz, or 0.15 m. If the velocity factor is 0.80, then the *physical length* of the transmission line that will achieve the desired *electrical length* is [(0.15 m)(v)]/F = [(0.15 m)(0.80)]/1 GHz = 0.12 m. The derivation of the rule of thumb is left as an exercise for the student. (*Hint:* It comes from the relationship between wavelength, frequency, and velocity of propagation for any form of wave.)

Certain practical considerations regarding velocity factor result from the fact that the physical and electrical lengths are not equal. For example, in a certain type of phased array antenna design radiating elements are spaced half-wavelengths apart and must be fed 180° (half-wave) out of phase with each other. The simplest

TABLE 4-1 TRANSMISSION LINE CHARACTERISTICS

TYPE OF LINE	Z_o (Ωs)	VEL.FACTOR (v)
1/2-in. TV parallel	300	0.95 line (air dielectric)
1-in. TV parallel	450	0.95 line (air dielectric)
TV twin-lead	300	0.82
UHF TV twin-lead	300	0.80
Polyethylene coaxial	*	0.66 cable
Polyethylene foam	*	0.79 coaxial cable
Air-space polyethylene	*	0.86 foam coaxial cable
Teflon®	*	0.70

* Various impedances depending on cable type.

interconnect is to use a half-wave transmission line between the 0° element and the 180° element. According to the standard wisdom, the transmission line will create the 180° phase delay required for the correct operation of the antenna. Unfortunately, because of the velocity factor, the physical length for a one-half electrical wavelength cable is shorter than the free space half-wave distance between elements. In other words, the cable will be too short to reach between radiating elements by the amount of the velocity factor!

Clearly, velocity factor is a topic that must be understood before transmission lines can be used in practical situations. Table 4-1 shows the velocity factors for several types of popular transmission line.

4-2.3 Transmission Line Noise

Transmission lines are capable of generating noise and spurious voltages that are interpreted by the system as valid signals. Several such sources exist. One source is coupling between noise currents flowing in the outer and inner conductors. Such currents are induced by nearby electromagnetic interference and other sources (for example, connection to a noisy ground plane). Although coaxial design reduces noise pickup compared with parallel line, the potential for EMI exists. Selection of high-grade line, with a high degree of shielding, reduces the problem.

Another source of noise is thermal noises in the resistances and conductances. This type of noise is proportional to resistance and temperature.

Noise is also created by mechanical movement of the cable. One species results from movement of the dielectric against the two conductors. This form of noise is caused by electrostatic discharges in much the same manner as the spark created by rubbing a piece of plastic against woolen cloth.

A second species of mechanically generated noise is piezoelectricity in the dielectric. Although more common in cheap cables, one should be aware of it. Mechanical deformation of the dielectric causes electrical potentials to be generated.

Both species of mechanically generated noise can be reduced or eliminated by proper mounting of the cable. Although rarely a problem at lower frequencies, such noise can be significant at microwave frequencies when signals are low.

4-2.4 Coaxial Cable Capacitance

A coaxial transmission line possesses a certain capacitance per unit of length. This capacitance is defined by

$$C = \frac{24\varepsilon}{\log \; D/d} \; \frac{\text{pF}}{\text{meter}} \qquad (4\text{-}8\text{A})$$

A long run of coaxial cable can build up a large capacitance. For example, a common type of coax is rated at 65 pF/m. A 150-m roll thus has a capacitance of (65 pF/m)(150 m), or 9750 pF. When charged with a high voltage, as is done in performing breakdown voltage tests at the factory, the *cable acts like a charged high-voltage capacitor*. Although rarely if ever lethal to humans, the stored voltage in new cable can deliver a nasty electrical shock and can irreparably damage electronic components.

4-2.5 Coaxial Cable Cutoff Frequency (F_c)

The normal mode in which a coaxial cable propagates a signal is as a transverse electromagnetic (TEM) wave, but others are possible, and usually undesirable. There is a maximum frequency above which TEM propagation becomes a problem and higher modes dominate. Coaxial cable should *not* be used above a frequency of

$$F = \frac{6.75}{(D + d)\sqrt{\varepsilon}} \; GHz \qquad (4\text{-}8\text{B})$$

where

$\quad$ F = TEM mode cutoff frequency
$\quad$ D = diameter of the outer conductor in inches
$\quad$ d = diameter of the inner conductor in inches
$\quad$ ε = dielectric constant

When maximum operating frequencies for cable are listed, it is the TEM mode that is cited. Beware of attenuation, however, when making selections for microwave frequencies. A particular cable may have a sufficiently high TEM-mode frequency, but still exhibit a high attenuation per unit length at X or Ku bands.

4-2.6 Transmission Line Responses

To understand the operation of transmission lines, we need to consider two cases: *step-function response* and the *steady-state ac response*. The step function case involves a single event when a voltage at the input of the line snaps from zero (or a steady value) to a new (or nonzero) value and remains there until all action dies out. This response tells us something of the behavior of pulses in the line, and in fact is used to describe the response to a single pulse stimulus. The steady-state ac response tells us something of the behavior of the line under stimulation by a sinusoidal RF signal.

4-3   STEP-FUNCTION RESPONSE OF A TRANSMISSION LINE

Figure 4-3 shows a parallel transmission line with characteristic impedance (Z_o) connected to a load impedance (Z_L). The generator at the input of the line consists of a voltage source (V) in series with a source impedance (Z_s) and a switch ($S1$). Assume for the present that all impedances are pure resistances (that is, $R + j0$). Also, assume that $Z_s = Z_o$.

When the switch is closed at time T_o (Fig. 4-4A), the voltage at the input of the line (V_{in}) jumps to $V/2$. When we discussed Fig. 4-2, you may have noticed that the LC circuit resembles a delay line circuit. As might be expected, therefore, the voltage wavefront propagates along the line at a velocity (v) of

$$v = \frac{1}{\sqrt{LC}} \tag{4-9}$$

where

> v = velocity in meters per second
> L = inductance in henrys
> C = capacitance in farads

At time $T1$ (Fig. 4-4B), the wavefront has propagated one-half the distance L, and by T_d it has propagated the entire length of the cable.

If the load is perfectly matched ($Z_L = Z_o$), then the load absorbs the wave and no component is reflected. But in a mismatched system (Z_L is not equal to Z_o), a portion of the wave is reflected back down the line toward the generator.

Figure 4-5 shows the rope analogy for reflected pulses in a transmission line. A taut rope (Fig. 4-5A) is tied to a rigid wall that does not absorb any of the energy in the pulse propagated down the rope. When the free end of the rope is given a vertical displacement (Fig. 4-5B), a wave is propagated down the rope at velocity

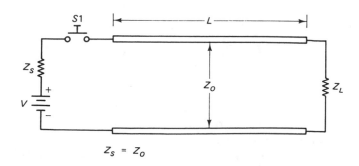

Figure 4-3   Model of a parallel transmission line.

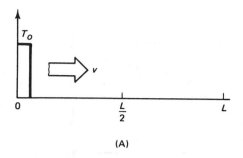

(A)

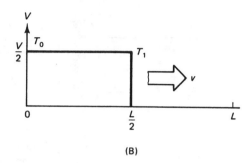

(B)

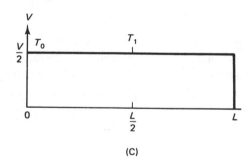

(C)

Figure 4-4 Propagation of an impulse along a transmission line.

v (Fig. 4-5C). When the pulse hits the wall (Fig. 4-5D), it is reflected (Fig. 4-5E) and propagates back down the rope toward the free end (Fig. 4-5F).

If a second pulse is propagated down the line before the first pulse dies out, there will be two pulses on the line at the same time (Fig. 4-6A). When the two pulses interfere, the resultant will be the algebraic sum of the two. If a pulse train is applied to the line, the interference pattern will set up *standing waves*, an example of which is shown in Fig. 4-6B.

4-3.1 Reflection Coefficient (*P*)

The *reflection coefficient* (*P*) of a circuit containing a transmission line and load impedance is a measure of how well the system is matched. The absolute value of the reflection coefficient varies from –1 to +1, depending on the magnitude of reflec-

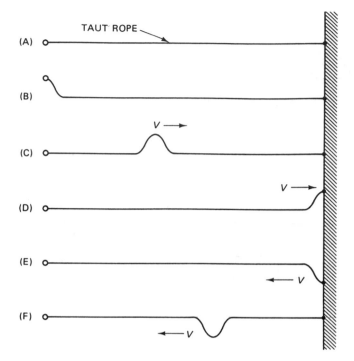

Figure 4-5   Rope analog for an impulse along a transmission line.

tion; $P = 0$ indicates a perfect match with no reflection, while -1 indicates a short-circuited load and $+1$ indicates an open circuit. To understand the reflection coefficient, let's start with a basic definition of the resistive load impedance $Z = R + j0$:

$$Z_L = \frac{V}{I} \qquad (4\text{-}10)$$

where

$\quad Z_L$ = load impedance $R + j0$
$\quad V$ = voltage across the load
$\quad I$ = current flowing in the load

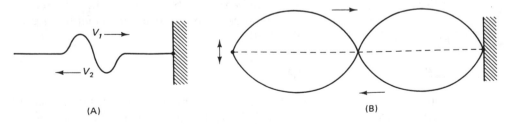

(A)   (B)

Figure 4-6   A) Combination of incident and reflected signals on a transmission line; B) standing waves occur when the impulse is replaced with a continuous signal.

Because there are both reflected and incident waves, we find that V and I are actually the sums of incident and reflected voltages and currents, respectively. Therefore,

$$Z_L = \frac{V}{I} = \frac{V_{inc} + V_{ref}}{I_{inc} + I_{ref}} \qquad (4\text{-}11)$$

where

V_{inc} = incident (that is, forward) voltage
V_{ref} = reflected voltage
I_{inc} = incident current
I_{ref} = reflected current

Because of Ohm's law, we may define the currents in terms of voltage, current, and the characteristic impedance of the line:

$$I_{Inc} = \frac{V_{ref}}{Z_o} \qquad (4\text{-}12)$$

and

$$I_{Inc} = \frac{-V_{ref}}{Z_o} \qquad (4\text{-}13)$$

The minus sign in Eq. (4-13) indicates that a direction reversal took place.

The two expressions for current, Eqs. (4-12) and (4-13), may be substituted into Eq. (4-11) to yield

$$Z_L = \frac{V_{inc} + V_{ref}}{(V_{inc}/Z_o) - (V_{ref}/Z_o)} \qquad (4\text{-}14)$$

The reflection coefficient (P) is defined as the ratio of reflected voltage to incident voltage (V_{ref}/V_{inc}), so by solving Eq. (4-14) for this ratio, we find

$$P = \frac{V_{ref}}{V_{inc}} \qquad (4\text{-}15)$$

and

$$P = \frac{Z_L - Z_o}{Z_L + Z_o} \qquad (4\text{-}16)$$

EXAMPLE 4-3

A 50-Ω transmission line is connected to a 30-Ω resistive load. Calculate the reflection coefficient, P.

   Solution

$$P = \frac{Z_L - Z_o}{Z_L + Z_o}$$

$$= \frac{50\,\Omega - 30\,\Omega}{50\,\Omega - 30\,\Omega} = \frac{20}{80} = 0.25$$

EXAMPLE 4-4

In Example 4-3, the incident voltage is 3 V rms. Calculate the reflected voltage.

   Solution

If

$$P = \frac{V_{ref}}{V_{inc}}$$

then

$$V_{ref} = PV_{inc}$$

$$= (0.25)(3\text{ V}) = 0.75\text{ V}$$

The *phase* of the reflected signal is determined by the relationship of load impedance and transmission line characteristic impedance. For resistive loads ($Z = R + j0$), if the ratio Z_L/Z_o is 1.0, then there is no reflection; if Z_L/Z_o is less than 1.0, then the reflected signal is 180° out of phase with the incident signal; if the ratio Z_L/Z_o is greater than 1.0, then the reflected signal is in phase with the incident signal. In summary:

RATIO	ANGLE OF REFLECTION
$Z_L/Z_o = 1$	No reflection
$Z_L/Z_o < 1$	180°
$Z_L/Z_o > 1$	0°

The step function (or pulse) response of the transmission line leads to a powerful means of analyzing the line and its load on an oscilloscope. Figure 4-7 shows the

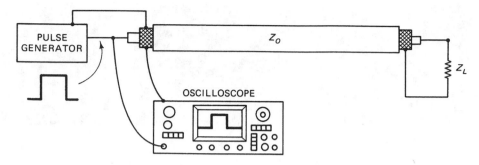

Figure 4-7   Equipment setup for time-domain reflectometry.

test setup for *time-domain reflectometry* (TDR) measurements. An oscilloscope and a pulse (or square wave) generator are connected in parallel across the input end of the transmission line. If a periodic waveform is supplied by the generator, the display on the oscilloscope will represent the sum of reflected and incident pulses. The duration of the pulse (that is, pulse width), or one-half the period of the square wave, is adjusted so that the returning reflected pulse arrives approximately in the center of the incident pulse.

Figure 4-8 shows a TDR display under several circumstances. Approximately 30 m of coaxial cable with a velocity factor of 0.66 was used in a test setup similar to Fig. 4-7. The pulse width was approximately 0.9 microseconds (us). The horizontal sweep time on the oscilloscope was adjusted to show only one pulse, which in this case represented one-half of a 550-kHz square wave (Fig. 4-8A).

The displayed trace in Fig. 4-8A shows the pattern when the load is matched to the line, or, in other words, $Z_L = Z_o$. A slight discontinuity exists on the high side of the pulse, and this represents a small reflected wave. Even though the load and line were supposedly matched, the connectors at the end of the line presented a slight impedance discontinuity that shows up on the oscilloscope as a reflected wave. In general, any discontinuity in the line, any damage to the line, any too-sharp bend or other anomaly causes a slight impedance variation, and hence a reflection.

Notice that the anomaly occurs approximately one-third of the 0.9-Ωs duration (or 0.3 Ωs) after the onset of the pulse. This fact tells us that the reflected wave arrives back at the source 0.3 Ωs after the incident wave leaves. Because this time period represents a round trip, we can conclude that the wave required 0.3 Ωs/2, or 0.15 Ωs to propagate the length of the line. Knowing that the velocity factor is 0.66 for this type of line, we can calculate its approximate length:

$$\text{Length} = c\,v\,T$$

$$= \frac{3 \times 10^8\,\text{m}}{\text{s}} \times (0.66) \times (1.5 \times 10^{-7}\,\text{s}) \qquad (4\text{-}17)$$

which agrees within experimental accuracy with the 30-m actual length prepared for the experiment ahead of time.

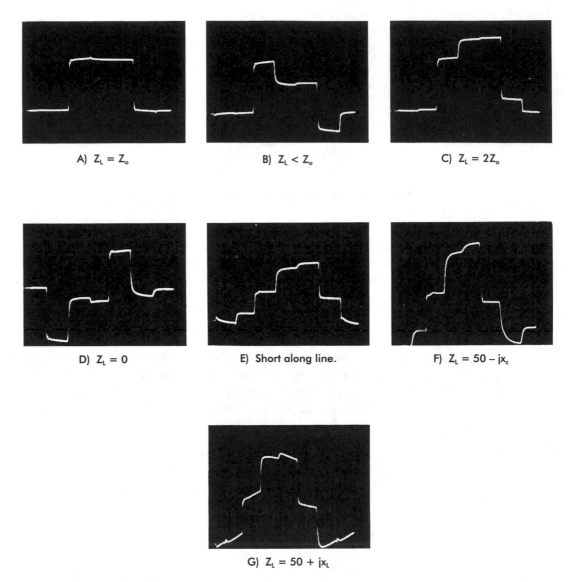

Figure 4-8   TDR waveforms expected for various loads.

Thus, the TDR setup (or a TDR instrument) can be used to measure the length of a transmission line. A general equation is

$$L_{meters} = \frac{cvT_d}{2}$$ (4-18)

where

L = length in meters
c = velocity of light (3×10^8 m/s)

v = velocity factor of the transmission line

T_d = round-trip time between the onset of the pulse and the first reflection

Figures 4-8B through G show the behavior of the transmission line with respect to the step function when the load impedance is mismatched to the transmission line (Z_L not equal to Z_o). In Figure 4-8B, we see what happens when the load impedance is less than the line impedance (in this case $0.5Z_o$). The reflected wave is inverted and sums with the incident wave along the top of the pulse. The reflection coefficient can be determined by examining the relative amplitudes of the two waves.

The opposite situation, in which Z_L is $2Z_o$, is shown in Fig. 4-8C. In this case, the reflected wave is in phase with the incident wave and so adds to the incident wave as shown. The cases for short-circuited load and open-circuited load are shown in Figs. 4-8D and E, respectively. The cases of reactive loads are shown in Figs. 4-8F and G. The waveform in Fig. 4-8F resulted from a capacitance in series with a 50-Ω (matched) resistance; the waveform in Fig. 4-8G resulted from a 50-Ω resistance in series with an inductance.

4-4   AC RESPONSE OF THE TRANSMISSION LINE

When a CW RF signal is applied to a transmission line, the excitation is sinusoidal (Fig. 4-9), so it becomes useful for us to investigate the steady-state ac response of the line. The term *steady state* implies a sine wave of constant amplitude, phase, and frequency. When ac is applied to the input of the line, it propagates along the line at a given velocity. The ac signal amplitude and phase will decay exponentially in the following manner:

$$V_R = Ve^{-\gamma l} \qquad\qquad (4\text{-}19)$$

where

V_R = voltage received at the far end of the line
V = applied voltage
l = length of the line
γ = *propagation constant* of the line

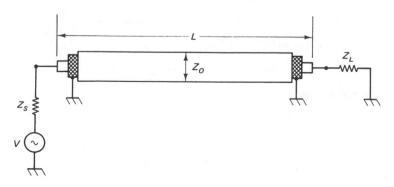

Figure 4-9 Circuit for showing ac response of a line.

The propagation constant (γ) is defined in various equivalent ways, each of which serves to illustrate its nature. For example, the propagation constant is proportional to the product of impedance and admittance characteristics of the line:

$$\gamma = \sqrt{ZY} \qquad (4\text{-}20)$$

or, since $Z = R + j\omega L$ and $Y = G + j\omega C$, we may write

$$\gamma = \sqrt{(R + j\omega L)(G + j\omega C)} \qquad (4\text{-}21)$$

We may also write an expression for the propagation constant in terms of the *line attenuation constant* (α) and *phase constant* (β):

$$\gamma = \alpha + j\beta \qquad (4\text{-}22)$$

If we can assume that susceptance dominates conductance in the admittance term, and reactance dominates resistance in the impedance term (both usually true at microwave frequencies), then we may neglect the R and G terms altogether and write

$$\gamma = j\omega\sqrt{LC} \qquad (4\text{-}23)$$

We may also reduce the phase constant (β) to

$$\beta = \omega\sqrt{LC} \qquad (4\text{-}24)$$

or

$$\beta = \omega Z_o C \quad \text{rad/m} \qquad (4\text{-}25)$$

and, of course, the characteristic impedance remains

$$Z_o = \sqrt{L/C} \qquad (4\text{-}26)$$

4-4.1 Special Cases

The impedance looking into a transmission line (Z) is the impedance presented to the source by the combination of load impedance and transmission line characteristic impedance. The following equations define the looking-in impedance experienced by a generator or source driving a transmission line.

The case where the load impedance and line characteristic impedance are matched is defined by

$$Z_L = R_L + j0 = Z_o$$

In other words, the load impedance is resistive and equal to the characteristic impedance of the transmission line. In this case, the line and load are matched, and the impedance looking in will be simply $Z = Z_L = Z_o$. In other cases, however, we find different situations where Z_L is not equal to Z_o.

1. Z_L *is not equal to* Z_o *in a random-length lossy line:*

$$Z = Z_o\left(\frac{Z_L + Z_o\tanh(\gamma l)}{Z_o + Z_L\tanh(\gamma l)}\right) \qquad (4\text{-}27)$$

where

   Z = impedance looking in in ohms
   Z_L = load impedance in ohms
   Z_o = line characteristic impedance in ohms
   l = length of the line in meters
   γ = propagation constant

2. Z_L *not equal to* Z_o *in lossless or very low-loss random-length line:*

$$Z = Z_o\left(\frac{Z_L + jZ_o\tanh(\beta l)}{Z_o + jZ_L\tanh(\beta l)}\right) \qquad (4\text{-}28)$$

Equations (4-27) and (4-28) serve for lines of any random length. For lines that are either integer multiples of half-wavelength or odd-integer (that is, 1, 3, 5, 7 . . .) multiples of quarter-wavelength, special solutions are found, and some of these solutions are very useful in practical situations. For example, consider the following.

3. *Half-wavelength lossy lines:*

$$Z = Z_o\left(\frac{Z_L + Z_o\tanh(\alpha l)}{Z_o + Z_L\tanh(\alpha l)}\right) \qquad (4\text{-}29)$$

EXAMPLE 4-5

A lossless 50-Ω (Z_o) transmission line is exactly one half-wavelength long and is terminated in a load impedance of $Z = 30 + j0$. Calculate the input impedance looking into the line (in a lossless line, $a = 0$).

Solution

$$Z = Z_o\left(\frac{Z_L + Z_o\tanh(\alpha\,l)}{Z_o + Z_L\tanh(\alpha\,l)}\right)$$

$$= (50\,\Omega) \times \left(\frac{30 + \left[50\,\tanh(0\pi)\right]}{50 + \left[30\,\tanh(0\pi)\right]}\right)$$

In Example 4-5, we discovered that the impedance looking into a lossless or very low-loss half-wavelength transmission line is the load impedance:

$$Z = Z_L \tag{4-30}$$

The fact that line input impedance equals load impedance is very useful in certain practical situations. For example, a resistive impedance is not changed by the line length. Therefore, when an impedance is inaccessible for measurement purposes, the impedance can be measured through a transmission line that is an integer multiple of a half-wavelength.

Our next special case involves a quarter-wavelength transmission line, and those that are *odd* integer multiples of quarter-wavelengths (*even* integer multiples of quarter-wavelength obey the half-wavelength criteria).

4. *Quarter-wavelength lossy lines:*

$$Z = Z_o\left(\frac{Z_L + Z_o\,\text{Coth}(\alpha\,l)}{Z_o + Z_L\,\text{Coth}(\alpha\,l)}\right) \tag{4-31}$$

5. *Quarter-wavelength lossless or very low-loss lines:*

$$Z = \frac{\sqrt{Z_o}}{Z_L} \tag{4-32}$$

From Eq. (4-32), we can discover an interesting property of the quarter-wavelength transmission line. First, divide each side of the equation by Z_o:

$$\frac{Z}{Z_o} = \frac{\sqrt{Z_o}}{Z_l Z_o} \tag{4-33}$$

$$\frac{Z}{Z_o} = \frac{Z_o}{Z_L} \qquad (4\text{-}34)$$

The ratio Z/Z_o shows an *inversion* of load impedance ratio Z_L/Z_o, or, stated another way,

$$\frac{Z}{Z_o} = \frac{1}{Z_L / Z_o} \qquad (4\text{-}35)$$

Again, from Eq. (4-32), we can deduce another truth about quarter-wavelength transmission lines. If

$$Z = \frac{\sqrt{Z_o}}{Z_L} \qquad (4\text{-}36)$$

then

$$Z\,Z_L = \sqrt{Z_o} \qquad (4\text{-}37)$$

which means that

$$Z_o = \sqrt{Z\,Z_L} \qquad (4\text{-}38)$$

Equation (4-38) shows that a quarter-wavelength transmission line can be used as an *impedance-matching network*. Called a *Q section*, the quarter-wavelength transmission line used for impedance matching requires a characteristic impedance Z_o if Z is the source impedance and Z_L is the load impedance.

EXAMPLE 4-6

A 50-Ω source must be matched to a load impedance of 36 Ω. Find the characteristic impedance required of a Q section matching network.

Solution

$$
\begin{aligned}
Z &= \sqrt{Z\,Z_L} \\
&= \sqrt{(50\,\Omega)(36\,\Omega)} \\
&= \sqrt{1800\,\Omega^2} = 42\,\Omega
\end{aligned}
$$

6. *Transmission line as a reactance:* Reconsider Eq. (4-28), which related impedance looking in to load impedance and line length:

$$Z = Z_o \left(\frac{Z_L + jZ_o\tan(\beta l)}{Z_o + jZ_L\tan(\beta l)} \right) \tag{4-39}$$

Now, for the case of a shorted line ($Z_L = 0$), the solution is

$$Z = Z_o \left(\frac{0 + jZ_o \tan(\beta l)}{Z_o + j(0) \tan(\beta l)} \right) \tag{4-40}$$

$$= Z_o \frac{jZ_o\tan(\beta l)}{Z_o} \tag{4-41}$$

$$= jZ_o \tan(\beta l) \tag{4-42}$$

Recall from Eq. (4-25) that

$$\beta = wZ_oC \tag{4-43}$$

Substituting Eq. (4-43) into Eq. (4-42) produces

$$Z = jZ_o \tan(\omega Z_oCl) \tag{4-44}$$

or

$$Z = jZ_o \tan(2\pi FZ_oCl) \tag{4-45}$$

Because the solution to Eqs. (4-44) and (4-45) is multiplied by the j operator, we know that the impedance is actually a reactance ($Z = 0 + jX$). It is possible to achieve almost any possible reactance (within certain practical limitations) by adjusting the length of the transmission line and shorting the load end. This fact leads us to a practical method for impedance matching.

Figure 4-10A shows a circuit in which an unmatched load is connected to a transmission line with characteristic impedance Z_o. The load impedance Z_L is of the form $Z = R \pm jX$ and in this case is equal to $50 - j20$. A complex impedance load can be matched to its source by interposing the complex conjugate of the impedance. For example, in the case where $Z = 50 - j20$, the matching impedance network will require an impedance of $50 + j20$ Ω. The two impedances combine to produce a result of 50 Ω. The situation of Fig. 4-10A shows a matching stub with a reactance equal in magnitude, but opposite sign, with respect to the reactive component of the load impedance. In this case, the stub has a reactance of $+j20$ Ω to cancel a reactance of $-j20$ Ω in the load.

A *quarter-wavelength shorted stub* is a special case of the stub concept that finds particular application in microwave circuits. *Waveguides* (Chapter 5) are based on the properties of the quarter-wavelength shorted stub. Figure 4-10B shows a quarter-wave stub and its current distribution. The current is maximum across the short, but wave cancellation forces it to zero at the terminals. Because $Z = V/I$, when I goes to zero the impedance goes infinite. Thus, a quarter-wavelength stub has an infinite impedance at its resonant frequency and acts as an insulator. The concept may be hard to swallow, but the stub is a "metal insulator." You will see this concept developed further in Chapter 5.

4-5 STANDING-WAVE RATIO

The reflection phenomenon was noted in Section 4-3 when we discussed the step function and single pulse response of a transmission line; the same phenomenon also applies when the transmission line is excited with an ac signal. When a transmission line is not matched to its load, some of the energy is absorbed by the load and some is reflected back down the line toward the source. The interference of incident (or *forward*) and reflected (or *reverse*) waves creates *standing waves* on the transmission line (refer again to Section 4-3).

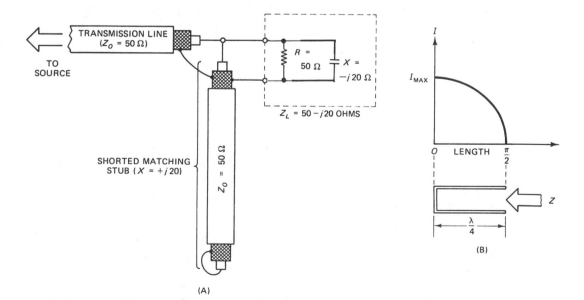

Figure 4-10 A) Coaxial matching stub; B) parallel line stub.

If the voltage or current is measured along the line, it will vary depending on the load, according to Fig. 4-11. Figure 4-11A shows the voltage versus length curve for a matched line, that is, where $Z_L = Z_o$. The line is said to be flat because the voltage (and current) is constant all along the line. But now consider Figs. 4-11B and 4-11C.

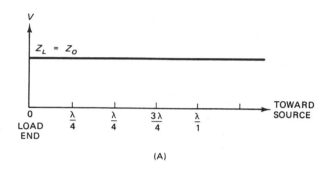

(A)

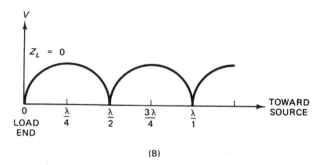

(B)

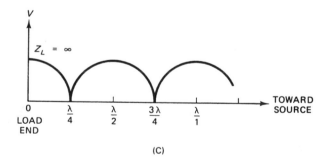

(C)

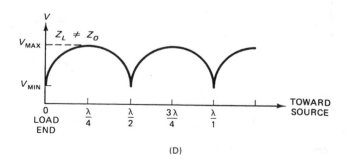

(D)

Figure 4-11 Impedance along a transmission line: A) flat line when $Z_L = Z_0$; B) shorted load; C) open load; D) load not equal to zero.

Figure 4-11B shows the voltage distribution over the length of the line when the load end of the line is *shorted* ($Z_L = 0$). At the load end the voltage is zero, which results from zero impedance. The same impedance and voltage situation is repeated every half-wavelength down the line from the load end toward the generator. Voltage *minima* are called *nodes*, while voltage *maxima* are called *antinodes*.

The pattern in Fig. 4-11C results when the line is unterminated (open), that is, $Z_L = \infty$. Note that the pattern is the same shape as Fig. 4-11B (shorted line), but phase shifted 90°. In both cases the reflection is 100%, but the phase of the reflected wave is opposite (Section 4-3).

Figure 4-11D shows the situation in which Z_L is not equal to Z_o, but is neither zero nor infinite. In this case, the nodes represent some finite voltage, V_{min}, rather than zero. The *standing-wave ratio* (SWR) reveals the relationship between load and line.

If the current along the line is measured, the pattern will resemble the patterns of Fig. 4-11. The SWR is then called *ISWR*, to indicate that it came from a current measurement. Similarly, if the SWR is derived from voltage measurements, it is called *VSWR*. Perhaps because voltage is easier to measure, VSWR is the term most commonly used in microwave work.

VSWR can be specified in any of several equivalent ways:

1. *From incident voltage (V_i) and reflected voltage (V_r):*

$$VSWR = \frac{V_i + V_r}{V_i + V_r} \qquad (4\text{-}46)$$

2. *From transmission line voltage measurements (Fig. 4-11D):*

$$VSWR = \frac{V_{max}}{V_{min}} \qquad (4\text{-}47)$$

3. *From load and line characteristic impedance:*

$$(Z_L > Z_o)\text{: } VSWR = Z_L/Z_o \qquad (4\text{-}48)$$

$$(Z_L < Z_o)\text{: } VSWR = Z_o/Z_L \qquad (4\text{-}49)$$

4. *From incident (P_i) and reflected (P_r) power:*

$$VSWR = \frac{1 + \sqrt{P_r / R_i}}{1 - \sqrt{R_r / P_i}} \qquad (4\text{-}50)$$

5. *From reflection coefficient (P):*

$$VSWR = \frac{1 + P}{1 - P} \qquad (4\text{-}51)$$

It is also possible to determine the reflection coefficient (P) from knowledge of VSWR:

$$P = \frac{VSWR - 1}{VSWR + 1} \qquad (4\text{-}52)$$

The relationship between reflection coefficient (P) and VSWR is shown in Fig. 4-12. VSWR is usually expressed as a *ratio*. For example, when Z_L is 100 Ω and Z_o is 50 Ω, the VSWR is $Z_L/Z_o = 100\ \Omega/50\ \Omega = 2$, which is usually expressed as VSWR = 2:1. VSWR can also be expressed in decibel form:

$$VSWR_{dB} = 20 \log (VSWR) \qquad (4\text{-}53)$$

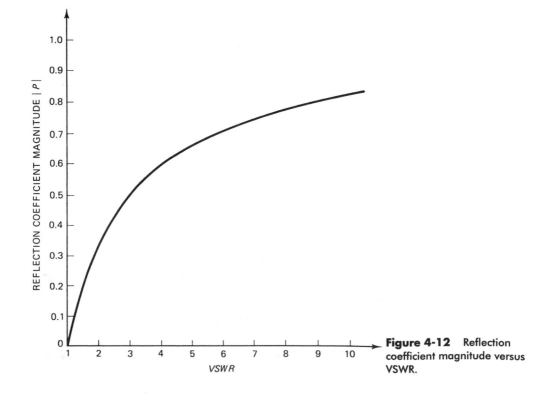

Figure 4-12   Reflection coefficient magnitude versus VSWR.

EXAMPLE 4-7

A transmission line is connected to a mismatched load. Calculate both the VSWR and VSWR decibel equivalent if the reflection coefficient (P) is 0.25.

Solution

(a)
$$\text{VSWR} = \frac{1+P}{1-P}$$
$$= \frac{1+0.25}{1-0.25}$$
$$= \frac{1.25}{0.75} = 1.67$$

(b)
$$\text{VSWR}_{dB} = 20 \log \text{VSWR}$$
$$= (20)(\log 1.67)$$
$$= (20)(0.22) = 4.4 \text{ dB}$$

The SWR is regarded as important in systems for several reasons. At the base of these reasons is the fact that the reflected wave represents energy lost to the load. For example, in an antenna system, less power is radiated if some of its input power is reflected back down the transmission line because the antenna feedpoint impedance does not match the transmission line characteristic impedance. In the next section, we will look at the problem of mismatch losses.

4-5.1 Mismatch (VSWR) Losses

The power reflected from a mismatched load represents a loss and will have implications that range from negligible to profound depending on the situation. For example, results might range from a slight loss of signal strength at a distant point from an antenna to destruction of the output device in a transmitter. The latter problem so plagued early solid-state transmitters that designers opted to include shut-down circuitry to sense high VSWR and turn down output power proportionally.

In microwave measurements, VSWR on the transmission lines that interconnect devices under test, instruments, and signal sources can cause erroneous readings and invalid measurements.

Determination of VSWR losses must take into account *two* VSWR situations. In Fig. 4-9, we have a transmission line of impedance Z_o interconnecting a load impedance (Z_L) and a source with an output impedance, Z_s. There is a potential for impedance mismatch at *both ends* of the line.

In the case where one end of the line is matched (either $Z_s = Z_s$), the *mismatch loss* due to SWR at the mismatched end is

$$ML = -10 \log \left(1 - \left(\frac{\text{VSWR} - 1}{\text{VSWR} + 1} \right)^2 \right) \qquad (4\text{-}54)$$

which from Eq. (4-52), you may recognize as

$$ML = -10 \log (1 - P^2) \hspace{3cm} (4\text{-}55)$$

EXAMPLE 4-8

A coaxial transmission line with a characteristic impedance of 50 Ω is connected to the 50-Ω output (Z_o) of a signal generator, and also to a 20-Ω load impedance (Z_L). Calculate the mismatch loss.

Solution

(a) First find the VSWR:

$$VSWR = Z_o/Z_L = 50 \ \Omega/20 \ \Omega = 2.5\text{:}1$$

(b) Mismatch loss:

$$ML = -10 \log \left(1 - \left(\frac{RSWR - 1}{RSWR + 1} \right)^2 \right)$$

$$= -10 \log \left(1 - \left(\frac{1.5}{3.5} \right)^2 \right)$$

$$= -10 \log [1 - (0.43)^2]$$

$$= -10 \log (0.815) = (-10)(0.089) = 0.89$$

When both ends of the line are mismatched, a different equation is required:

$$ML = 20 \log \left[1 \pm (P_1 \times P_2) \right] \hspace{3cm} (4\text{-}56)$$

where

P_1 = reflection coefficient at the source end of the line, $(VSWR_1 - 1)/(VSWR_1 + 1)$
P_2 = reflection coefficient at the load end of the line, $(VSWR_2 - 1)/(VSWR_2 + 1)$

Note that the solution to Eq. (4-56) has two values: $[1 + (P_1 P_2)]$ and $[1 - (P_1 P_2)]$.

The preceding equations reflect the mismatch loss solution for low-loss or lossless transmission lines. Although a close approximation, there are situations where they are insufficient, that is, when the line is lossy. While not very important at low frequencies, loss becomes higher at microwave frequencies. Interference between incident and reflected waves produces increased current at certain antinodes, which increases ohmic losses, and increased voltage at certain antinodes, which increases dielectric losses. It is the latter that increases with frequency. Equation (4-57) relates reflection coefficient and line losses to determine total loss on a given line.

$$\text{Loss} = 10 \log\left(\frac{n^2 - p^2}{n - np^2}\right) \qquad (4\text{-}57)$$

where

loss = total line loss in decibels
P = reflection coefficient
n = quantity $10^{(A/10)}$
A = total attenuation presented by the line, in decibels, when the line is properly matched ($Z_L = Z_o$)

4-6 SUMMARY

1. Transmission lines are conduits for RF signals. They are equivalent to complex *RLC* circuits and function in a manner similar to delay lines. Three basic forms of transmission line are *parallel line, coaxial cable,* and *strip line.*

2. The characteristic impedance of a transmission line is the impedance set by the stray inductance and capacitance of the line. Also determining the characteristic impedance is the dielectric constant of the insulator between the conductors. When a transmission line is terminated in a resistive impedance equal to its characteristic impedance, the line is said to be matched and no signal is reflected.

3. A transmission line will exhibit a looking-in impedance that is a function of its length, characteristic impedance, and load impedance. Because of this, lines can be used for impedance matching.

4. Several special cases are found when dealing with transmission lines: (a) matched lines ($Z_L = Z_o$) reflect the characteristic impedance to the input, regardless of length; (b) half-wavelength lines reflect the load impedance to the input; (c) quarter-wavelength lines reflect an impedance that is a function of load and characteristic impedance; (d) shorted lines reflect a reactive impedance that is a function of line length and characteristic impedance. In the case of the quarter-wavelength shorted line, the impedance is infinite, but a phase reversal takes place.

5. When a transmission line is terminated in any impedance other than its characteristic impedance, at least some of the incident signal is not absorbed by the load and is reflected back down the line toward the generator. This reflection phenomenon can be used to evaluate the line in a measurement method called *time-domain reflectometry.* The interference between reflected and incident waves produces *standing waves* on the line.

6. The *standing-wave ratio* is a measure of the standing waves, and hence the degree of mismatch on the line. SWR can be expressed either as a ratio or in decibel form. Standing waves on a transmission line give rise to mismatch losses, which can vary from negligible to severe depending on the situation, especially the natural line losses of the transmission line.

4-7   RECAPITULATION

Now return to the objectives and prequiz questions at the beginning of the chapter and see how well you can answer them. If you cannot answer certain questions, place a check by each and review the appropriate parts of the text. Next, try to answer the following questions and work the problems using the same procedure.

QUESTIONS AND PROBLEMS

1. _____ transmission line consists of a pair of parallel conductors separated by a dielectric material.
2. _____ transmission line consists of a pair of concentric conductors separated by a dielectric material.
3. List three different forms of coaxial cable.
4. A _____ transmission line consists of a rectangular conductor on a printed circuit board over a ground plane.
5. Another term for characteristic impedance is _____ impedance.
6. Calculate the characteristic impedance of a lossless transmission line if it has 0.5-μH/m inductance and 65-pF/m capacitance.
7. Calculate the surge impedance of a transmission line if it exhibits 0.75-μH/m inductance and 37-pF/m capacitance.
8. What is the characteristic impedance of a transmission line that exhibits 3.75 nH/ft and 1.5 pF/ft?
9. Find the dielectric constant of a material in which an electromagnetic wave propagates at a velocity of 0.85c.
10. Find the velocity factor of a line if the dielectric constant is 2.45.
11. Find the velocity factor of a line if the dielectric constant is 1.75.
12. A parallel line uses dry air as the dielectric ($\varepsilon = 1.006$). Calculate the characteristic impedance if 0.1-in. conductors are spaced 1.00 inch apart.
13. Calculate the characteristic impedance of the parallel line in question 12 if the line is placed inside a perfect vacuum. Compare the two answers and draw a practical conclusion.
14. A coaxial cable consists of an anhydrous air dielectric inside a 2-in. pipe. Calculate the characteristic impedance of the line for

    (a) a 0.25-in. center conductor, (b) a 0.45-in. center conductor.
15. A stripline transmission line is constructed on a double-sided printed circuit board that has a dielectric constant of 5.5 at microwave frequencies. Calculate the characteristic impedance of the line if it is 0.25 in. wide and the board is 0.1 in. thick.
16. Using the same board as in question 15, calculate the width of a stripline required to match a 50-Ω system impedance.
17. A 90-Ω transmission line is needed on a microwave transmission line. If the board is 2 mm thick and has a dielectric constant of 2.75, what width is needed to achieve the needed impedance?
18. On a transmission line, a measurement is made of the reflected and incident voltages.

If the reflected voltage is 0.2 V peak and the incident voltage is 1.78 V peak, find the reflection coefficient.

19. Calculate the reflection coefficient if the forward voltage is 0.9 and the reflected voltage is 0.10.

20. Calculate the reflected voltage if the incident voltage is 3.5 and the reflection coefficient is 0.34.

21. A load impedance of $50 + j0$ is connected to a 50-Ω transmission line. Calculate the reflection coefficient.

22. A load impedance of 12 Ω is connected to a 50-Ω transmission line. Calculate the reflection coefficient.

23. A load impedance of 95 Ω is connected to a 50-Ω transmission line. Calculate the reflection coefficient.

24. Calculate the reflection coefficient of a shorted 50-Ω transmission line.

25. A transmission line with a 75-Ω characteristic impedance is connected to a 6-Ω resistive load impedance. Calculate

    (a) the reflection coefficient, (b) the VSWR, (c) the angle of the reflected wave relative to the incident wave.

26. A pulse generator and oscilloscope are connected across the input end of a slightly mismatched transmission line. The pulse has a duration of 1.1 μs and a reflection "pip" is noted 0.37 μs after the onset of the incident pulse. Assuming that Teflon® dielectric is used ($\mu = 0.7$), find the length of the line.

27. A reflection arrives back at the input end of a transmission line 210 ns after it started. If the line has a velocity factor of 0.66, calculate the length of the line.

28. Calculate the velocity factor of a transmission line if the reflected pulse arrives back at the input end 0.33 μs after the incident pulse leading edge, and the line is known to be 29 m in length.

29. A 50-ft transmission line is tested, and it is found that the reflected pulse requires 0.15 μs for a round trip. Calculate the velocity factor.

30. A transmission line is measured and found to have a velocity factor of 0.80. The dielectric is probably _____ .

31. Calculate the phase constant of a transmission line operated at 1 GHz if the line has a capacitance per unit length of 45 pF/m and a 50-Ω characteristic impedance.

32. Calculate the inductance per meter of the transmission line in question 31.

33. A Q section matching stub is used to match the 200-Ω input impedance of a high-pass filter to a 50-Ω load. Calculate the characteristic impedance required of the line.

34. A microwave transistor has an input impedance of $20 + j0$ Ω. Calculate the characteristic impedance of a Q section matching line required to match it to a 50-Ω system impedance.

35. In a 50-Ω system operated at 3.3 GHz, it is necessary to match an impedance of $50 - j20$ using a shorted stub in parallel with the load. Calculate its length if the capacitance is 65 pF/m.

36. Calculate the VSWR if the incident voltage is 3.5 and the reflected voltage is 1.25. Express the answer both as a ratio and in decibel form.

37. The voltage along a transmission line is measured. It is found to vary as a function of line length. Is the load impedance matched or mismatched?

38. The maximum voltage along a transmission line is 2.75 and the minimum voltage is 0.50. Find the VSWR.

39. A 50-Ω transmission line is connected to a 100-Ω load. Calculate the VSWR.

40. A 50-Ω transmission line is connected to a 50-Ω load. Calculate the VSWR.

41. A 75-Ω transmission line is connected to a 40-Ω load. Calculate the VSWR.

42. A wattmeter is capable of reading both forward and reflected power. Calculate the VSWR if the forward power is 10.5 watts (W) and the reflected power is 2.25 W.

43. Calculate the reflection coefficients for the following VSWRs:

    (a) 2.5:1, (b) 1.5:1, (c) 5.5:1, (d) 7.75 dB.

44. A line shows a reflection coefficient of 0.4. Calculate the VSWR.

45. A line shows a reflection coefficient of 0.2. Calculate the VSWR.

46. Express the following VSWRs in decibel form:

    (a) 2.5:1, (b) 4:1, (c) 1.25:1.

47. Express the following VSWRs in ratio form:

    (a) 3 dB, (b) 4.5 dB, (c) 2.5 db, (d) 10 dB.

48. A system has a mismatch on one end of 3.5:1. Calculate the mismatch loss (assuming lossless transmission line).

49. Calculate the mismatch loss for a 2:1 VSWR.

50. Calculate the mismatch loss for a lossless line that sees a reflection coefficient of 0.25.

51. A 50-Ω transmission line is connected to a $25 - j0$-Ω impedance. Calculate the mismatch loss.

52. A transmission line is mismatched at both ends. Assuming that very-low-loss line is used, calculate the mismatch loss if the source end has a reflection coefficient of 0.1 and the load end a reflection coefficient of 0.25.

53. A transmission line is rated to show a loss of 5.5 dB/100 ft at microwave frequencies. Calculate the mismatch loss for a 1-m line that has a 4.5:1 VSWR.

54. In problem 53, the transmission line is shortened to 1 ft. Calculate the mismatch loss.

KEY EQUATIONS

1. Characteristic impedance of a lossy line:

$$Z_0 = \sqrt{\frac{R + j\omega L}{G + j\omega C}}$$

2. Characteristic impedance of a lossless line:

$$Z_0 = \sqrt{\frac{L}{C}}$$

3. Dielectric constant as a function of velocity:

$$\varepsilon = \frac{1}{v^2}$$

4. Characteristic impedance of transmission lines:
   (a) Parallel line:

$$Z_o = \frac{276}{\sqrt{\varepsilon}} \log \frac{2S}{d}$$

(b) Coaxial line:

$$Z_o = \frac{138}{\sqrt{\varepsilon}} \log \frac{D}{d}$$

(c) Shielded parallel line:

$$Z_o = \frac{276}{\sqrt{\varepsilon}} \log \left(2A \frac{1 - B^2}{1 + B^2} \right)$$

where:

$$A = \frac{s}{d}$$

$$B = \frac{s}{D}$$

(d) Stripline:

$$Z_o = \frac{377}{\sqrt{\varepsilon}} \frac{T}{W}$$

5. Width of a stripline as a function of impedance:

$$W = \frac{377T}{Z_o \sqrt{\varepsilon}}$$

6. Velocity of a wavefront along a transmission line:

$$v = \frac{1}{\sqrt{LC}}$$

7. Transmission line impedance as a function of voltage and current:

$$Z_L = \frac{V_{inc} + V_{ref}}{I_{inc} + I_{ref}}$$

8. Reflection coefficient:

$$P = \frac{V_{ref}}{V_{inc}}$$

and

$$P = \frac{Z_L - Z_o}{Z_L + Z_o}$$

9. Length of a transmission line as a function of reflection transit time:

$$L_{meters} = \frac{cvT_d}{2}$$

10. Voltage at the end of a transmission line:

$$V_r = Ve^{-\gamma l}$$

11. Propagation constant (γ):

$$\gamma = \sqrt{ZY}$$

$$\gamma = \sqrt{(R + j\omega L)(G + j\omega C)}$$

$$\gamma = a + jB$$

$$\gamma = j\omega\sqrt{LC}$$

12. Phase constant (β):

$$\beta = \omega\sqrt{LC}$$

$$\beta = \omega Z_o C \text{ rad/m}$$

13. Impedance looking in to a transmission line:
(a) Z_L is not equal to Z_0 in a random-length lossy line:

$$Z = Z_o \frac{Z_L + Z_o \tan(\gamma l)}{Z_o + Z_L \tan(\gamma l)}$$

(b) Z_L not equal to Z_0 in lossless or very-low-loss, random-length line:

$$Z = Z_o \frac{Z_L + jZ_o \tan(\beta l)}{Z_o + jZ_L \tan(\beta l)}$$

(c) Half-wavelength lossy lines:

$$Z = Z_o \frac{Z_L + Z_o \tanh(\alpha l)}{Z_o + Z_L \tanh(\alpha l)}$$

(d) Half-wavelength lossless lines:

$$Z = Z_L$$

(e) Quarter-wavelength lossy lines:

$$Z = Z_o \frac{Z_L + Z_o \, \text{Coth} \, (\alpha l)}{Z_o + Z_L \, \text{Coth} \, (\alpha l)}$$

(f) Quarter-wavelength lossless or very-low-loss lines:

$$Z = \frac{\sqrt{Z_o}}{Z_L}$$

(g) Reactance of a shorted line:

$$Z = jZ_o \tan (\beta l)$$

$$Z = jZ_o \tan (\omega Z_o Cl)$$

$$Z = jZ_o \tan (2\pi F Z_o Cl)$$

14. Impedance inversion in a quarter-wavelength shorted stub:

$$\frac{Z}{Z_o} = \frac{Z_o}{Z_L}$$

$$\frac{Z}{Z_o} = \frac{1}{Z_L / Z_o}$$

15. Characteristic impedance for a quarter-wavelength Q section:

$$Z_o = \sqrt{ZZ_L}$$

16. Transmission line as a reactance:

$$Z = jZ_o \tan (\beta l)$$

17. VSWR:
    (a) From incident voltage (V_i) and reflected voltage (V_r):

$$\text{VSWR} = \frac{V_i + V_r}{V_i - V_r}$$

(b) From transmission line voltage measurements (Fig. 4-11D):

$$\text{VSWR} = \frac{V_{max}}{V_{min}}$$

(c) From load and line characteristic impedance:

$$Z_L > Z_o: \quad \text{VSWR} = \frac{Z_L}{Z_o}$$

$$Z_L < Z_o: \quad \text{VSWR} = \frac{Z_o}{Z_L}$$

(d) From incident (P_i) and reflected (P_r) power:

$$\text{VSWR} = \frac{1 + \sqrt{P_r / P_i}}{1 - \sqrt{P_r / P_i}}$$

(e) From reflection coefficient (P):

$$\text{VSWR} = \frac{1 + P}{1 - P}$$

(f) Reflection coefficient (P) from knowledge of VSWR:

$$P = \frac{\text{VSWR} - 1}{\text{VSWR} + 1}$$

CHAPTER 5

Waveguides

OBJECTIVES

1. Understand the theory of operation for waveguides.
2. Be able to describe the TE and TM modes of propagation in waveguides.
3. Learn the limitations and constraints on waveguide applications.
4. Know the various components of waveguide systems.

5-1   PREQUIZ

These questions test your prior knowledge of the material in this chapter. Try answering them before you read the chapter. Look for the answers (especially those you answered incorrectly) as you read the text. After you have finished studying the chapter, try answering these questions again and those at the end of the chapter.

1. The TE_{10} mode is called the _____ mode for a rectangular waveguide.
2. The _____ mode of propagation has the electric field transverse (that is, orthogonal) to the direction of wave propagation.
3. A _____ _____ uses a pyramid- or wedge-shaped block of absorptive material at the terminating end of the waveguide.
4. A vertical radiating element protruding into the waveguide interior and positioned a quarter-wavelength from the end cap produces a _____ mode wave propagating in the direction away from the end cap.

5-2   INTRODUCTION

The microwave spectrum covers frequencies from about 0.4 to 300 GHz, with wavelengths in free space ranging from 75 cm down to 1 mm. Transmission lines

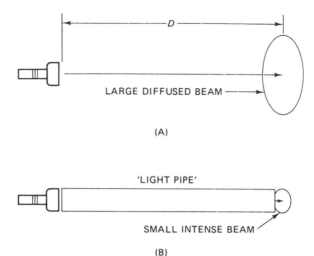

LARGE DIFFUSED BEAM

(A)

'LIGHT PIPE'

SMALL INTENSE BEAM

(B)

Figure 5-1 Flashlight analogy: in free space the beam spreads according to the inverse square law (intensity = $1/D^2$).

can be used at frequencies from dc to about 50 or 60 GHz, although above about 5 GHz only short runs are practical because attenuation increases dramatically as frequency increases. There are three types of losses in conventional transmission lines: *ohmic*, *dielectric*, and *radiation*. Ohmic losses are caused by the current flowing in the resistance of the conductors making up the transmission lines. Because of the skin effect, which increases resistance at higher frequencies, these losses tend to increase in the microwave region. Dielectric losses are caused by the electric field acting on the molecules of the insulator and thereby causing heating through molecular agitation. Radiation losses represent loss of energy as an electromagnetic wave propagates away from the surface of the transmission line conductor.

Losses on long runs of coaxial transmission line (the type most commonly used) give designers cause for concern even in as low as the 0.4- to 5-GHz region. Also, because of the increased losses, power-handling capability decreases at higher frequencies. Therefore, at higher microwave frequencies, or where long runs make coax attenuation losses unacceptable, or where high power levels would overheat the coax, *waveguides* are used in lieu of transmission lines.

What is a waveguide? Consider the light pipe analogy depicted in Fig. 5-1. A flashlight serves as our RF source, which, given that light is also an electromagnetic wave, is not altogether unreasonable. In Fig. 5-1A, the source radiates into free space, and spreads out as a function of distance. The intensity per unit area at the destination (a wall) falls off as a function of distance (D) according to the *inverse square law* ($1/D^2$).

But now consider the transmission scheme in Fig. 5-1B. The light wave still propagates over distance D, but is now confined to the interior of a mirrored pipe. Almost all of the energy (less tiny losses) coupled to the input end is delivered to the output end, where the intensity is practically undiminished. While not perfect, the light pipe analogy neatly summarizes on a simple level the operation of microwave waveguides.

Thus, we can consider the waveguide as an RF pipe without seeming too serenely detached from reality. Similarly, fiber-optic technology is waveguidelike at optical (IR and visible) wavelengths. In fact, the analogy between fiber optics

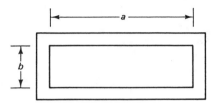

Figure 5-2 End view of rectangular waveguide.

and waveguide can withstand more rigorous comparison than the simplistic light pipe analogy.

The internal walls of the waveguide are not mirrored surfaces, as in our optical analogy, but are rather electrical conductors. Most waveguides are made of aluminum, brass, or copper. In order to reduce ohmic losses, some waveguides have their internal surfaces electroplated with either gold or silver, both of which have lower resistivities than the other metals mentioned.

Waveguides are hollow metal pipes and may have either circular or rectangular cross sections (although the rectangular are, by far, the most common). Figure 5-2 shows an end view of the rectangular waveguide. The dimension *a* is the wider dimension, and *b* is the narrower. These letters are considered the standard form of notation for waveguide dimensions and will be used in the equations developed in this chapter.

5-3 DEVELOPMENT OF THE RECTANGULAR WAVEGUIDE FROM PARALLEL TRANSMISSION LINES

One way of visualizing how a waveguide works is to develop the theory of waveguides from the theory of elementary parallel transmission lines (see Chapter 4). Figure 5-3A shows the basic parallel transmission line, which was introduced in Chapter 4. The line consists of two parallel conductors separated by an air dielec-

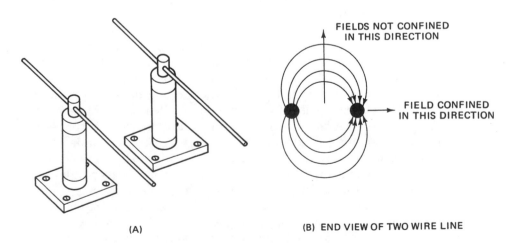

(A)　　　　　　　　(B) END VIEW OF TWO WIRE LINE

Figure 5-3 A) Parallel lines; B) electrical field between the conductors.

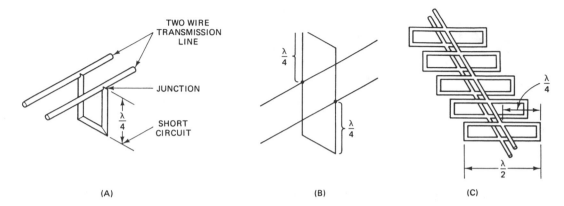

Figure 5-4   A) Quarter-wavelength shorted stub on parallel line; B) two lines at the same point; C) analog to waveguide.

tric. Because air will not support the conductors, ceramic or other material insulators are used as supports.

There are several reasons why the parallel line *per se* is not used at microwave frequencies. Skin effect increases ohmic losses to a point that is unacceptable. We also find that the insulators supporting the two conductors are significantly more lossy at microwave frequencies than at lower frequencies. Finally, radiation losses increase dramatically. Figure 5-3B shows the electric fields surrounding the conductors. The fields add algebraically either constructively or destructively, resulting in pinching of the resultant field along one axis and bulging along the other. This geometry increases radiation losses at microwave frequencies.

Now let's revisit the quarter-wavelength shorted stub introduced in Chapter 4. Recall that the looking-in impedance of such a stub is infinite. When placed in parallel across a transmission line (Fig. 5-4A), the stub acts like an insulator. In other words, at its resonant frequency the stub is a *metallic insulator*, and can be used to physically support the transmission line.

Again, because the impedance is infinite, we can connect two quarter-wavelength stubs in parallel with each other across the same points on the transmission line (Fig. 5-4B) without loading down the line impedance. This arrangement effectively forms a half-wavelength pair. The impedance is still infinite, so no harm is done. Likewise, we can parallel a large number of center-fed half-wavelength pairs along the line, as might be the case when a long line is supported at multiple points. The waveguide is analogous to an infinite number of center-fed *half-wave pairs* of quarter-wave shorted stubs connected across the line. The result is the continuous metal pipe structure of common rectangular waveguide.

At first glance, relating rectangular waveguide to quarter-wavelength shorted stubs seems to fall down except at the exact resonant frequency. It turns out, however, that the analogy also holds up at other frequencies, as long as the frequency is higher than a certain cutoff frequency. The waveguide thus acts like a high-pass filter. There is also a practical upper frequency limit. In general, waveguides support a bandwidth of 30% to 40% of cutoff frequency. As shown in Fig. 5-5, the center line of the waveguide, which represents the points where the conductors are in the par-

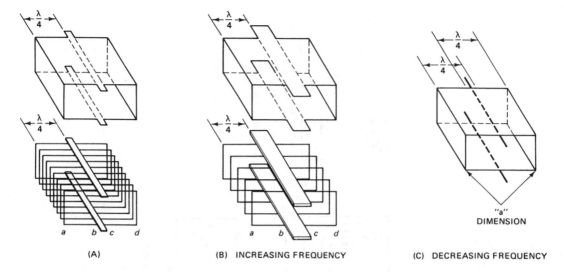

Figure 5-5 The analogy of quarter-wavelength shorted stubs continued, to show what happens as frequency changes.

allel line analogy, becomes a *shorting bar* between segments, and that bar widens or narrows according to operating frequency. Thus, the active region is still a quarter-wavelength shorted stub.

Below the cutoff frequency, the structure disappears entirely, and the waveguide acts like a parallel transmission line with a low-impedance inductive reactance shorted across the conductors. When modeled as a pair of quarter-wavelength stubs, the *a* dimension of the waveguide is a half-wavelength long. The cutoff frequency is defined as the frequency at which the *a* dimension is less than a half-wavelength.

5-4   PROPAGATION MODES IN WAVEGUIDE

The signal in a microwave waveguide propagates as an electromagnetic wave, not as a current. Even in a transmission line, the signal propagates as a wave because the current in motion down the line gives rise to electric and magnetic fields, which behave as an electromagnetic field. The specific type of field found in transmission lines, however, is a *transverse electromagnetic (TEM) field*. The term *transverse* implies things at right angles to each other, so the electric and magnetic fields are perpendicular to the direction of travel. In addition to the word *transverse*, these right angle waves are said to be *normal* or *orthogonal* to the direction of travel, which are three different ways of saying the same thing: right-angleness.

5-4.1 Boundary Conditions

The TEM wave will not propagate in a waveguide because certain *boundary conditions* apply. While the wave in the waveguide propagates through the air or inert

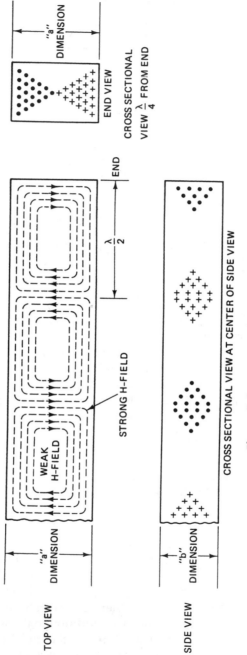

Figure 5-6 Magnetic fields inside a waveguide.

gas dielectric in a manner similar to free-space propagation, the phenomenon is bounded by the walls of the waveguide, which implies that certain conditions must be met. The boundary conditions for waveguides are as follows:

1. The electric field must be orthogonal to the conductor in order to exist at the surface of that conductor.
2. The magnetic field must not be orthogonal to the surface of the waveguide.

In order to satisfy these boundary conditions, the waveguide gives rise to two types of propagation modes: *transverse electric mode* (TE mode) and *transverse magnetic mode* (TM mode). The TEM mode violates the boundary conditions because the magnetic field is not parallel to the surface and so does not occur in waveguides.

The transverse electric field requirement means that the E field must be perpendicular to the conductor wall of the waveguide. This requirement is met by use of a proper coupling scheme (see Section 5-8) at the input end of the waveguide. A vertically polarized coupling radiator will provide the necessary transverse field.

One boundary condition requires that the magnetic (H) field must not be orthogonal to the conductor surface. Because it is at right angles to the E field, it will meet this requirement (see Fig. 5-6). The planes formed by the magnetic field are parallel to both the direction of propagation and the wide dimension surface.

As the wave propagates away from the input radiator, it resolves into two components that are not along the axis of propagation and are not orthogonal to the walls. The component along the waveguide axis violates the boundary conditions and so is rapidly attenuated. For the sake of simplicity, only one component is shown in Fig. 5-7. Three cases are shown in Fig. 5-7: high, medium, and low frequency. Note that the angle of incidence with the waveguide wall increases as frequency drops. The angle rises toward 90° as the cutoff frequency is approached from above. Below the cutoff frequency, the angle is 90°, so the wave bounces back and forth between the walls without propagating.

5-4.2 Coordinate System and Dominant Mode in Waveguides

Figure 5-8 shows the coordinate system used to denote dimensions and directions in microwave discussions. The a and b dimensions of the waveguide correspond to the X and Y axes of a Cartesian coordinate system, while the Z axis is the direction of wave propagation.

In describing the various modes of propagation, we use a shorthand notation as follows:

$$Tx_{m,n}$$

where

x = E for transverse electric mode and M for transverse magnetic mode
m = number of half-wavelengths along the X axis (the a dimension)
n = number of half-wavelengths along the Y axis (the b dimension)

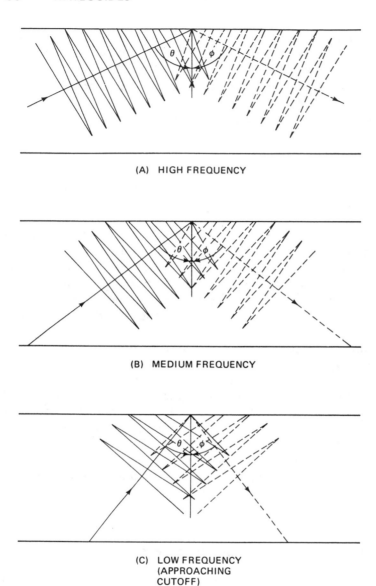

(A) HIGH FREQUENCY

(B) MEDIUM FREQUENCY

(C) LOW FREQUENCY
     (APPROACHING
     CUTOFF)

θ = $\angle$ OF INCIDENCE
ϕ = $\angle$ OF REFLECTION

Figure 5-7 Propagation along a waveguide at various frequencies.

The TE_{10} mode is called the *dominant mode* and is the best mode for low-attenuation propagation in the Z axis. The nomenclature TE_{10} indicates that there is one half-wavelength in the a dimension and zero half-wavelengths in the b dimension. The dominant mode exists at the lowest frequency at which the waveguide is half-wavelength.

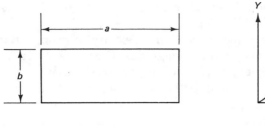
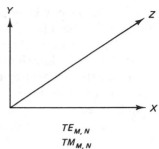

Figure 5-8   A) Rectangular waveguide showing dimensional conventions; B) vector relationships for mode convention.

5-4.3 Velocity and Wavelength in Waveguides

Figure 5-9 shows the geometry for two wave components simplified for sake of illustration. There are three different wave velocities to consider with respect to waveguides: *free-space velocity* (*c*), *group velocity* (V_g), and *phase velocity* (V_p). The

$$V_p = \frac{c}{\sin a}$$

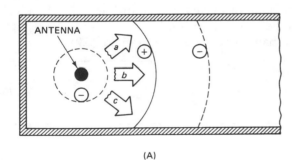

(A)

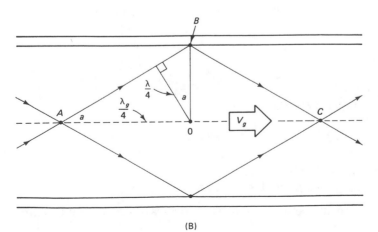

(B)

Figure 5-9   Waveguide antenna: A) physical structure; B) propagation diagram.

free-space velocity is the velocity of propagation in unbounded free space, that is, the speed of light ($c = 3 \times 10^8$ m/s).

The *group velocity* is the straight-line velocity of propagation of the wave down the center line (Z axis) of the waveguides. The value of V_g is always less than c because the actual path length taken as the wave bounces back and forth is longer than the straight-line path (that is, path ABC is longer than path AC). The relationship between c and V_g is

$$V_g = c \sin (a) \tag{5-1}$$

where

V_g = group velocity in meters per second (m/s)
c = free-space velocity (3×10^8 m/s)
a = angle of incidence in the waveguide

The *phase velocity* is the velocity of propagation of the spot on the waveguide wall where the wave impinges (for example, point B in Fig. 5-9). This velocity is actually faster than both the group velocity and the speed of light. The relationship between phase and group velocities can be seen in the beach analogy. Consider an ocean beach in which the waves arrive from offshore at an angle other than 90°. In other words, the arriving wavefronts are not parallel to the shore. The arriving waves have a group velocity, V_g. But as a wave hits the shore, it will strike a point down the beach first, and the point of strike races up the beach at phase velocity, V_p, that is much faster than the group velocity. In a microwave waveguide, the phase velocity can be greater than c, as can be seen from Eq. (5.2):

$$V_p = \frac{c}{\sin a} \tag{5-2}$$

EXAMPLE 5-1

Calculate the group and phase velocities for an angle of incidence of 33°.

Solution

(a) Group velocity:
$$\begin{aligned}
V_g &= c \sin a \\
&= (3 \times 10^8)(\sin 33) \\
&= (3 \times 10^8)(0.5446) = 1.6 \times 10^8 \text{ m/s}
\end{aligned}$$

(b) Phase velocity:
$$\begin{aligned}
V_p &= c/\sin a \\
&= (3 \times 10^8 \text{ m/s})/\sin 33 \\
&= (3 \times 10^8 \text{ m/s})/(0.5446) = 5.51 \times 10^8 \text{ m/s}
\end{aligned}$$

We can also write a relationship between all three velocities by combining Eqs. (5-1) and (5-2), resulting in

$$c = \sqrt{V_p V_g} \qquad (5\text{-}3)$$

In any wave phenomenon the product of *frequency* and *wavelength* is the *velocity*. Thus, for a TEM wave in unbounded free space, we know that

$$c = F \lambda_o \qquad (5\text{-}4)$$

Because the frequency (F) is fixed by the generator, only the wavelength can change when the velocity changes. In a microwave waveguide, we can relate phase velocity to wavelength as the wave is propagated in the waveguide (λ):

$$V_p = \frac{\lambda \times c}{\lambda_o} \qquad (5\text{-}5)$$

where

V_p = phase velocity in meters per second (m/s)
c = free-space velocity $(3 \times 10^8 \text{ m/s})$
λ = wavelength in the waveguide in meters (m)
λ_o = wavelength in free space (c/F) in meters; see Eq. (5-4)

Equation (5.5) can be rearranged to find the wavelength in the waveguide:

$$\lambda = \frac{V_p \lambda_o}{c} \qquad (5\text{-}6)$$

EXAMPLE 5-2

A 5.6-GHz microwave signal is propagated in a waveguide. Assume that the internal angle of incidence to the waveguide surface is 42°. Calculate (a) phase velocity, (b) wavelength in unbounded free space, and (c) wavelength of the signal in the waveguide.
(a) Phase velocity:

$$
\begin{aligned}
V_p \;\; &= \frac{c}{\sin\ a} \\[2mm]
&= \frac{3 \times 10^8 \text{ m/s}}{\sin\ 42} \\[2mm]
&= \frac{3 \times 10^8 \text{ m/s}}{0.6691} = 4.5 \times 10^8 \text{ m/s}
\end{aligned}
$$

(b) Wavelength in free space:

$$\lambda_o = \frac{c}{F}$$

$$= \frac{3 \times 10^8 \text{ m/s}}{5.6 \times 10^9 \text{ Hz}} = 0.054 \text{ m}$$

(c) Wavelength in waveguide:

$$\lambda = \frac{V_p \lambda_o}{c}$$

$$= \frac{(4.5 \times 10^8 \text{ m/s})(0.054 \text{ m})}{3 \times 10^8 \text{ m/s}}$$

$$= 0.08 \text{ m}$$

Comparing, we find that the free-space wavelength is 0.054 m, while the wavelength inside of the waveguide increases to 0.08 m.

5-4.4 Cutoff Frequency (F_c)

The propagation of signals in a waveguide depends in part on the operating frequency of the applied signal. As discussed earlier, the angle of incidence made by the plane wave to the waveguide wall is a function of frequency. As the frequency drops, the angle of incidence increases towards 90°.

   The propagation of waves depends on the angle of incidence and the associated reflection phenomenon. Indeed, both phase and group velocities are functions of the angle of incidence. When the frequency drops to a point where the angle of incidence is 90°, then group velocity is zero and the concept of phase velocity is meaningless.

   We can define a general mode equation based on our system of notation:

$$\frac{1}{\lambda_c^2} = \left(\frac{m}{2a} \right)^2 + \left(\frac{n}{2b} \right)^2 \tag{5-7}$$

where

   λ_c = longest wavelength that will propagate
   a, b = waveguide dimensions (see Fig. 5-2)
   m, n = integers that define the number of half-wavelengths that will fit in
          the a and b dimensions, respectively

Evaluating Eq. (5-7) reveals that the longest TE-mode signal that will propagate in the dominant mode (TE$_{10}$) is given by

$$\lambda_c = 2a \tag{5-8}$$

from which we can write an expression for the cutoff frequency:

$$F_c = \frac{c}{2a} \tag{5-9}$$

where

F_c = lowest frequency that will propagate, in hertz
c = speed of light (3×10^8 m/s)
a = wide waveguide dimension

EXAMPLE 5-3

A rectangular waveguide has dimensions of 3×5 cm. Calculate the TE_{10} mode cutoff frequency.

Solution

$$F_c = \frac{c}{2a}$$

$$= \frac{3 \times 10^8 \text{ m/s}}{2[5 \text{ cm} \times (1 \text{ m}/100 \text{ cm})]}$$

$$= \frac{3 \times 10^8 \text{ m/s}}{(2)(0.05 \text{ m})} = 3 \text{ GHz}$$

Equation (5-7) assumes that the dielectric inside the waveguide is air. A more generalized form, which can accommodate other dielectrics, is

$$F_c = \frac{1}{\sqrt{\mu\varepsilon}} \sqrt{\left(\frac{m}{a}\right)^2 + \left(\frac{n}{b}\right)^2} \tag{5-10}$$

where

ε = dielectric constant
μ = permeability constant

For air dielectrics, $\mu = \mu_o$ and $\varepsilon = \varepsilon_o$, from which

$$c = \frac{1}{\sqrt{\varepsilon_o \mu_o}} \tag{5-11}$$

To determine the cutoff wavelength, we can rearrange Eq.(5-10) to the form

$$\lambda_c = \frac{2}{\sqrt{\left(\dfrac{m}{a}\right)^2 + \left(\dfrac{n}{b}\right)^2}} \qquad (5\text{-}12)$$

Another expression for the air-filled waveguide calculates the actual wavelength in the waveguide from knowledge of the free-space wavelength and actual operating frequency:

$$\lambda_g = \frac{\lambda_o}{\sqrt{1 - \left(\dfrac{F_c}{F}\right)^2}} \qquad (5\text{-}13)$$

where

λ_g = wavelength in the waveguide
λ_o = wavelength in free space
F_c = waveguide cutoff frequency
F = operating frequency

Transverse magnetic modes also propagate in waveguides, but the base TM_{10} mode is excluded by the boundary conditions. Thus, the TM_{11} mode is the lowest magnetic mode that will propagate.

5-5   WAVEGUIDE IMPEDANCE

All forms of transmission line, including waveguide, exhibit a characteristic impedance, although in the case of waveguide it is a little difficult to pin down conceptually. This concept was developed for ordinary transmission lines in Chapter 4. For waveguide, the characteristic impedance is approximately equal to the ratio of the electric and magnetic fields (E/H). The impedance of the waveguide is a function of waveguide characteristic impedance (Z_o) and the wavelength in the waveguide:

$$Z = \frac{Z_o \lambda_g}{\lambda_o} \qquad (5\text{-}14)$$

Or, for rectangular waveguide with constants taken into consideration,

$$Z = \frac{120\pi\lambda_g}{\lambda_o} \qquad (5\text{-}15)$$

The *propagation constant* (β) for rectangular waveguide is a function of both cutoff frequency and operating frequency:

$$\beta = \omega\sqrt{\varepsilon\mu}\sqrt{1 - \left(\frac{F_c}{F}\right)^2} \qquad (5\text{-}16)$$

from which we can express the TE-mode impedance as

$$Z_{TE} = \frac{\sqrt{\varepsilon\mu}}{\sqrt{1 - \left(\frac{F_c}{F}\right)^2}} \qquad (5\text{-}17)$$

and the TM-mode impedance as

$$Z_{TM} = 377\sqrt{1 - \left(\frac{F_c}{F}\right)^2} \qquad (5\text{-}18)$$

5-6   WAVEGUIDE TERMINATIONS

When an electromagnetic wave propagates down a waveguide, it must eventually reach the end of the guide. If the end is open, then the wave will propagate into free space. The horn radiator antenna discussed in Chapter 9 is an example of an unterminated waveguide. If the waveguide terminates in a metallic wall, then the wave reflects back down the waveguide from where it came. The interference between incident and reflected waves forms standing waves (see Chapter 4). Such waves are stationary in space but vary in the time domain.

To prevent standing waves, or more properly the reflections that give rise to standing waves, the waveguide must be *terminated* in a matching impedance. When a properly designed antenna is used to terminate the waveguide, it forms the matched load required to prevent reflections. Otherwise, a *dummy load* must be provided. Figure 5-10 shows several types of dummy load.

The classic termination is shown in Fig. 5-11A. The resistor making up the dummy load is a mixture of sand and graphite. When the fields of the propagated

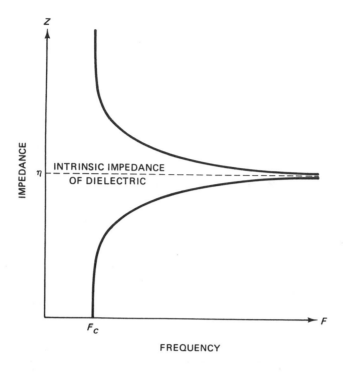

Figure 5-10 Impedance versus frequency diagram.

wave enter the load, they cause currents to flow, which in turn causes heating. Thus, the RF power dissipates in the sand-graphite rather than being reflected back down the waveguide.

A second dummy load is shown in Fig. 5-11B. The resistor element is a carbonized rod critically placed at the center of the electric field. The E field causes currents to flow, resulting in I^2R losses that dissipate the power.

Two bulk loads, similar to the graphite-sand chamber, are shown in Figs. 5-11C, D, and E. A bulk material such as graphite or a carbonized synthetic material is used. These loads are used in much the same way as the sand load; that is, currents are set up and I^2R losses dissipate the power.

The resistive vane load is shown in Fig. 5-11F. The plane of the element is orthogonal to the magnetic lines of force. When the magnetic lines cut across the vane, currents are induced, which gives rise to the I^2R losses. Very little RF energy reaches the metallic end of the waveguide, so there is little reflected energy and a low VSWR.

In some situations it is not desirable to terminate the waveguide in a dummy load. Several reflective terminations are shown in Fig. 5-12. Perhaps the simplest form is the permanent end plate shown in Fig. 5-12A. The metal cover must be welded or otherwise affixed through a very-low-resistance joint. At the substantial power levels typically handled in transmitter waveguides, even small resistances can be important.

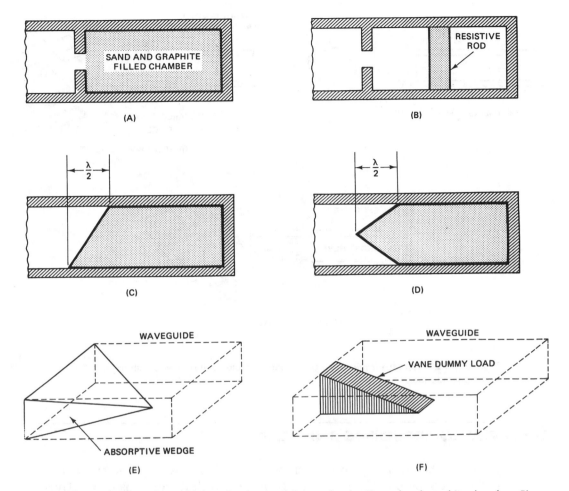

Figure 5-11   Waveguide terminations and dummy loads: A) sand and graphite chamber; B) resistive rod; C) one-sided taper; D) dual taper; E) absorptive wedge; F) vane style.

The end plate shown in Fig. 5-12B uses a quarter-wavelength cup to reduce the effect of joint resistances. The cup places the contact joint at a point that is a quarter-wavelength from the end. This point is a *minimum current node*, so I^2R losses in the contact resistance become less important.

The adjustable short circuit is shown in Fig. 5-12C. The walls of the waveguide and the surface of the plunger form a half-wavelength channel. Because the metallic end of the channel is a short circuit, the impedance reflected back to the front of the plunger is zero ohms, or nearly so. Thus, a *virtual short* exists at the points shown. By this means, the contact or joint resistance problem is overcome.

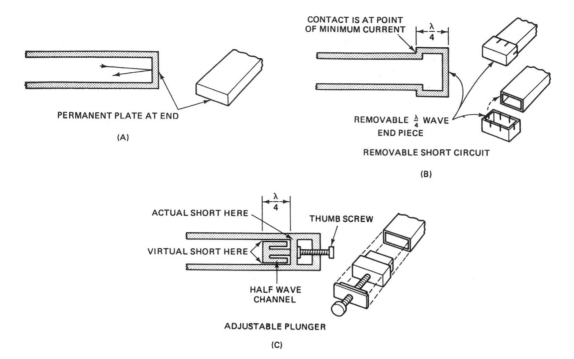

Figure 5-12   End terminations: A) permanent end plate; B) removable short circuit; C) adjustable plunger.

5-7   WAVEGUIDE JOINTS AND BENDS

Joints and bends in any form of transmission line or waveguide are seen as impedance discontinuities and so are points at which disruptions occur. Thus, improperly formed bends and joints are substantial contributors to a poor VSWR. In general, bends, twists, joints, or abrupt changes in waveguide dimension can deteriorate the VSWR by giving rise to reflections.

Extensive runs of waveguide are sometimes difficult to make in a straight line. Although some installations do permit a straight waveguide, many others require directional change. This possibility is especially likely on shipboard installations. Figure 5-13A shows the proper ways to bend waveguide around a corner. In each case, the radius of the bend must be at least two wavelengths, at the lowest frequency that will be propagated in the system.

The twist shown in Fig. 5-13B is used to rotate the polarity of the E and H fields by 90°. This type of section is sometimes used in antenna arrays for phasing the elements. As in the case of the bend, the twist must be made over a distance of at least two wavelengths.

When an abrupt 90° transition is needed, it is better to use two successive 45° bends spaced one quarter-wavelength apart (see Fig. 5-13C). The theory behind this kind of bend is to cause interference of the direct reflection of one bend with the inverted reflection of the other end. The resultant relationship between the fields is reconstructed as if no reflections had taken place.

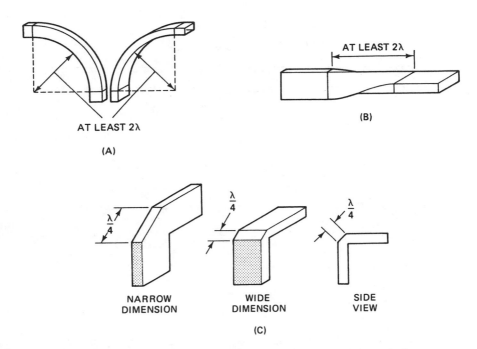

Figure 5-13 Waveguide transitions: A) smooth corner turn; B) twist in line; C) alternate turn method.

Joints are necessary in practical waveguides because it is not possible to construct a single length of practical size for all situations. Three types of common joints are used: *permanent*, *semipermanent*, and *rotating*.

To make a *permanent joint*, the two waveguide ends must be machined extremely flat so that they can be butt-fitted together. A weld or brazed seam bonds the two sections together. Because such a surface represents a tremendous discontinuity, reflections and VSWR will result unless the interior surfaces are milled flat and then polished to a mirrorlike finish.

A *semipermanent joint* allows the joint to be disassembled for repair and maintenance, as well as allowing easier on-site assembly. The most common example of this class is the *choke joint* shown in Fig. 5-14.

One surface of the choke joint is machined flat and is a simple butt-end planar flange. The other surface is the mate to the planar flange, but has a quarter-wavelength circular slot cut at a distance of one quarter-wavelength from the waveguide aperture. The two flanges are shown in side view in Fig. 5-14A, while the slotted end view is shown in Fig. 5-14B. The method for fitting the two ends together is shown in the oblique view in Fig. 5-14C.

Rotating joints are used in cases where the antenna has to point in different directions at different times. Perhaps the most common example of such an application is the radar antenna.

The simplest form of rotating joint is shown in Fig. 5-15. The key to its operation is that the selected mode is symmetrical about the rotating axis. For this reason, a circular waveguide operating in the TM_{01} mode is selected. In this rotating choke joint, the actual waveguide rotates, but the internal fields do not (thereby minimiz-

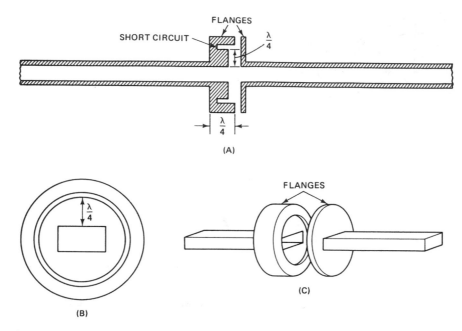

Figure 5-14 Waveguide connection flanges: A) side view; B) end view; C) oblique view showing two waveguides mating.

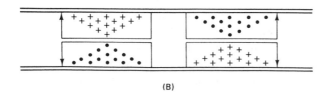

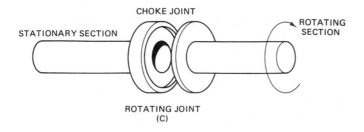

Figure 5-15
Rotating waveguide joint:
a) crossectional view showing electrical field direction;
b) section view showing magnetic field direction;
c) rotating joint mechanism

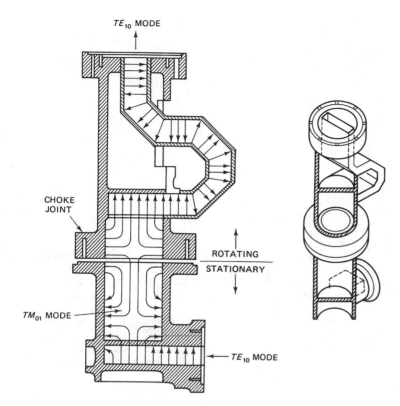

TE_{10} MODE

CHOKE
JOINT

ROTATING

STATIONARY

TM_{01} MODE

TE_{10} MODE

Figure 5-16
Rotating waveguide joint
showing two circular sections
joining sections of rectangular
waveguide.

ing reflections). Because most waveguide is rectangular, however, a somewhat more complex system is needed. Figure 5-16 shows a rotating joint consisting of two circular waveguide sections inserted between segments of rectangular waveguide. On each end of the joint there is a rectangular-to-circular transition section.

In Fig. 5-16, the rectangular input waveguide operates in the TE_{10} mode that is most efficient for rectangular waveguide. The *E* field lines of force couple with the circular segment, thereby setting up a TM_{01} mode wave. The TM_{01} mode has the required symmetry to permit coupling across the junction, where it meets another transition zone and is reconverted to TE_{10} mode.

5-8 WAVEGUIDE COUPLING METHODS

Except possibly for the case where an oscillator exists inside a waveguide, it is necessary to have some form of input or output coupling in a waveguide system. Three basic types of coupling are used in microwave waveguide: *capacitive* or *probe*, *inductive* or *loop*, and *aperture* or *slot*.

Capacitive (also called probe) coupling is shown in Fig. 5-17. This type of coupling uses a vertical radiator inserted into one end of the waveguide. Typically, the probe is quarter-wavelength in a fixed-frequency system. The probe is analogous to

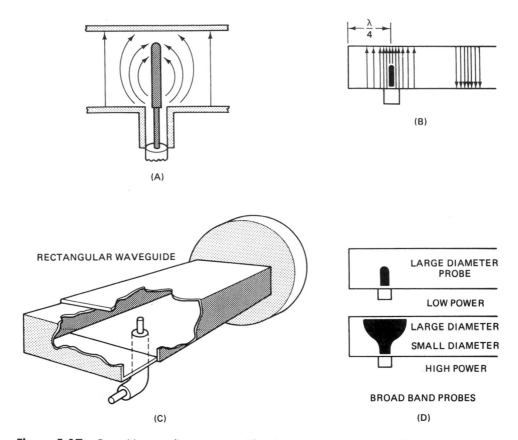

Figure 5-17   Capacitive coupling to waveguide: a) input antenna structure; b) position of antenna element; c) tunable waveguide with adjustable end cap; d) broad band probes.

the vertical antennas used at lower frequencies. A characteristic of this type of radiator is that the E field is parallel to the waveguide top and bottom surfaces. This arrangement satisfies the first boundary condition for the dominant TE_{10} mode.

The radiator is placed at a point that is a quarter-wavelength from the rear wall (Fig. 5-17B). By traversing the quarter-wavelength distance (90° phase shift), being reflected from the rear wall (180° phase shift), and then retraversing the quarter-wavelength distance (another 90° phase shift), the wave undergoes a total phase shift of one complete cycle, or 360°. Thus, the reflected wave arrives back at the radiator in phase to reinforce the outgoing wave. Hence, none of the excitation energy is lost.

Some waveguides have an adjustable end cap (Fig. 5-17C) to accommodate multiple frequencies. The end cap position is varied to accommodate the different wavelength signals.

Figure 5-17D shows high- and low-power broadband probes that are typically not quarter-wavelength except at one particular frequency. Broadbanding is accomplished by attention to the diameter-to-length ratio. The *degree of coupling* can be varied in any of several ways: the *length* of the probe can be varied, the *position* of the probe in the E field can be changed, or *shielding* can be used to partially shade the radiator element.

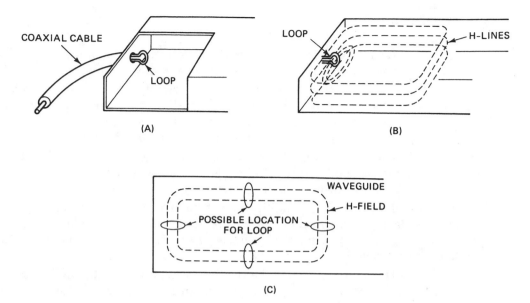

Figure 5-18   a) Coupling loop method for connecting to waveguide; b) H-field around loop; c) permissable positions of loop.

Inductive or loop coupling is shown in Fig. 5-18. A small loop of wire (or other conductor) is placed such that the number of magnetic flux lines is maximized. This form of coupling is popular on microwave receiver antennas to make a waveguide-to-coaxial cable transition. In some cases, the loop is formed by the pigtail lead of a detector diode that, when combined with a local oscillator, downconverts the microwave signal to an IF frequency in the 30- to 300-MHz region.

Aperture or slot coupling is shown in Fig. 5-19. This type of coupling is used to couple together two sections of waveguide, as on an antenna feed system. Slots can be designed to couple electric, magnetic, or electromagnetic fields. In Fig. 5-19, slot A is placed at a point where the *E* field peaks and so allows electrical field coupling. Similarly, slot B is at a point where the *H* field peaks and so allows magnetic field coupling. Finally, we have slot C, which allows electromagnetic field coupling.

Slots can also be characterized according to whether they are *radiating* or *nonradiating*. A nonradiating slot is cut at a point that does not interrupt the flow of currents in the waveguide walls. The radiating slot, on the other hand, does interrupt currents flowing in the walls. A radiating slot is the basis for several forms of antenna, which are discussed in Chapter 9.

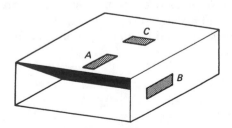

Figure 5-19
Aperture or slot coupling.

5-9 SUMMARY

1. A waveguide is a pipe in which an electromagnetic wave can propagate as either transverse electric or transverse magnetic waves.

2. The dimensions of the waveguide determine the longest wavelength, and hence lowest frequency, that will propagate. This wavelength is that for which one half-wavelength fits in the long dimension of the rectangular waveguide aperture. In general, a waveguide supports a bandwidth of 30% to 40%.

3. The dominant mode of propagation in a waveguide is called the TE_{10} mode, that is, the mode for which there is one half-wavelength in the long (a) dimension and zero half-wavelength in the short (b) dimension.

4. Two boundary conditions must be met for a wave to propagate in a waveguide: (1) *the electric field must be orthogonal to the conductor in order to exist at the surface of that conductor*, and (2) *the magnetic field must not be orthogonal to the surface of the waveguide.*

5. Three velocities must be considered with respect to waveguides: *free-space velocity* (speed of light, c), *group velocity*, and *phase velocity*. Of these, the group velocity is less than c, while phase velocity can be greater than c.

6. The waveguide impedance is directly proportional to the characteristic impedance of the waveguide (E/H), the wavelength of the signal in the waveguide, and inversely proportional to the wavelength of the signal in free space.

7. Waveguides must be terminated in a dissipating load such as an antenna or a dummy load such as an absorptive wedge.

8. Joints in waveguide runs must be made smooth and well polished or else reflections and impedance discontinuities will exist. Rotating joints tend to be made with circular waveguide choke joints fitted between sections of rectangular-to-circular transition elements.

9. Bends in waveguide must be made gently, over at least two wavelengths. Abrupt 90° bends are better made in two successive 45° bends.

10. Three forms of waveguide coupling are commonly used: *capacitive* or *probe*, *inductive* or *loop*, and *aperture* or *slot*. Slots can be radiating or nonradiating, depending on whether or not they interrupt a current antinode.

5-10 RECAPITULATION

Now return to the objectives and prequiz questions at the beginning of the chapter and see how well you can answer them. If you cannot answer certain questions, place a check by each and review the appropriate parts of the text. Next, try to answer the following questions and work the problems using the same procedure.

QUESTIONS AND PROBLEMS

1. List three types of loss encountered in regular transmission lines: _____, _____, and _____.

2. At microwave frequencies, the _____ effect causes ohmic losses to increase dramatically.

3. The rectangular waveguide can be modeled as an extension of the _____-wave stub from parallel transmission line theory.

4. Bandwidths of ____ % to ____ % are typically available on waveguides.

5. The TEM mode (will/will not) propagate efficiently in a waveguide. In this mode, the electric and magnetic fields are _____ to the direction of travel.

6. State the two boundary conditions regulating propagation in the waveguide.

7. List two modes that will propagate efficiently in waveguide.

8. The TE_{10} mode is also called the _____ mode. This mode occurs at the _____ frequency at which a wave will propagate.

9. A wave is such that two wavelengths will fit into the a dimension of a rectangular waveguide, and zero wavelengths fit into the b dimension. Write the standard notation for the transverse electric version of this mode.

10. List three different velocities pertaining to waveguides.

11. A signal forms an incident angle of 37° with the surface of the waveguide. Calculate (a) group velocity, (b) phase velocity.

12. A signal forms an incident angle of 45° with the surface of the waveguide. Calculate both group and phase velocities.

13. What is the group velocity of a signal that forms a 90° incident angle with the surface of the waveguide?

14. Write the first six modes (m and $n = 0, 1, 2$) for a rectangular waveguide with an air dielectric.

15. In problem 14, the a dimension is 60 mm and the b dimension is 28 mm. Calculate the cutoff wavelength for the dominant mode.

16. A waveguide has the following dimensions: 4.5×1.75 cm. Calculate (a) cutoff wavelength, (b) cutoff frequency. Assume the TE_{10} mode.

17. A 5.6-GHz signal is injected into a waveguide that has a cutoff frequency of 4.5 GHz. Calculate the wavelength of the signal in the waveguide.

18. A 10.75-GHz signal is injected into a waveguide that has a cutoff frequency of 8.8 GHz. Calculate the wavelength of the signal in the waveguide.

19. In questions 17 and 18, we use a rectangular waveguide with an air dielectric. Calculate the impedance of the waveguide.

20. To prevent reflections in the end of a waveguide that is not connected to a radiator antenna, we need to use a _____ consisting of an absorptive material.

KEY EQUATIONS

1. Group velocity in a waveguide:

$$V_g = c \sin a$$

2. Phase velocity in a waveguide:

$$V_p = \frac{c}{\sin a}$$

3. Relationship between phase and group velocity:

$$c = \sqrt{V_p V_g}$$

4. Relationship between frequency and free-space wavelength:

$$c = F \lambda_o$$

5. Phase velocity as a function of wavelength:

$$V_p = \frac{\lambda_c}{\lambda_o}$$

6. Wavelength in a waveguide:

$$\lambda = \frac{V_p \lambda_o}{c}$$

7. General mode equation for air dielectric:

$$\frac{1}{(\lambda_c)^2} = \left(\frac{m}{2a}\right)^2 + \left(\frac{n}{2b}\right)^2$$

8. Cutoff wavelength:

$$\lambda = 2a$$

9. Cutoff frequency:

$$F_c = \frac{c}{2a}$$

10. General mode equation for all dielectrics:

$$F_c = \frac{1}{2\sqrt{\mu\varepsilon}} \sqrt{\left(\frac{m}{a}\right)^2 + \left(\frac{n}{b}\right)^2}$$

11. Cutoff wavelength:

$$\lambda_c = \frac{2}{\sqrt{(m/a)^2 (n/b)^2}}$$

12. Wavelength in a waveguide:

$$\lambda_g = \frac{\lambda_o}{\sqrt{1 - [(F_c / F)]^2}}$$

13. Waveguide impedance (general):

$$Z = \frac{Z_o \lambda_g}{\lambda_o}$$

14. Waveguide impedance for a rectangular waveguide and air dielectric:

$$Z = \frac{120\pi\lambda_g}{\lambda_o}$$

15. Propagation constant as a function of frequency:

$$\beta = \omega\sqrt{\varepsilon\mu}\sqrt{1 - \left(\frac{F_c}{F}\right)^2}$$

16. TE-mode impedance:

$$Z_{TE} = \frac{\sqrt{\mu/\varepsilon}}{\sqrt{1 - \left(\frac{(F_c}{F)}\right)^2}}$$

17. TM-mode impedance:

$$Z_{TM} = 377\sqrt{1 - \left(\frac{F_c}{F}\right)^2}$$

CHAPTER 6

The Smith Chart

OBJECTIVES

1. Understand the structure of the Smith chart.
2. Know the features and capabilities of the Smith chart.
3. Be able to interpret a Smith chart plot.
4. Learn to draw Smith chart plots.

6-1   PREQUIZ

These questions test your prior knowledge of the material in this chapter. Try answering them before you read the chapter. Look for the answers (especially those you answered incorrectly) as you read the text. After you have finished studying the chapter, try answering these questions again and those at the end of the chapter.

1. A Smith chart consists of a series of _____ circles that depict complex impedances.
2. The horizontal line through the center of the Smith chart represents pure _____.
3. The circles on the Smith chart that are bisected by the horizontal line through the center of the Smith chart and are tangent to the far-right end of the line represent constant _____ points on the graph.
4. A circle with its center in the center of the chart has a radius that passes through the 0.30 point on the horizontal line. This is a _____ circle.

6-2   THE SMITH CHART

The mathematics of transmission lines and certain other devices becomes cumbersome at times, especially when dealing with complex impedances and nonstandard situations. In 1939, Phillip H. Smith published a graphical device for solving these problems, followed in 1945 by an improved version of the chart. This graphic aid, somewhat modified over time, is still in constant use in microwave electronics and other fields where complex impedances and transmission line problems are found. The *Smith chart* is indeed a powerful tool for the RF designer.

The modern Smith chart is shown in Fig. 6-1 and consists of a series of overlapping orthogonal circles (circles that intersect each other at right angles). In this chapter, we will dissect the Smith chart so that the origin and use of these circles are apparent. The set of orthogonal circles makes up the basic structure of the Smith chart.

6-2.1 The Normalized Impedance Line

A baseline is highlighted in Fig. 6-2 and bisects the Smith chart outer circle. This line is called the *pure resistance line* and forms the reference for measurements made on the chart. Recall that a complex impedance contains both resistance and reactance and is expressed in the mathematical form:

$$Z = R \pm jX \qquad (6\text{-}1)$$

where

Z = complex impedance
R = resistive component of the impedance
X = reactive component of the impedance

The pure resistance line represents the situation where $X = 0$, and the impedance is therefore equal to the resistive component only. To make the Smith chart universal, the impedances along the pure resistance line are *normalized* with reference to system impedance (for example, Z_o in transmission lines); for most microwave RF systems, the system impedance is standardized at 50 Ω. In order to normalize the actual impedance, divide it by the system impedance. For example, if the load impedance of a transmission line is Z_L, and the characteristic impedance of the line is Z_o, then $Z = Z_L/Z_o$. In other words,

$$Z = \frac{R \pm jX}{Z_o} \qquad (6\text{-}2)$$

The pure resistance line is structured such that the system standard impedance is in the center of the chart and has a normalized value of 1.0 (see point A in Fig. 6-2). This value derives from the fact that $Z_o/Z_o = 1.0$.

To the left of the 1.0 point are decimal fraction values used to denote impedances less than the system impedance. For example, in a 50-Ω transmission line system with a 25-Ω load impedance, the normalized value of impedance is 25 Ω/50 Ω or 0.50. This point is marked B in Fig. 6-2. Similarly, points to the right of 1.0 are greater than 1 and denote impedances that are higher than the system impedance.

NAME	TITLE	DWG. NO.
SMITH CHART FORM 82-BSPR (9-66)	KAY ELECTRIC COMPANY, PINE BROOK, N.J., © 1966. PRINTED IN U.S.A.	DATE

IMPEDANCE OR ADMITTANCE COORDINATES

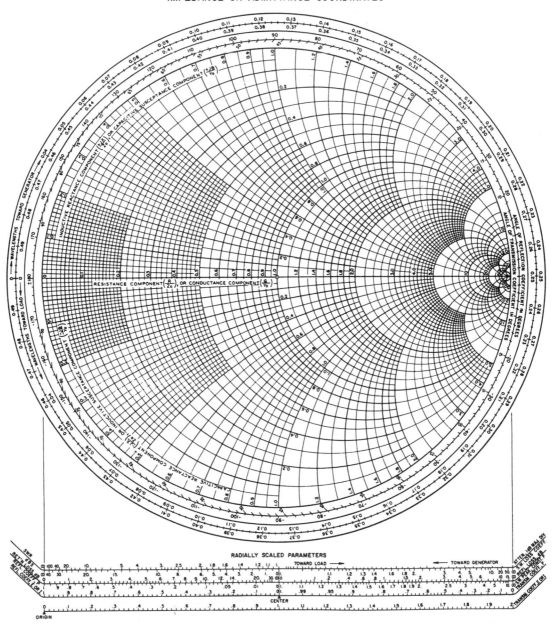

RADIALLY SCALED PARAMETERS

Figure 6-1 The Smith chart.

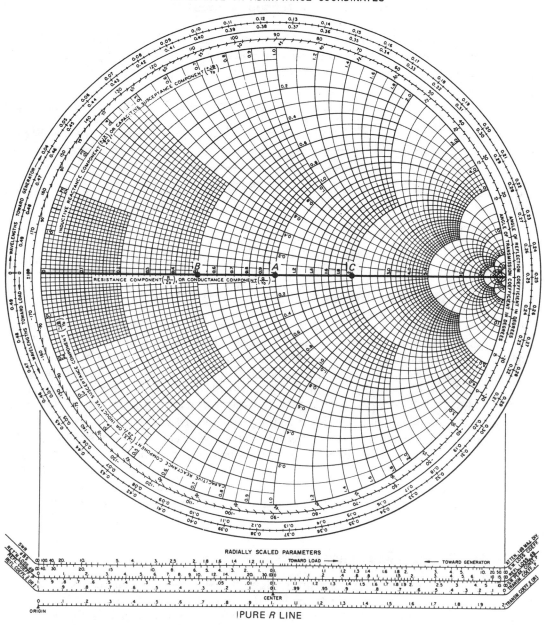

Figure 6-2   Pure resistance line.

For example, in a 50-Ω system connected to a 100-Ω resistive load, the normalized impedance is 100 Ω/50 Ω, or 2.0; this value is shown as point C in Fig. 6-2. By using normalized impedances, we can use the Smith chart for almost any practical combination of system and load and/or source impedances, whether resistive, reactive, or complex.

Reconversion of the normalized impedance to actual impedance values is done by multiplying the normalized impedance by the system impedance. For example, if the resistive component of a normalized impedance is 0.45, then the actual impedance is

$$Z = Z_{normal}(Z_o) \tag{6-3}$$

$$= 0.45 \ (50 \ \Omega) \tag{6-4}$$

$$= 22.5 \ \Omega \tag{6-5}$$

6-2.2 The Constant Resistance Circles

The *isoresistance circles*, also called the *constant resistance circles*, represent points of equal resistance. Several of these circles are shown highlighted in Fig. 6-3. These circles are all tangent to the point at the right-hand extreme of the pure resistance line and are bisected by that line. When you construct complex impedances (for which X = nonzero) on the Smith chart, the points on these circles will all have the same resistive component. Circle A, for example, passes through the center of the chart and so has a normalized constant resistance of 1.0. Note that impedances that are pure resistances ($Z = R + j0$) will fall at the intersection of a constant resistance circle and the pure resistance line, while complex impedances (that is, X not equal to zero) will appear at all other points on the circle. In Fig. 6-2, circle A passes through the center of the chart and represents all points on the chart with a normalized resistance of 1.0. This particular circle is sometimes called the *unity resistance circle*.

6-2.3 The Constant Reactance Circles

Constant reactance circles are highlighted in Fig. 6-4. The circles (or circle segments) *above* the pure resistance line (Fig. 6-4A) represent the *inductive reactance* (+X),while those circles (or segments) *below* the pure resistance line (Fig. 6-4B) represent *capacitive reactance* (-X). In both cases, circle A represents a normalized reactance of 0.80.

One of the outer circles (circle A in Fig. 6-4C) is called the *pure reactance circle*. Points along circle A represent reactance only, in other words, an impedance of $Z = 0 \pm jX$ ($R = 0$).

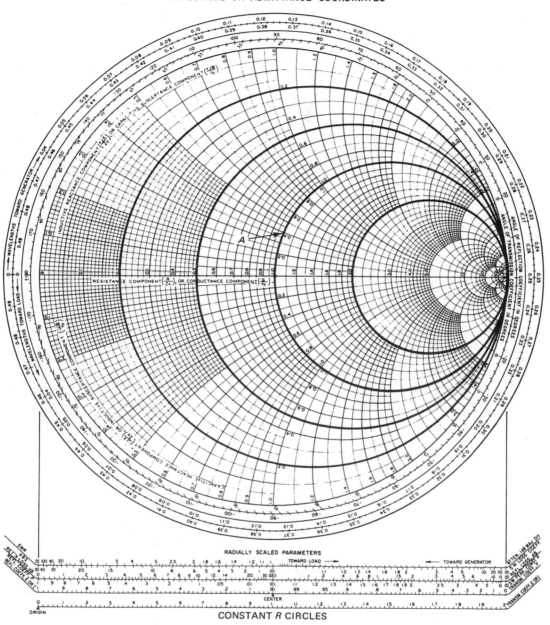

IMPEDANCE OR ADMITTANCE COORDINATES

Figure 6-3   Isoresistance circles.

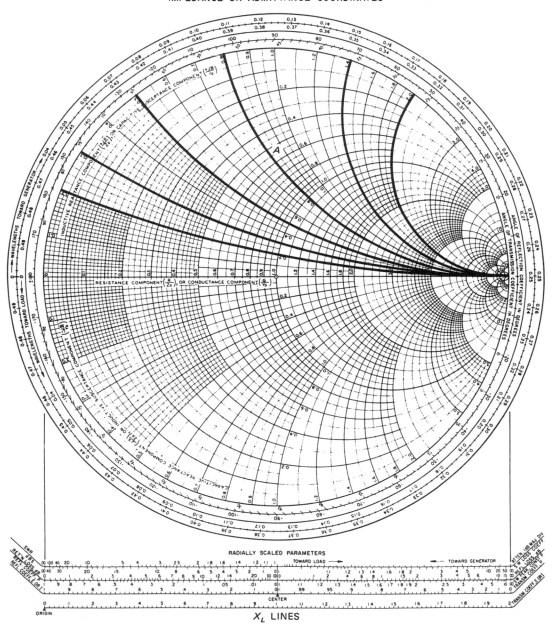

IMPEDANCE OR ADMITTANCE COORDINATES

Figure 6-4 Constant reactance circles: A) inductance reactance (+X) lines; B) capacitance resistance (-X); C) pure reactance circle; D) plotting admittance and impedance on a Smith chart.

Figure 6-4 (*continued*)

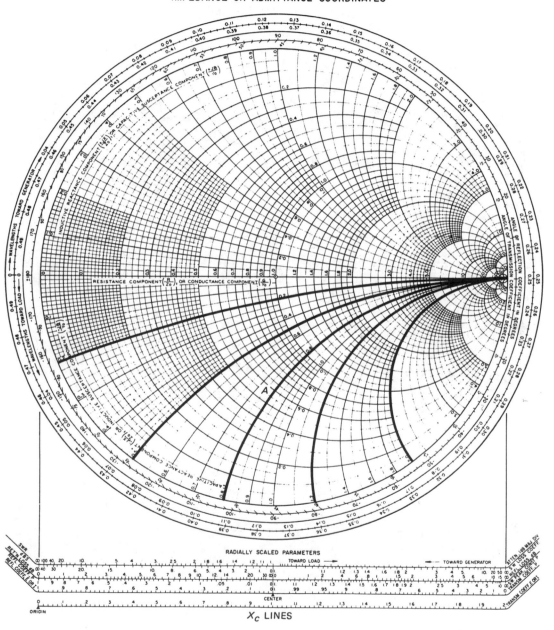

IMPEDANCE OR ADMITTANCE COORDINATES

RADIALLY SCALED PARAMETERS

X_C LINES

Figure 6-4b

Continued

Figure 6-4 *(continued)*

NAME	TITLE	DWG. NO.
SMITH CHART FORM 82-BSPR(9-66)	KAY ELECTRIC COMPANY, PINE BROOK, N.J., © 1966. PRINTED IN U.S.A.	DATE

IMPEDANCE OR ADMITTANCE COORDINATES

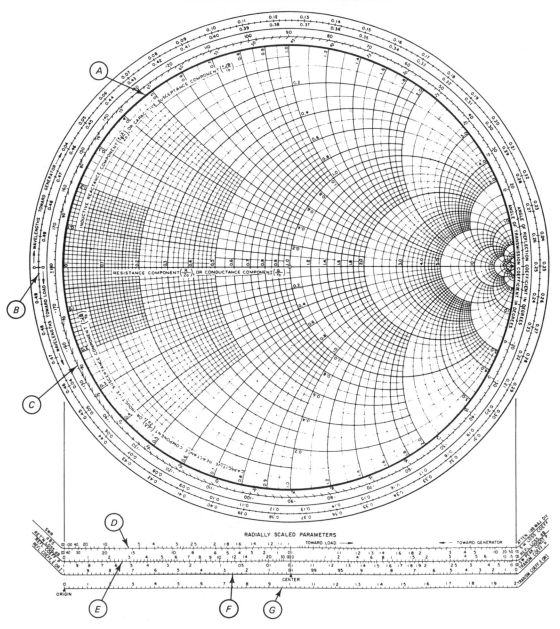

Figure 6-4c

Figure 6-4   (continued)

NAME	TITLE		DWG. NO
SMITH CHART FORM 82-BSPR (9-66)	KAY ELECTRIC COMPANY, PINE BROOK, N.J., © 1966. PRINTED IN U.S.A.		DATE

IMPEDANCE OR ADMITTANCE COORDINATES

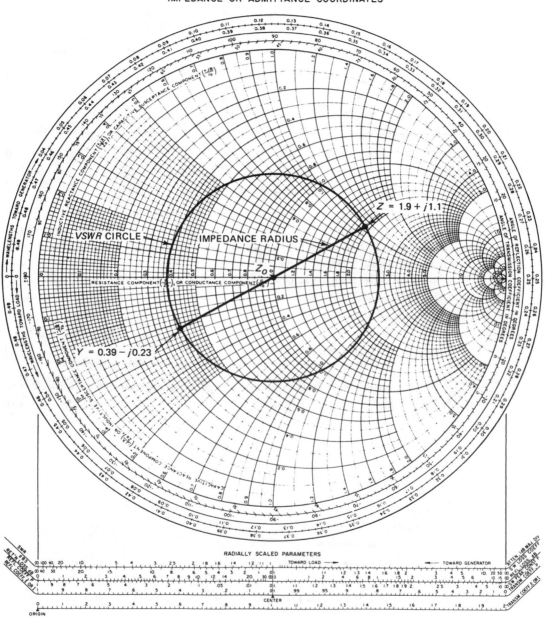

RADIALLY SCALED PARAMETERS

Figure 6-4d

Figure 6-4D shows how to plot impedance and admittance on the Smith chart. Consider an example in which system impedance Z_o is 50 Ω and the load impedance is $Z_L = 95 + j55$ Ω. This load impedance is normalized to

$$Z = \frac{Z_L}{Z_o} \qquad (6\text{-}6)$$

$$Z = \frac{95 + j55 \, \Omega}{50 \, \Omega} \qquad (6\text{-}7)$$

$$Z = 1.9 + j1.1 \qquad (6\text{-}8)$$

An *impedance radius* is constructed by drawing a line from the point represented by the normalized load impedance, 1.9 + j1.1, to the point represented by the normalized system impedance (1.0) in the center of the chart. A circle is constructed from this radius and is called the *VSWR circle*.

Admittance is the reciprocal of impedance and so is found from

$$Y = \frac{1}{Z} \qquad (6\text{-}9)$$

Because impedances in transmission lines are rarely purely resistive, but rather contain a reactive component also, impedances are expressed using complex notation:

$$Z = R \pm jX \qquad (6\text{-}10)$$

where

Z = complex impedance
R = resistive component
X = reactive component

To find the *complex admittance*, we take the reciprocal of the complex impedance by multiplying the simple reciprocal by the complex conjugate of the impedance. For example, when the normalized impedance is 1.9 + j1.1, the normalized admittance will be

$$Y = \frac{1}{Z} \qquad (6\text{-}11)$$

$$Y = \left(\frac{1}{1.9 + j1.1} \right) \times \left(\frac{1.9 - j1.1}{1.9 - j1.1} \right) \tag{6-12}$$

$$Y = \frac{1.9 - j1.1}{3.6 + 1.2} \tag{6-13}$$

$$= 0.39 - j0.23 \tag{6-14}$$

One delight of the Smith chart is that this calculation is reduced to a quick graphical interpretation! Simply extend the impedance radius through the 1.0 center point until it intersects the VSWR circle again. This point of intersection represents the normalized admittance of the load.

6-2.4 Outer Circle Parameters

The standard Smith chart shown in Fig. 6-4C contains three concentric calibrated circles on the outer perimeter of the chart. Circle *A* is the pure reactance circle and has already been discussed. The other two circles define wavelength distance relative to either the load or generator (*B*) end of the transmission line and either the transmission or reflection coefficient angle in degrees (*C*).

There are two scales on the *wavelengths* circle (*B* in Fig 6-4C) and both have their zero origin on the left-hand extreme of the pure resistance line. Both scales represent *one half-wavelength for one entire revolution* and are calibrated from 0 through 0.50 such that these two points are identical with each other on the circle. In other words, starting at the zero point and traveling 360° around the circle brings one back to zero, which represents one half-wavelength, or 0.5λ.

Although both wavelength scales are of the same magnitude (0 to 0.50), they are opposite in direction. The outer scale is calibrated clockwise and represents *wavelengths toward the generator*; the inner scale is calibrated counterclockwise and represents *wavelengths toward the load*. These two scales are complementary at all points. Thus, 0.12 on the outer scale corresponds to 0.50 minus 0.12 or 0.38 on the inner scale.

The *angle of transmission coefficient* and *angle of reflection coefficient* scales are shown in circle *C* in Fig. 6-4C. These scales are relative phase angles between reflected and incident waves. Recall from transmission line theory (see Chapter 4) that a short circuit at the load end of the line reflects the signal back toward the generator 180° out of phase with the incident signal; an open line (infinite impedance) reflects the signal back to the generator in phase (that is, 0°) with the incident signal. These facts are shown on the Smith chart by the fact that both scales start at 0° on the right-hand end of the pure resistance line, which corresponds to an infinite resistance, and go halfway around the circle to 180° at the 0 end of the pure resistance line. Note that the upper half-circle is calibrated from 0° to +180°, and the bottom half-circle is calibrated from 0° to +180°, reflecting inductive or capacitive reactance situations, respectively.

6-2.5 Radially Scaled Parameters

There are six scales laid out on five lines (*D* through *G* in Fig. 6-4C and in expanded form in Fig. 6-5) at the bottom of the Smith chart. These scales are called the *radially scaled parameters*, and are both very important and often overlooked. With these scales, we can determine such factors as VSWR (both as a ratio and in decibels), return loss in decibels, voltage or current reflection coefficient, and power reflection coefficient.

The *reflection coefficient* (*P*) is defined as the *ratio of the reflected signal to the incident signal*. For voltage or current,

$$P = \frac{E_{ref}}{E_{inc}} \qquad (6\text{-}15)$$

and

$$P = \frac{I_{ref}}{I_{inc}} \qquad (6\text{-}16)$$

Power is proportional to the square of voltage or current, so

$$P_{pwr} = P^2 \qquad (6\text{-}17)$$

or

$$P_{pwr} = \frac{P_{ref}}{P_{inc}} \qquad (6\text{-}18)$$

EXAMPLE 6-1

Ten watts of microwave RF power are applied to a lossless transmission line, of which 2.8 W are reflected from the mismatched load. Calculate the reflection coefficient.

$$P_{pwr} = \frac{P_{ref}}{P_{inc}} \qquad (6\text{-}19)$$

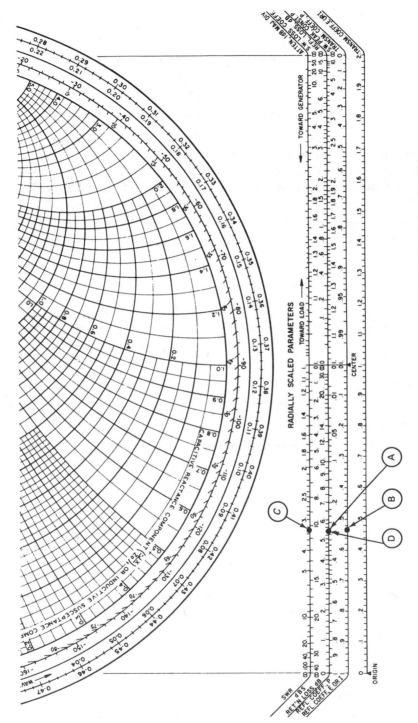

Figure 6-5 Radially scaled parameters on slide rule chart below the circles.

$$= \frac{2.8\,\text{W}}{10\,\text{W}} \tag{6-20}$$

$$= 0.28 \tag{6-21}$$

The voltage reflection coefficient (P) is found by taking the square root of the power reflection coefficient, so in this example it is equal to 0.529. These points are plotted at A and B in Fig. 6-5.

The *standing-wave ratio* (SWR) can be defined in terms of the reflection coefficient:

$$\text{VSWR} = \frac{1 + P}{1 - P} \tag{6-22}$$

or

$$\text{VSWR} = \frac{1 + \sqrt{P_{\text{pwr}}}}{1 - \sqrt{P_{\text{pwr}}}} \tag{6-23}$$

or, in our example,

$$\text{VSWR} = \frac{1 + \sqrt{0.28}}{1 - \sqrt{0.28}} \tag{6-24}$$

$$\text{VSWR} = \frac{1 + 0.529}{1 - 0.529} \tag{6-25}$$

$$\frac{1.529}{0.471} = 3.25{:}1 \tag{6-26}$$

or, in decibel form,

$$\text{VSWR}_{\text{dB}} = 20 \log (\text{VSWR}) \tag{6-27}$$

$$\text{VSWR}_{\text{dB}} = 20 \log (20) \qquad\qquad (6\text{-}28)$$

$$\text{VSWR}_{\text{dB}} = (20)\,(0.510) = 10.2 \text{ dB} \qquad\qquad (6\text{-}29)$$

These points are plotted at C in Fig. 6-5. Shortly, we will work an example to show how these factors are calculated in a transmission line problem from a known complex load impedance.

    Transmission loss is a measure of the one-way loss of power in a transmission line because of reflection from the load. *Return loss* represents the two-way loss and so is exactly twice the transmission loss. Return loss is found from

$$= \text{Loss}_{\text{ret}} = 10 \log (P_{\text{pwr}}) \qquad\qquad (6\text{-}30)$$

and, for our example in which $P_{\text{pwr}} = 0.28$,

$$= \text{Loss}_{\text{ret}} = 10 \log (0.28) \qquad\qquad (6\text{-}31)$$

$$= (10)\,(\text{-}0.553) = -5.53 \text{ dB} \qquad\qquad (6\text{-}32)$$

This point is shown as D in Fig. 6-5.

    The *transmission loss coefficient* can be calculated from

$$\text{TLC} = \frac{1 + P_{\text{pwr}}}{1 - P_{\text{pwr}}} \qquad\qquad (6\text{-}33)$$

or, for our example,

$$\text{TLC} = \frac{1 + 0.28}{1 - 0.28} \qquad\qquad (6\text{-}34)$$

$$= \frac{1.28}{0.72} = 1.78 \qquad\qquad (6\text{-}35)$$

The TLC is a correction factor that is used to calculate the attenuation due to mismatched impedance in a lossy, as opposed to an ideal lossless, line. The TLC is found from laying out the impedance radius on the *loss coefficient scale* on the radially scaled parameters at the bottom of the chart.

6-3   SMITH CHART APPLICATIONS

One of the best ways to demonstrate the usefulness of the Smith chart is by practical example. In the sections to follow, we will look at two general cases: *transmission line problems* and *stub matching systems*.

6-3.1 Transmission Line Problems

Figure 6-6 shows a 50-Ω transmission line connected to a complex load impedance Z_L of 36 + j40 Ω. The transmission line has a velocity factor (v) of 0.80, which means that the wave propagates along the line at 8/10 the speed of light (c = 300,000,000 m/s). The length of the transmission line is 28 cm. The generator (V_{in}) is operated at a frequency of 4.5 GHz and produces a power output of 1.5 W. Let's see what we can tell from the Smith chart (Fig. 6-7).

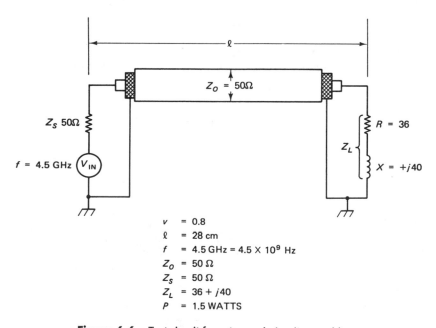

$$
\begin{aligned}
v &= 0.8 \\
\ell &= 28 \text{ cm} \\
f &= 4.5\,\text{GHz} = 4.5 \times 10^9 \text{ Hz} \\
Z_0 &= 50\ \Omega \\
Z_S &= 50\ \Omega \\
Z_L &= 36 + j40 \\
P &= 1.5 \text{ WATTS}
\end{aligned}
$$

Figure 6-6   Test circuit for a transmission line problem.

First, normalize the load impedance. This is done by dividing the load impedance by the system impedance (in this case, $Z_o = 50\ \Omega$):

$$Z = \frac{36 = j40\ \Omega}{50\ \Omega} \tag{6-36}$$

$$= 0.72 + j0.8 \tag{6-37}$$

The resistive component of impedance Z is located along the 0.72 pure resistance circle (see Fig. 6-7). Similarly, the reactive component of impedance Z is located by traversing the 0.72 constant resistance circle until the $+j0.8$ constant reactance circle is intersected. This point graphically represents the normalized load impedance $Z = 0.72 + j0.80$. A VSWR circle is constructed with an impedance radius equal to the line between 1.0 (in the center of the chart) and the $0.72 + j0.8$ point.

At a frequency of 4.5 GHz, the length of a wave propagating in the transmission line, assuming a velocity factor of 0.80, is

$$\lambda_{\text{line}} = \frac{cv}{F_{\text{Hz}}} \tag{6-38}$$

$$\lambda_{\text{line}} = \frac{(3 \times 10^8\ \text{m/s})\,(0.80)}{4.5 \times 10^9\ \text{Hz}} \tag{6-39}$$

$$\lambda_{\text{line}} = \frac{2.4 \times 10^8\ \text{m/s}}{4.5 \times 10^9\ \text{Hz}} \tag{6-40}$$

$$\lambda_{\text{line}} = 0.053\ \text{m} \times \frac{100\ \text{cm}}{\text{m}} = 5.3\ \text{cm} \tag{6-41}$$

One wavelength is 5.3 cm, so a half-wavelength is 5.3 cm/2, or 2.65 cm. The 28 cm line is 28-cm/5.3 cm, or 5.28 wavelengths long. A line drawn from the center (1.0) to the load impedance is extended to the outer circle and intersects the circle at 0.1325. Because one complete revolution around this circle represents one half-wavelength, 5.28 wavelengths from this point represent ten revolutions plus 0.28 more. The residual 0.28 wavelength is added to 0.1325 to form a value of (0.1325 +

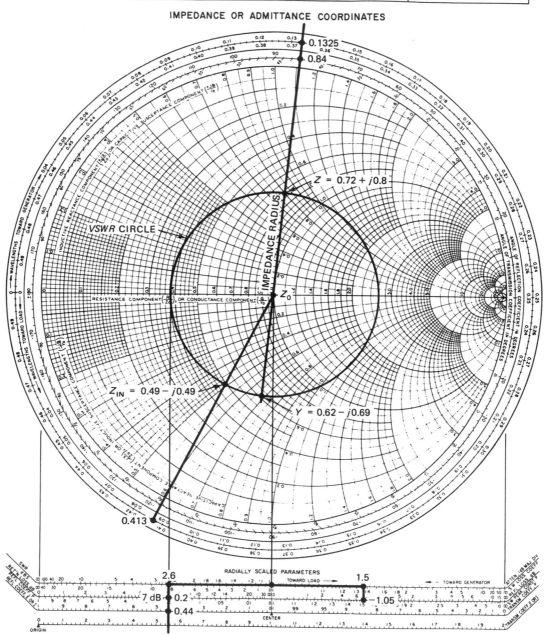

Figure 6-7 Results from test circuit.

0.28), or 0.413. The point 0.413 is located on the circle and is marked. A line is then drawn from 0.413 to the center of the circle, and it intersects the VSWR circle at 0.49 − j0.49, which represents the input impedance (Z_{in}) looking into the line.

To find the actual impedance represented by the normalized input impedance, we have to *denormalize* the Smith chart impedance by multiplying the result by Z_o:

$$Z_{in} = 0.49 - j0.49 \ (50 \ \Omega) \tag{6-42}$$

$$= 24.5 - j24.5 \ \Omega \tag{6-43}$$

It is this impedance that must be matched at the generator by a conjugate matching network.

The admittance represented by the load impedance is the reciprocal of the load impedance and is found by extending the impedance radius through the center of the VSWR circle until it intersects the circle again. This point is found and represents the admittance $Y = 0.62 - j0.69$. Confirming the solution mathematically,

$$Y = \frac{1}{Z} \tag{6-44}$$

$$Y = \frac{1}{0.72 + j0.80} \times \frac{0.72 - j0.80}{0.72 - j0.80} \tag{6-45}$$

$$Y = \frac{0.72 + j0.80}{1.16} = 0.62 - j0.69 \tag{6-46}$$

The VSWR is found by transferring the *impedance radius* of the VSWR circle to the radial scales below. The radius (0.72 − 0.8) is laid out on the VSWR scale (topmost of the radially scaled parameters) with a pair of dividers from the center mark, and we find that the VSWR is approximately 2.6:1. The decibel form of VSWR is 8.3 dB (next scale down from VSWR), and this is confirmed by

$$VSWR_{dB} = 20 \log (VSWR) \tag{6-47}$$

$$= (20) \log (2.7) \tag{6-48}$$

$$= (20) (0.431) = 8.3 \text{ dB} \tag{6-49}$$

The transmission loss coefficient is found in a manner similar to the VSWR, using the radially scaled parameter scales. In practice, once we have found the VSWR we need only drop a perpendicular line from the 2.6:1 VSWR line across the other scales. In this case, the line intersects the voltage reflection coefficient at 0.44. The power reflection coefficient (P_{pwr}) is found from the scale above and is equal to P^2. The perpendicular line intersects the power reflection coefficient line at 0.20.

The angle of reflection coefficient is found from the outer circles of the Smith chart. The line from the center to the load impedance ($Z = 0.72 + j0.80$) is extended to the *angle of reflection coefficient in degrees* circle and intersects it at approximately 84°. The reflection coefficient is therefore 0.44/84°.

The transmission loss coefficient (TLC) is found from the radially scaled parameter scales also. In this case, the impedance radius is laid out on the *loss coefficient* scale, where it is found to be 1.5. This value is confirmed from

$$\text{TLC} = \frac{1 + P_{pwr}}{1 - P_{pwr}} \tag{6-50}$$

$$= \frac{1 + 0.20}{1 - 0.21} \tag{6-51}$$

$$= (1.20/0.79) = 1.5 \tag{6-52}$$

The *return loss* is also found by dropping the perpendicular from the VSWR point to the return loss *dB* line, and the value is found to be approximately 7 dB, which is confirmed by

$$\text{Loss}_{ret} = 10 \log (P_{pwr}) \text{ dB} \tag{6-53}$$

$$= 10 \log (0.21) \text{ dB} \tag{6-54}$$

$$= (10) \, (-0.677) \text{ dB} \qquad (6\text{-}55)$$

$$= 6.77 \text{ dB} \qquad (6\text{-}56)$$

The reflection loss is the amount of RF power reflected back down the transmission line from the load. The difference between incident power supplied by the generator (1.5 W in our example), is the *absorbed power* (P_a), or in the case of an antenna, the radiated power. The reflection loss is found graphically by dropping a perpendicular from the TLC point (or by laying out the impedance radius on the Reflection Loss dB scale), and in our example (Fig. 6-7) is −1.05 dB. We can check our calculations. The return loss was −7 dB, so we know that

$$-7 \text{ dB} = 10 \log\!\left(\frac{P_{\text{ref}}}{P_{\text{inc}}}\right) \qquad (6\text{-}57)$$

$$-7 = 10 \log\!\left(\frac{P_{\text{ref}}}{1.5 \text{ W}}\right) \qquad (6\text{-}58)$$

$$\frac{-7}{10} = \log\!\left(\frac{P_{\text{ref}}}{1.5 \text{ W}}\right) \qquad (6\text{-}59)$$

$$10^{(-7/10)} = \frac{P_{\text{ref}}}{1.5 \text{ W}} \qquad (6\text{-}60)$$

$$0.2 = \frac{P_{\text{ref}}}{1.5 \text{ W}} \qquad (6\text{-}61)$$

$$(0.2)(1.5 \text{ W}) = P_{\text{ref}} \qquad (6\text{-}62)$$

$$= 0.3 \text{ W} = P_{\text{ref}} \qquad (6\text{-}63)$$

The power absorbed by the load (P_a) is the difference between incident power (P_{inc}) and reflected power (P_{ref}). If 0.3 W is reflected, then that means the absorbed power is 1.5 – 0.3, or 1.2 W. The reflection loss is –1.05 dB, and can be checked from

$$-1.05 \text{ dB} = 10 \log\left(\frac{P_a}{P_{\text{inc}}}\right) \qquad (6\text{-}64)$$

$$\frac{-1.05 \text{ dB}}{10} = \frac{P_a}{1.5 \text{ W}} \qquad (6\text{-}65)$$

$$10^{(-1.05/10)} = \frac{P_a}{1.5 \text{ W}} \qquad (6\text{-}66)$$

$$0.785 = \frac{P_a}{1.5 \text{ W}} \qquad (6\text{-}67)$$

$$(1.5 \text{ W}) \times (0.785) = P_a \qquad (6\text{-}68)$$

$$1.2 \text{ W} = P_a \qquad (6\text{-}69)$$

Now let's review what we have learned from our Smith chart. Recall that we input 1.5 W of 4.5-GHz microwave RF signal into a 50-Ω transmission line that was 28 cm long. The load connected to the transmission line has an impedance of 36 + j40. From the Smith chart, we discovered the following:

Admittance (load):	0.62 – j0.69
VSWR:2.6:	1
VSWR(dB):	8.3 dB
Reflection coefficient (E):	0.44
Reflection coefficient (P):	0.2

Reflection coefficient angle:	84°
Return loss:	-7 dB
Reflection loss:	-1.05 dB
Transmission loss coefficient	:1.5

Note that in all cases the mathematical interpretation corresponds to the graphical interpretation of the problem within the limits of accuracy of the graphical method.

6-3.2 Stub Matching Systems

A properly designed matching system will provide a conjugate match to a complex impedance. Some sort of matching system or network is needed any time the load impedance (Z_L) is not equal to the characteristic impedance (Z_o) of the transmission line. In a transmission line system, it is possible to use a *shorted stub* connected in parallel with the line at a critical distance back from the mismatched load in order to effect a match. The stub is merely a section of transmission line shorted at the end not connected to the main transmission line. The reactance (hence also susceptance) of a shorted line can vary from –1 to +1, depending on length, so we can use a line of critical length $L2$ to cancel the reactive component of the load impedance. Because the stub is connected in parallel with the line, it is a bit easier to work with admittance parameters rather than impedance.

Consider the example of Fig. 6-8 in which the load impedance is $Z = 100 + j60$, which is normalized to $2.0 + j1.2$. This impedance is plotted on the Smith chart in

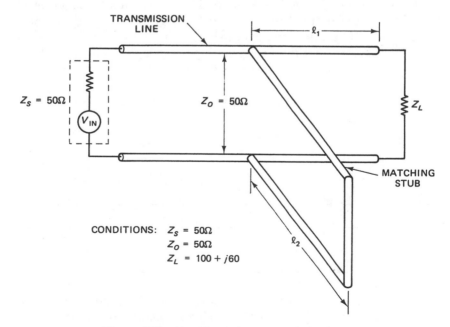

Figure 6-8 Matching stub on transmission line.

Fig. 6-9, and a VSWR circle is constructed. The admittance is found on the chart at point $Y = 0.37 - j0.22$.

To provide a properly designed matching stub, we need to find two lengths. $L1$ is the length (relative to wavelength) from the load toward the generator (see $L1$ in Fig. 6-8); $L2$ is the length of the stub itself.

The first step in finding a solution to the problem is to find the points where the unit conductance line (1.0 at the chart center) intersects the VSWR circle; two such points are shown in Fig. 6-9: $1.0 + j1.1$ and $1.0 - j1.1$. We select one of these (choose $1.0 + j1.1$) and extend a line from the center 1.0 point through the $1.0 + j1.1$ point to the outer circle (Wavelengths toward Generator). Similarly, a line is drawn from the center through the admittance point $0.37 - 0.22$ to the outer circle. These two lines intersect the outer circle at the points 0.165 and 0.461. The distance of the stub back toward the generator is found from

$$L1 = 0.165 + (0.500 - 0.461) \, \lambda \qquad \text{(6-70)}$$

$$= 0.165 + 0.039 \, \lambda \qquad \text{(6-71)}$$

$$= 0.204 \, \lambda \qquad \text{(6-72)}$$

The next step is to find the length of the stub required. This is done by finding two points on the Smith chart. First, locate the point where admittance is infinite (far right side of the pure conductance line); second, locate the point where admittance is $0 - j1.1$ (note that the susceptance portion is the same as that found where the unit conductance circle crossed the VSWR circle). Because the conductance component of this new point is 0, the point will lie on the $-j1.1$ circle at the intersection with the outer circle. Now draw lines from the center of the chart through each of these points to the outer circle. These lines intersect the outer circle at 0.368 and 0.250. The length of the stub is found from

$$L2 = (0.368 - 0.250) \, \lambda \qquad \text{(6-73)}$$

$$= 0.118 \, \lambda \qquad \text{(6-74)}$$

From the preceding analysis, we can see that the impedance $Z = 100 + j60$ can be matched by placing a stub of a length 0.118λ at a distance 0.204λ back from the load.

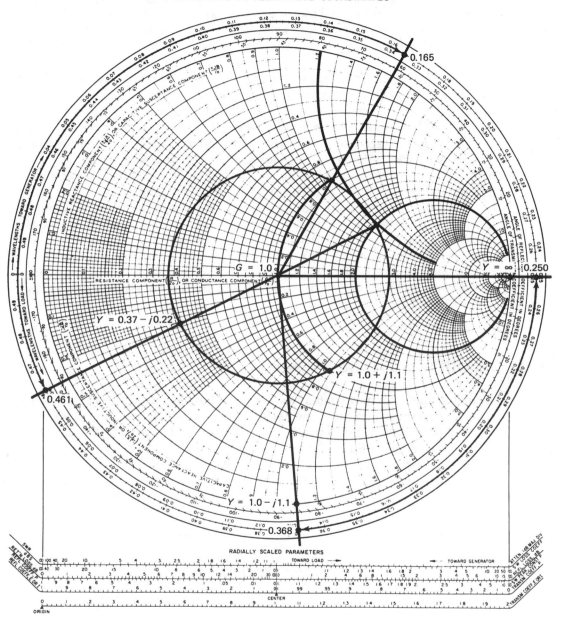

Figure 6-9   Smith chart solution to problem.

6-4 SMITH CHART IN LOSSY CIRCUITS

Thus far in our discussion of the Smith chart we have dealt with situations in which loss is either zero (ideal transmission lines) or so small as to be negligible. In situations where there is appreciable loss in the circuit or line, however, we see a slightly modified situation. The VSWR circle in that case is actually a spiral, rather than a circle.

Figure 6-10 shows a typical situation. Assume that the transmission line is 0.60λ long and is connected to a normalized load impedance of $Z = 1.2 + j1.2$. An ideal VSWR circle is constructed on the impedance radius represented by 1.2 + $j1.2$. A line (A) is drawn from the point where this circle intersects the pure resistance baseline (B) perpendicularly to the Attenuation 1-dB/Major Division line on the radially scaled parameters. A distance representing the loss (3 dB) is stepped off on this scale. A second perpendicular line is drawn from the −3-dB point back to the pure resistance line (C). The point where line C intersects the pure resistance line becomes the radius for a new circle that contains the actual input impedance of the line. The length of the line is 0.60λ, so we must step back (0.60 − 0.50)λ or 0.1λ. This point is located on the Wavelengths toward Generator outer circle. A line is drawn from this point to the 1.0 center point. The point where this new line intersects the new circle is the actual input impedance (Z_{in}). The intersection occurs at 0.76 + $j0.4$, which when denormalized represents an input impedance of 38 + $j20$ Ω.

6-5 FREQUENCY ON THE SMITH CHART

A complex network may contain resistive, inductive reactance, and capacitive reactance components. Because the reactance component of such impedances is a function of frequency, the network or component tends also to be frequency sensitive. We can use the Smith chart to plot the performance of such a network with respect to various frequencies. Consider the load impedance connected to a 50-Ω transmission line in Fig. 6-11. In this case, the resistance is in series with a 2.2-pF capacitor, which will exhibit a different reactance at each frequency. The impedance of this network is

$$Z = R - j\left(\frac{1}{\lambda C}\right) \tag{6-75}$$

or

$$Z = 50 - j\left(\frac{1}{2\pi FC}\right) \tag{6-76}$$

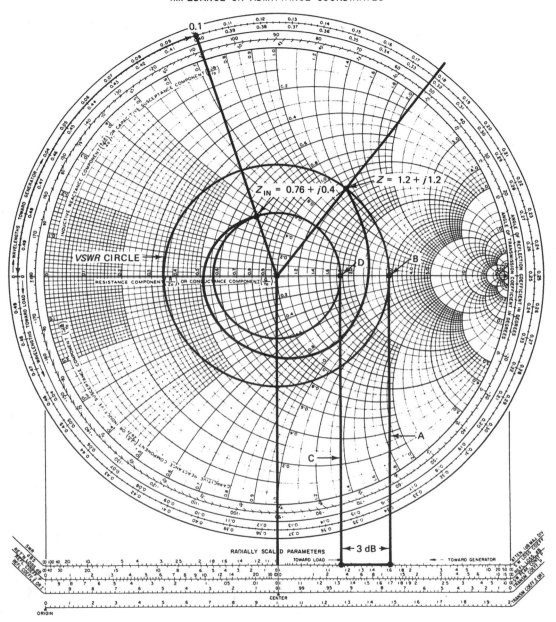

NAME	TITLE	DWG. NO
SMITH CHART FORM 82-BSPR(9-66)	KAY ELECTRIC COMPANY, PINE BROOK, N.J. ©1966. PRINTED IN U.S.A.	DATE

IMPEDANCE OR ADMITTANCE COORDINATES

$Z_{IN} = 0.76 + j0.4$

$Z = 1.2 + j1.2$

VSWR CIRCLE

RADIALLY SCALED PARAMETERS

Figure 6-10   Smith chart solution for lossy circuit.

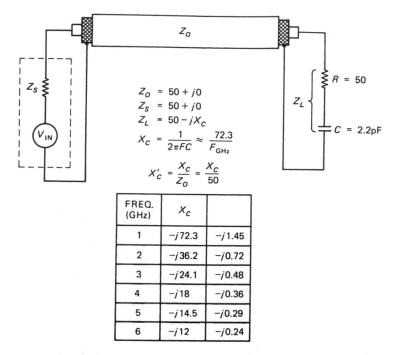

Figure 6-11 Test circuit for reactive load.

and, in normalized form,

$$Z' = 1.0 - \frac{j}{(2\pi FC) \times 50} \qquad (6\text{-}77)$$

$$= 1.0 - \frac{j}{6.9 \times 10^{-10}\,F} \qquad (6\text{-}78)$$

or, converted to gigahertz,

$$Z' = 1.0 - \frac{j72.3}{F_{GHz}} \qquad (6\text{-}79)$$

The normalized impedances for the sweep of frequencies from 1 to 6 GHz is therefore

$$Z = 1.0 - j1.45 \qquad (6\text{-}80)$$

$$= 1.0 - j0.72 \qquad\qquad (6\text{-}81)$$

$$= 1.0 - j0.48 \qquad\qquad (6\text{-}82)$$

$$= 1.0 - j0.36 \qquad\qquad (6\text{-}83)$$

$$= 1.0 - j0.29 \qquad\qquad (6\text{-}84)$$

$$= 1.0 - j0.24 \qquad\qquad (6\text{-}85)$$

These points are plotted on the Smith chart in Fig. 6-12. For complex networks in which both inductive and capacitive reactances exist, take the difference between the two reactances; that is, $X = X_L - X_C$.

6-6   SUMMARY

1.   The Smith chart is a means to perform graphical solutions to complex transmission line problems.

2.   The Smith chart consists of a series of orthogonal circles arranged around a central base line. The base line represents pure resistance, while one set of circles represents inductive reactance and another represents capacitive reactance; pure reactances are graphed along an outer circle that is bisected by the pure resistance line. The circles that are tangent to the right side and bisected by the pure resistance line are constant resistance circles.

3.   Typical parameters that can be graphed or found on the Smith chart in transmission line problems are complex impedance, complex admittance, VSWR (both as a ratio and in decibel form), voltage reflection coefficient, power reflection coefficient, reflection coefficient angle, transmission coefficient angle, return loss (dB), reflection loss (dB), and transmission loss coefficient.

4.   The Smith chart can be used for a variety of problems, including transmission line, impedance-matching stubs, tuned circles, and others involving complex impedances and/or frequency.

6-7   RECAPITULATION

Now return to the objectives and prequiz questions at the beginning of the chapter and see how well you can answer them. If you cannot answer certain questions,

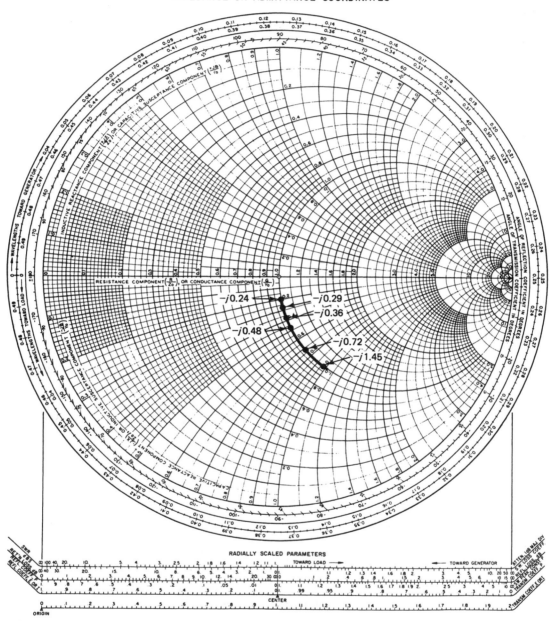

IMPEDANCE OR ADMITTANCE COORDINATES

Figure 6-12 Smith chart solution.

place a check mark by each and review the appropriate parts of the text. Next, try to answer the following questions and work the problems using the same procedure.

QUESTIONS AND PROBLEMS

When answering the questions and working these problems, use the blank Smith chart in Fig. 6-13 or obtain chart paper elsewhere.

1. On Fig. 6-13, locate and identify the pure resistance line.

2. On Fig. 6-13, locate and identify the constant resistance circle that represents a load resistance of 50 Ω when the system impedance is also 50 Ω.

3. Impedances are represented in _____ form on the Smith chart in order to accommodate a wide variety of system impedances and situations.

4. If the system impedance (transmission line characteristic impedance) is 50 Ω, the normalized impedance for a 25-Ω resistive load will appear at the _____ point on the pure resistance line.

5. A 100-Ω resistive load is plotted at the _____ point on the pure resistance line.

6. Locate and identify the $+j1.2$ reactance line on the Smith chart. Similarly, locate the $-j1.2$ reactance line.

7. A complex impedance of $Z = 40 + j25$ Ω is connected to a 50-Ω transmission line. Calculate the *normalized* impedance for this load that will be plotted on the Smith chart.

8. On a Smith chart, find and mark the point that represents a normalized impedance of $Z = 0.80 - j1.2$.

9. On a Smith chart, find and mark the point that represents an impedance of $40 + j60$ Ω if the system impedance is 50 Ω.

10. A load of $30 + j40$ Ω is connected to a 30-cm transmission line that has a velocity factor of 0.80. A 4-GHz RF source inputs 4 W to this line. Using a Smith chart, find admittance (load); VSWR; VSWR (dB); reflection coefficient (E); reflection coefficient (P); reflection coefficient angle; return loss; reflection loss; transmission loss coefficient.

11. A load of $100 + j60$ Ω is connected to a 15-cm transmission line that has a velocity factor of 0.75. An 8.5-GHz RF source inputs 4 W to this line. Using a Smith chart, find admittance (load); VSWR; VSWR (dB); reflection coefficient (E); reflection coefficient (P); reflection coefficient angle; return loss; reflection loss; transmission loss coefficient.

12. Calculate the admittance represented by a load impedance of $80 + j55$ Ω. Find and mark this point on a Smith chart. Find both the graphical and mathematical solutions.

13. Calculate the admittance represented by a load impedance of $25 - j30$ Ω. Find and mark this point on a Smith chart. Find both the graphical and mathematical solutions.

14. A complex load impedance is normalized to $0.7 + j0.9$. Find the VSWR represented by this load using the graphical method.

15. Find the VSWR when a microwave transmitter inputs 1.5 W to a transmission line, and 150 W is reflected from the load.

16. In problem 15, calculate the power and voltage reflection coefficients.

17. A complex load impedance of $75 + j55$ Ω is connected to a long, 50-Ω transmission line. Calculate the distance from the load and length of a shorted stub used to impedance match this system. The lengths should be in wavelengths.

NAME	TITLE	DWG. NO.
SMITH CHART FORM 82-BSPR (9-66)	KAY ELECTRIC COMPANY, PINE BROOK, N.J., ©1966. PRINTED IN U.S.A.	DATE

IMPEDANCE OR ADMITTANCE COORDINATES

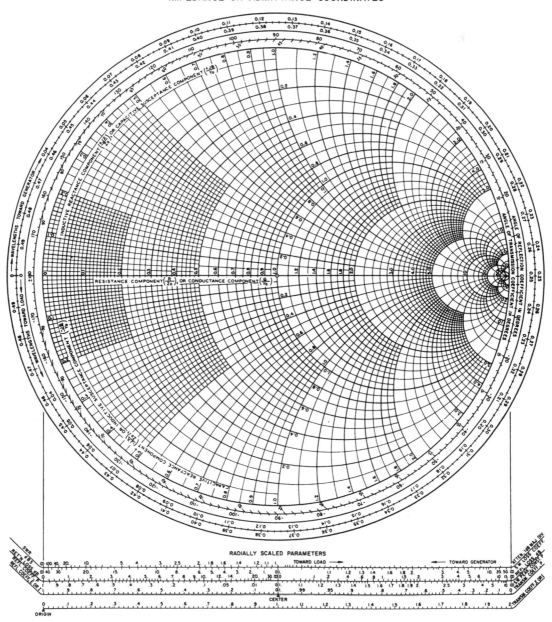

Figure 6-13   Smith chart for problems.

18. A 100-Ω resistor is connected in series with a 0.5-μH inductor. Plot the impedance of this network on a Smith chart for the frequencies 1, 2, 3, 4, 5, 6, and 7 GHz.

19. A 50-Ω resistor is connected in parallel with a 2.7-pF capacitor. Draw the Smith chart plot for a frequency sweep from 1 to 10 GHz.

KEY EQUATIONS

1. Complex impedance:

$$Z = R \pm jX$$

2. Normalized impedance:

$$Z = \frac{R \pm jX}{Z_o}$$

3. Voltage reflection coefficient:

$$P = \frac{E_{ref}}{E_{inc}}$$

4. Current reflection coefficient:

$$P = \frac{I_{ref}}{I_{inc}}$$

5. Power reflection coefficient:

$$P_{pwr} = P^2$$

or

$$P_{pwr} = \frac{P_{ref}}{P_{inc}}$$

6. VSWR as a function of reflection coefficient:

$$VSWR = \frac{1 + P}{1 - P}$$

or

$$VSWR = \frac{1 + (P_{pwr})^{1/2}}{1 - (P_{pwr})^{1/2}}$$

7. Return loss as a function of VSWR:

$$Loss_{ret} = 10 \log (P_{pwr})$$

8. Transmission loss coefficient:

$$TLC = \frac{1 + P_{pwr}}{1 - P_{pwr}}$$

CHAPTER 7

Microwave Components

OBJECTIVES

1. Learn the different types of components available.
2. Understand the applications of various microwave components.
3. Describe the characteristics of microwave components.
4. Learn the limitations of microwave components.

7-1 PREQUIZ

These questions test your prior knowledge of the material in this chapter. Try answering them before you read the chapter. Look for the answers (especially those you answered incorrectly) as you read the text. After you have finished studying the chapter, try answering these questions again and those at the end of the chapter.

1. Describe three types of microwave attenuator.
2. A post completely shorting together the two *a* dimension surfaces of a waveguide is a _____ reactance.
3. _____ components (for example, resistors, capacitors, and inductors) are used at microwave frequencies.
4. VSWR and power measurements are sometimes made with the aid of a _____ coupler.

7-2 INTRODUCTION

This chapter is a potpourri of topics because there are many different forms and types of microwave component. The intent of this chapter is to provide the student with insight into the various types of microwave components that are available and might be found in common systems.

A significant fact that emerges from the study of microwave devices is that concepts are often reduced to their most fundamental form. In Section 7-3, for example, we will study reactance devices. Although we will take a look at the microwave version of traditional forms of inductive and capacitive reactance, there are other forms that operate on different principles. For example, in waveguides a post or screw that shorts together the two wide-dimension surfaces of the waveguide form an inductive reactance. Students of lower-frequency electronics sometimes come to view inductors as mere coils of wire that have certain magnetic properties and capacitors as parallel conductors separated by an insulator. But at a more basic level both are merely devices for changing the phase relationship between voltage and current, differing primarily in the *direction* of the phase shift.

7-3   REACTANCE DEVICES

A *reactor* in an electronic circuit is a device that operates on an electric or magnetic field in such a way that the phase relationship between voltage and current changes. In a pure resistance circuit, the voltage and current are in phase with each other. An inductive reactance *opposes changes in current*, so the current tends to lag behind voltage. At lower frequencies, inductive reactance is created by coils of wire. Similarly, a capacitive reactance *opposes changes in applied voltage*. Thus, in a capacitive reactance circuit, the voltage tends to lag behind current. Equal inductive and capacitive reactances tend to cancel each other by providing equal but opposite effects on the applied waveform.

In microwave circuits, we find both the traditional forms of reactor and special types that are unique to the microwave world. *Chip components* and *microstrip components* are microwave examples of traditional reactors, which are analogous to lower-frequency devices. Because component values are very small, however, and strays therefore predominate, such components take on unusual appearances in their microwave versions. An inductor, for example, might be a single straight or curved section of printed circuit track.

An inductance can be formed by the action of the magnetic field around a current-carrying conductor. In lower-frequency electronics, larger values of inductance are created by coiling the conductor to concentrate the magnetic field. But at microwave frequencies, straight sections of conductor possess sufficient inductance. Figure 7-1 shows both series (Fig. 7-1A) and shunt (Fig. 7-1B) printed circuit inductances. In the series case, the relatively broad, low-inductance, printed circuit track is narrowed to form the required value of inductance, which is a function of both width and length. Similarly, a narrow conductor dropped from a conductor to the ground plane forms a shunt inductor.

Figure 7-2 shows a printed circuit capacitor used at microwave frequencies. The series capacitor is formed by interrupting a conductor, leaving a dielectric section between. The shunt capacitor is formed by widening the track at a point, as shown in Fig. 7-2B.

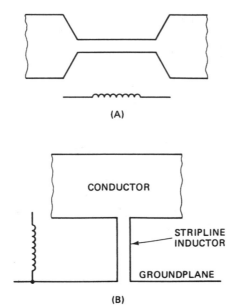

(A)

CONDUCTOR

STRIPLINE
INDUCTOR

GROUNDPLANE

(B)

Figure 7-1
Microwave printed circuit
inductors: A) series; B) shunt.

Figure 7-3 shows *iris reactors* used in rectangular waveguides. In all three cases shown, the iris interrupts either the magnetic or electric fields. An inductive iris is shown in Fig. 7-3A, which consists of a vertical vane between the *a* dimension surfaces of the waveguide. This iris parallels the electric field and cuts across the magnetic field. Induced currents in the vane can thus flow between the surfaces under the influence of the *E*-field potential difference. These currents in turn set up additional magnetic flux lines, thereby acting as an inductive reactance shunted across the waveguide.

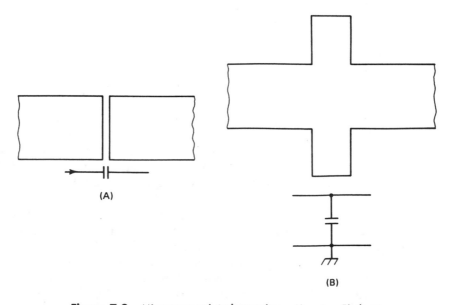

(A)

(B)

Figure 7-2 Microwave printed capacitors: A) series; B) shunt.

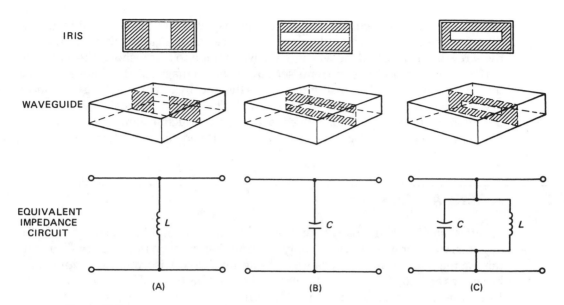

Figure 7-3 Microwave waveguide *LC* components: A) inductor; B) capacitor; C) *LC* tuned circuit.

The capacitive reactance iris is shown in Fig. 7-3B. In this case, the direction of the iris is rotated 90° such that it runs between the *b* dimension surfaces of the waveguide. The capacitive iris operates by concentrating the *E*-field flux lines in the waveguide.

An *LC* resonant tank circuit can be formed by a hybrid of inductive and capacitive reactances (Fig. 7-3C). In this type of vane, the iris extends to neither *a* nor *b* surfaces, so both modes are effective, but neither mode dominates.

A post or screw can also serve as a reactive element (Fig. 7-4). The only significant difference between posts and screws is that posts are fixed and screws are

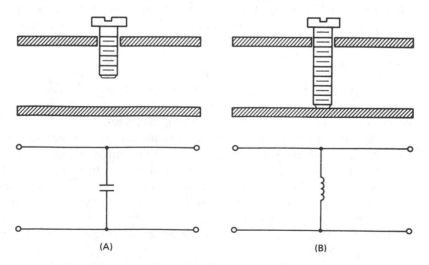

Figure 7-4 Adjustable waveguide components: A) capacitive setting; B) inductive setting.

adjustable. When the screw is advanced only partway into the waveguide, it concentrates the *E* field and acts like a capacitive reactance. On the other hand, when the screw is advanced all the way into the waveguide so that it shorts the two wide surfaces together, induced currents set up additional magnetic flux lines and force it to behave like an inductor shunted across the line. By advancing the screw most of the way into the waveguide, but not so that it touches the opposite surface, a situation is created that permits both capacitive and inductive elements to exist, but neither predominates. The screw acts like an *LC* tuned circuit in those cases.

7-4   ATTENUATORS

An *attenuator* is a device used to reduce the strength or amplitude of a signal. In RF systems, the attenuator must not only perform that function but do so while maintaining the characteristic impedance of the system. Otherwise, the attenuator would be seen as an impedance discontinuity, and reflections (hence VSWR) would result. In addition, some attenuators must be precision devices and provide an exact amount of signal reduction. Thus, a precise 3-dB attenuator must cut signal power in half. Still other attenuators are adjustable, sometimes crudely so and other times with great precision. In this section, we will deal with microwave attenuators and their uses.

Some applications of attenuators are almost obvious, reducing signal levels, for example. When measuring the output level of a transmitter, for example, it is common practice to use a low-power RF wattmeter and then reduce the output power of the transmitter through a chain of precision attenuators in cascade. If a 3-dB attenuator, properly matched to system impedance, is placed between the transmitter and the RF wattmeter, then the meter reading must be multiplied by 2 to arrive at the correct measurement.

There are also other uses for attenuators. Antenna and amplifier gains are often measured by an *equal deflection method*. A receiver is used to monitor a distant signal received through both a gain antenna and a reference antenna (usually either a monopole or dipole). The reference signal produces an output reference level. The gain antenna is then connected and a new (higher) output level is obtained. Calibrated attenuators are then placed in the antenna line until the output level drops to the level produced by the reference antenna. The gain of the antenna can be stated in terms of the number of decibels of attenuation required to achieve equal deflections of the receiver output (or S) meter.

Another attenuator application is to stabilize circuits that are sensitive to load impedance variations (or errors). Filters, for example, are dependent on seeing a specified resistive impedance at both input and output terminals. Incorrect impedances distort the filter passband characteristic, often dramatically. Oscillators often experience frequency variation if the output load varies. As a result, some designers place a resistive attenuator in the output line to swamp out impedance variations. Similarly, a microwave amplifier with improperly terminated input lines may oscillate.

Figure 7-5 shows several attenuator circuits of more or less usefulness at microwave frequencies. The simplest, and least useful, form is the in-line resistor

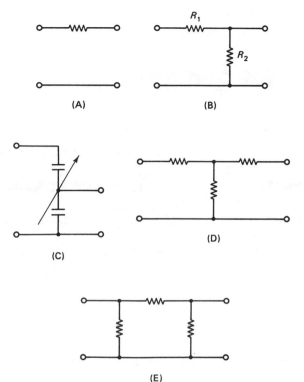

Figure 7-5
Attenuator circuits: A) series in-
line resistor; B) resistor voltage
divider; C) capacitor voltage
divider; D) T pad; E) pi pad.

shown in Fig. 7-5A. This attenuator is a resistance in series with the line and so absorbs power and thereby provides attenuation.

The second form is the inaptly named *minimum loss* attenuator of Fig. 7-5B. This attenuator is primarily used in voltage signal systems where impedance matching is not the issue. The attenuation ratio is determined by the voltage-divider equation $[VR2/(R1 + R2)]$. A capacitive reactance variant of this minimum loss design is sometimes used to build RF variable attenuators, even at microwave frequencies.

Constant input and output impedance are a necessary requirement of attenuators used in most RF circuits. As a result, T pads (Fig. 7-5D) and pi pads (Fig. 7-5E) are favored. These circuits are used extensively in microwave attenuators, especially those exhibiting fixed attenuation ratios.

Figure 7-6 shows three waveguide attenuators. A simple in-line resistor attenuator is shown in Fig. 7-6A. The attenuator element is a resistive material. When the wave passes the attenuator, it induces currents that are energy lost to the wave. Thus, attenuation occurs by heating of the resistive element.

Similar mechanisms work in the T pad (Fig. 7-6B) and pi pad (Fig. 7-6C) versions, but the structures of the resistive element are different.

Variable attenuators are shown in Fig. 7-7. The *movable vane attenuator* of Fig. 7-7A provides coarse values of attenuation. It works by placing more or less of the rotary resistive flap or vane in the path. The attenuator shown in Fig. 7-7B can be either coarse or precision, depending on the mechanism used to vary the position

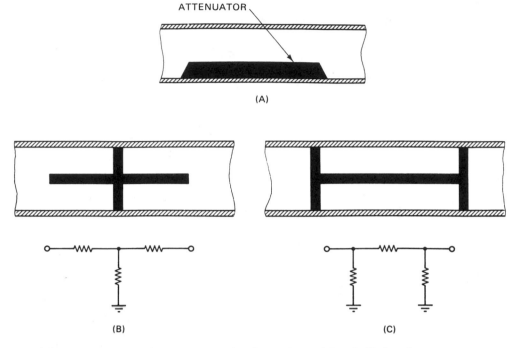

Figure 7-6   Waveguide attenuators: A) in-line resistor; B) T pad; C) pi pad.

of the resistive element within the waveguide. In Fig. 7-7B, the element is controlled by a micrometer dial, so the attenuator is identified as a precision model.

Figure 7-8 shows a fixed attenuator such as might be found in hybrid microelectronic, printed circuit, and other applications where stripline transmission lines are typically used. The resistor elements are vapor deposited or otherwise affixed

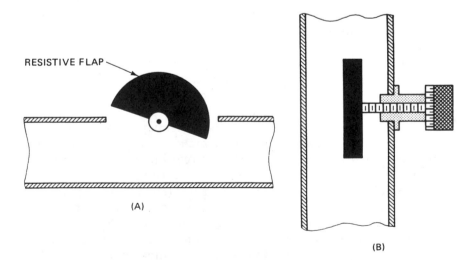

Figure 7-7   Adjustable attenuators: A) resistive vane; B) adjustable disk.

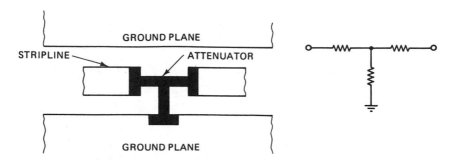

Figure 7-8   Printed circuit strip-line attenuator.

to the substrate at a break in the line. This method is also used to form fixed in-line attenuators. In such cases, a transmission line segment containing the attenuator element is constructed inside a fitting that has a male coaxial connector on one end and a female matching connector on the other end.

7-5   RESONANT CAVITIES

At frequencies lower than microwaves, resonant circuits can be made using inductive and capacitive elements. At microwave frequencies, however, the values of the LC components at resonance are too small for practical implementation. Thus, at microwave frequencies other methods for providing resonant circuits must be used. One popular solution is the *resonant cavity*, also called the *tuned cavity*.

The resonant cavity is a metal-walled chamber made of low-resistivity material enclosing a good dielectric (for example, vacuum, dry air, or dry nitrogen). It is analogous to a section of waveguide with both ends shorted, and with a means for injecting energy into or extracting energy from the cavity.

The cavity supports both TE and TM modes of propagation (Fig. 7-9). The electromagnetic wave is confined by the walls of the cavity. Energy stored in the magnetic field represents the inductance, while energy stored in the electric field represents the capacitance. Thus, both elements of the LC tuned resonant tank circuit are present in the cavity.

The resonant modes occur in a cavity at those frequencies for which the d dimension along the Z axis (see Fig. 7-9A) is a half-wavelength. The alphanumeric system for designating modes in a cavity is the same as for waveguide, except that a third integer (p) is added to the subscript to indicate the number of half-wavelengths that fit into the d dimension:

$$Tx_{m,n,p}$$

where

$x = M$ for magnetic dominant modes and E for electric dominant modes
$m =$ number of half-wavelengths in the a dimension
$n =$ number of half-wavelengths in the b dimension
$p =$ number of half-wavelengths in the d dimension

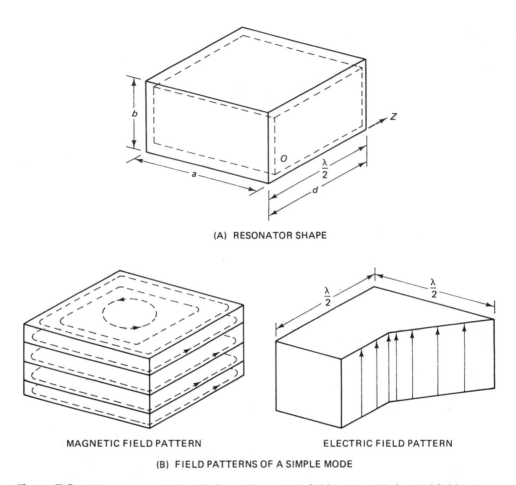

(A) RESONATOR SHAPE

MAGNETIC FIELD PATTERN   ELECTRIC FIELD PATTERN

(B) FIELD PATTERNS OF A SIMPLE MODE

Figure 7-9   Microwave resonator: A) shape; B) magnetic field pattern; C) electrical field pattern.

For the usual case where $b < a < d$, the TE_{101} mode is the dominant mode. In the dominant mode for the impressed signal (at the resonant frequency), the standing waves are at their maximum amplitude, and the peak amount of energy is stored; both E and H fields are equal in this case.

The electromagnetic field in the cavity satisfies Maxwell's equations within the following boundary conditions:

1. The tangential electric field goes to zero.
2. The orthogonal magnetic field goes to zero.

The general mode dependent equation for resonant frequency of the cavity is

$$F_r = \frac{1}{2\sqrt{\mu\varepsilon}} \sqrt{\left(\frac{m}{a}\right)^2 + \left(\frac{n}{b}\right)^2 + \left(\frac{p}{d}\right)^2} \qquad (7\text{-}1)$$

Because for air $\mu = \mu_o$ and $\varepsilon = \varepsilon_o$, and the quantity $(\mu_o\varepsilon_o)^{-1/2} = c$, we can write

$$F_r = \frac{c}{2}\sqrt{\left(\frac{m}{a}\right)^2 + \left(\frac{n}{b}\right)^2 + \left(\frac{p}{d}\right)^2} \qquad (7\text{-}2)$$

EXAMPLE 7-1

Find the resonant frequency in the dominant TE_{101} mode for a cavity of the following dimensions: $a = 6$ cm (0.06 m), $b = 2.5$ cm (0.025 m), and $d = 10$ cm (0.10 m).

Solution

$$F_r = \frac{c}{2}\sqrt{\left(\frac{m}{a}\right)^2 + \left(\frac{n}{b}\right)^2 + \left(\frac{p}{d}\right)^2}$$

$$= \frac{c}{2}\sqrt{\left(16.7^2 + 0 + (10)^2\right)}$$

$$= \frac{c}{2}\sqrt{279 + 100}$$

$$= \frac{3 \times 10^8 \text{ m/s}}{2} \times 19.5 = 2.925 \text{ GHz}$$

The Q of any resonant tank circuit is defined as the ratio of energy stored to energy dissipated per cycle

$$Q_o = \frac{2\pi U_s}{U_d} \qquad (7\text{-}3)$$

where

  Q_o = quality factor
  U_s = total energy stored in the cavity
  U_d = energy dissipated per cycle

Stated another (and often more useful) way, Q is the ratio of center resonant frequency to the bandwidth:

$$Q_o = \frac{F_r}{BW_{3 \text{ dB}}} \qquad (7\text{-}4)$$

where

F_r = resonant frequency
$BW_{3\,dB}$ is the bandwidth between –3-dB points on the passband response curve

The Q of the resonant cavity tends to be very much higher than the Q of LC tank circuits. Values on the order of 10,000 are not uncommon.

As is also true of LC tank circuits, input and output coupling tend to reduce Q, creating the *loaded* Q concept. The coefficient of coupling (K) determines the relationship between loaded Q and unloaded Q (Q versus Q_o) according to

$$Q = \frac{Q_o}{1 + K} \qquad (7\text{-}5)$$

Three general cases are recognized: *critical coupling*, *subcritical coupling* (also called *undercoupling*), and *overcritical coupling* (also called *overcoupling*). In the case of critical coupling, the value of K is 1, so

$$Q = \frac{Q_o}{1 + 1} \qquad (7\text{-}6)$$

$$Q = \frac{Q_o}{2} \qquad (7\text{-}7)$$

In the case of overcritical coupling, K is greater than 1 and is equal to the SWR (P). The loaded Q is found from

$$Q = \frac{Q_o}{1 + P} \qquad (7\text{-}8)$$

For undercritical coupled cases, K is less than 1 and is equal to $1/P$, so the value of loaded Q is

$$Q = \frac{P\,Q_o}{1 + P} \qquad (7\text{-}9)$$

Our discussion thus far has centered on rectangular resonant cavities, so now let's briefly take a look at three other types. Figure 7-10A shows a *cylindrical cavity*. This

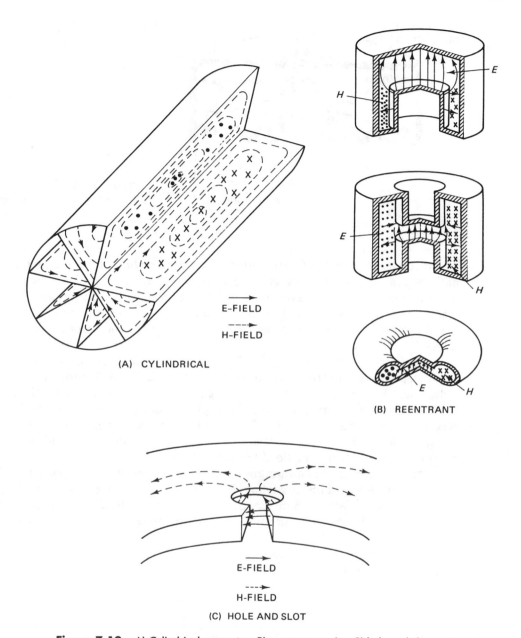

(A) CYLINDRICAL

E-FIELD
- - - ►
H-FIELD

(B) REENTRANT

E-FIELD
- - - ►
H-FIELD

(C) HOLE AND SLOT

Figure 7-10 A) Cylindrical resonator; B) reentrant cavity; C) hole and slot cavity.

type of cavity is made from cylindrical waveguide that has both ends shorted. Figure 7-10B shows a *reentrant cavity*. This type of cavity is easily identified by the fact that the walls project into the cavity interior. Some microwave tubes use reentrant cavities where electrons must be injected into and interact with the cavity. The *hole and slot cavity* shown in Fig. 7-10C is used in cavity magnetron oscillators.

Cavity input and output coupling methods are the same as for waveguide: probe, loop, and slot. In addition, reentrant cavities have electron injection coupling (Fig. 7-11).

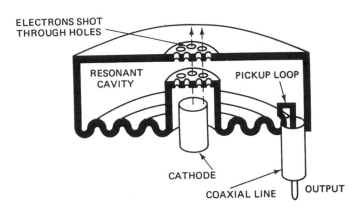

Figure 7-11
Pickup loop in a resonant cavity.

The resonant frequency of the cavity can be varied or *tuned* by three methods: *inductance*, *capacitance*, and *volume*. Recall from our discussion of waveguide that an inductance can be made using a metal post or screw protruding into the space. The issue is how well the object concentrates or disperses magnetic flux. In Fig. 7-12A, we see inductive tuning of a resonant cavity in which an adjustable screw is used.

The version shown in Fig. 7-12B is the *blade* or *vane* method. The inductance value depends on the position of the flat side of the blade with respect to the magnetic lines of force. A rotation of 90° changes the inductance over its full range.

A relatively complex version of capacitive tuning is shown in Fig. 7-13. The resonant cavity uses a flexible wall so that the distance across the gap where the electric field is concentrated is varied. A rather delicate mechanism adjusts the capacitance by distending the flexible wall of the cavity.

Perhaps the most common means for varying the frequency of the cavity is to vary the volume of the cavity, as shown in Fig. 7-14. Recall from Eq. (7.2) that the dimensions affect resonant frequency. In Fig. 7-14, the *d* dimension is varied. As was true in the shorted end caps used in waveguide, quarter-wavelength slots are sometimes cut into the tuning disk to produce a virtual short that is independent of the imperfections inherent in a screw-type adjustment mechanism for the disk.

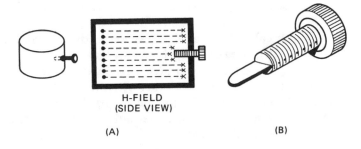

H-FIELD
(SIDE VIEW)

(A)   (B)

Figure 7-12
A) Tuning the cavity; B) blade turner.

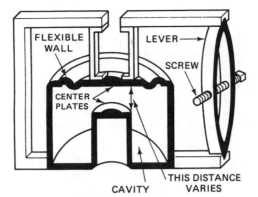

Figure 7-13
Flexible wall tuning.

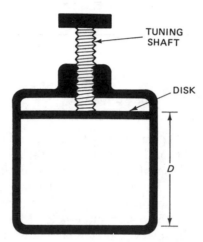

Figure 7-14
Adjustable disk tuning.

7-6   DIRECTIONAL COUPLERS

A *directional coupler* is a transmission line device that couples together two circuits in one direction, while providing a high degree of isolation in the opposite direction. Figure 7-15 shows a model for a typical directional coupler.

In Fig. 7-15, two sections of transmission line are closely coupled across a space, *S*. The respective ends of the lines form four ports, which are labeled 1, 2, 3, and 4. The main line connects ports 1 and 2, while the couple line connects ports 3 and 4.

The length of the coupled line is critical and must be an odd multiple of quarter-wavelength at the center of the band of operation. This requirement tells us that typical directional couplers tend to be narrow-band devices. The family of acceptable lengths is defined by

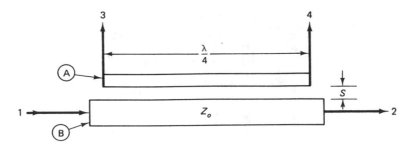

Figure 7-15   Directional coupler made from strip line.

$$L = \frac{M\lambda}{4} \qquad\qquad (7\text{-}10)$$

where

L = length of the line
λ = wavelength in the same units as L
M = an odd integer $(1, 3, 5, 7, \ldots)$

The main line segment (A) connects ports 1 and 2 and is a low-loss section of line. The power at the output (port 2) is the input power (port 1) less the *insertion loss* of the device. This loss consists of normal transmission line losses, plus any energy lost as a result of being transferred to the coupled section of line (B).

At the input end of the coupler, the impedance of the coupled section is low and should be matched to the system impedance. Maximum energy transfer takes place at this point. Recall, however, that the impedance at the far end of a quarter-wavelength line will be maximum. Therefore, minimum power is transferred between the main line and the coupled line segments.

Port 3, which receives power, is called the *coupled port*, while port 4 is called the *isolated port*. The signal available at these ports is measured in decibels relative to the port 1 signal. For example, a certain directional coupler might list insertion loss of −0.2 dB, a coupled port loss of −6 dB, and an isolated port loss of −60 dB. These specifications mean that the signal at port 2 is 0.2 dB below the port 1 signal, the port 3 signal is 6 dB below the port 1 signal, and the port signal is 60 dB below the port 1 signal.

The type of directional coupler shown in Fig. 7-15 is usually constructed of stripline sections, two examples of which are shown in Fig. 7-16. The version in Fig. 7-16A is *side coupled*, while that in Fig. 7-16B is *broadside coupled*.

A waveguide directional coupler is shown in Fig. 7-17. The lower section is the main waveguide, while the upper is the coupled guide. Energy is coupled through the slots from the main to the coupled guide. Because the slots are a quarter-wavelength apart, the energy in the coupled guide will cancel in one direction and reinforce in the other direction.

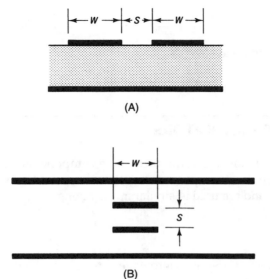

(A)

(B)

Figure 7-16
A) Side-coupled version;
B) broadside-coupled.

Consider a wave propagating from port 1 to port 2. When the wave passes slot *a*, energy is radiated into the coupled guide, where it radiates in both directions. The main guide wave continues to propagate toward slot *b*. Part of the wave couples through slot *b* into the other guide. As before, the coupled wave propagates in both directions in the other guide. The portion that propagates towards port 4 is in phase with slot *a* energy and thus reinforces the signal. But the portion that propagates from slot *b* back toward slot *a* is phase shifted 180° (that is, 90° in the main guide and 90° in the coupled guide). Thus, the port 3 signals from slots *a* and *b* are out of phase by 180° and cancel each other. We can label port 1 the input, port 2 the output, port 3 the isolated port, and port 4 the coupled port.

The spacing between slots *a* and *b* is critical because it is necessary to effect a 180° phase shift in the *a-b-b-a* path. The slot spacing should satisfy the following equation

$$S = \frac{(2N + 1)\lambda_g}{4} \qquad (7\text{-}11)$$

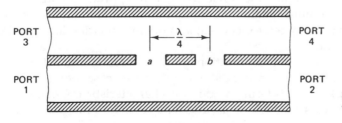

Figure 7-17
Waveguide directional coupler.

where

S = spacing
λ_g = group velocity wavelength
N = an integer

7-7 ISOLATORS AND CIRCULATORS

In this section, we will look at several microwave components: *isolators*, *circulators*, and *hybrids*. Grouping these components together is convenient because they work on similar principles and are used in similar applications.

7-7.1 Isolators

An *isolator* is a unilateral, two-port microwave component. That is, power can flow from input to output, but not from output backward toward the input. Thus, the isolator can be used to reduce the adverse effects on a signal source of varying load impedance. Some oscillators will "pull" in operating frequency when load impedance changes, and an isolator prevents that problem. Two common forms of isolator are used: *Faraday* and *terminated circulator*.

The Faraday device works because of a property of ferrite materials. When an electromagnetic wave passes through the material, its polarity is reversed. When the input wave passes through the ferrite polarity shifter, the E field is rotated to become perpendicular to a resistive vane. This signal causes little loss in the vane and so propagates unattenuated. The reflected signal, however, now has its E field parallel to the attenuator vane and so causes maximum loss. A typical Faraday isolator has 1 dB or less forward attenuation and 30 dB or more reverse attenuation. In other words, it is a unilateral (nonreciprocal) device.

7-7.2 Circulators

A *circulator* is similar in concept to the isolator, except that it is a multiport device. Like the isolator, the circulator is a unilateral device. That is, power flows in one direction only.

Figure 7-18A shows the schematic symbol for the basic three-port circulator. Power applied to any port always flows to the next adjacent port in the direction of rotation (shown by the arrow). Although both right- and left-hand circulator directions are possible, the right-hand devices (Fig. 7-18A) predominate.

Circulators can be built of either waveguide or stripline transmission lines. In both cases, a ferrite polarity rotator is used, and operation is similar to that of the two-port simple isolator.

Figures 7-18B and C show applications of circulators. In Fig. 7-18B, we see a three-port circulator used as a two-port isolator. In this case, power flows from port 1 to port 2, while port 3 is terminated in the characteristic impedance.

Figure 7-18C shows a circulator used as a *duplexer* to connect a receiver and transmitter to a common antenna. These circuits are used in radar and communi-

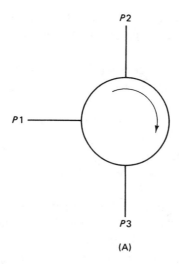

(A)

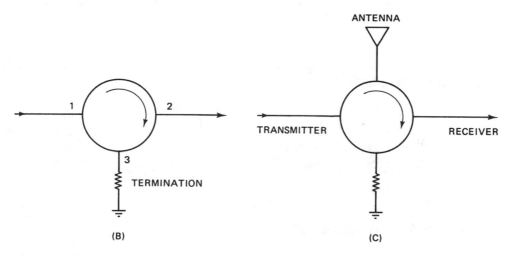

(B)

(C)

Figure 7-18 A) Three-port circulator; B) three-port circulator used as a two-port isolator; C) circulator used as a duplexer to connect both receiver and transmitter to a common antenna.

cations systems. Normally, transmitters and receivers cannot be connected to the same antenna because of the vast power difference of transmitter and receiver signals. But with the isolation provided by the circulator, the two devices can be served by the same antenna, without resorting to lossy, unreliable mechanical switching.

7-8 SUMMARY

1. Microwave components often differ from lower-frequency components because the basic physics underlying the phenomena are at a fundamental level.

2. Reactances (inductive or capacitive) can be formed of stripline components, by irises in waveguide, and by stubs, posts, or screws in waveguide. Chip components are also used.

3. Attenuators are devices that absorb power and thus have a negative gain. Attenuator elements are formed by resistors, conductive vanes in a signal path, or capacitor voltage dividers.

4. A resonant cavity is a metal-walled chamber that contains a region of good dielectric material. It is analogous to a waveguide with both ends shorted.

5. Directional couplers are transmission line devices that couple together two circuits in one direction, while providing a great degree of isolation in the opposite direction.

6. An isolator is a unilateral, two-port microwave device. It passes power from input to output with low attenuation, but provides high attenuation to signals from output backward to the input.

7. A circulator is a three or more port device that passes signals in one direction to the next adjacent port. Circulators are often used to couple an antenna to both the receiver and the transmitter in the same system.

7-9   RECAPITULATION

Now return to the objectives and prequiz questions at the beginning of the chapter and see how well you can answer them. If you cannot answer certain questions, place a check mark by each and review the appropriate parts of the text. Next, try to answer the following questions and work the problems using the same procedure.

QUESTIONS AND PROBLEMS

1. List three methods for producing an inductive reactance in a microwave circuit. Make at least one of them for nonwaveguide applications.

2. An *LC* tuned tank circuit can be formed in a rectangular waveguide with a partial _____ in both *a* and *b* directions that is neither fully inductive nor fully capacitive.

3. List three applications for attenuators.

4. A _____ vane is an example of a variable attenuator.

5. A _____ _____ is a metal-walled chamber that encloses a good dielectric.

6. A tuned (resonant) cavity supports both ____ and ___ modes of propagation.

7. For the case where a waveguide's dimensions are $b < a < d$, the _____ mode is the dominant mode.

8. The electromagnetic fields in a resonant cavity will satisfy Maxwell's equations with what two boundary conditions?

9. A resonant cavity operates in the TM_{101} mode. Assuming an air dielectric, calculate the resonant frequency if the following dimensions exist: $a = 3$ cm, $b = 1.5$ cm, and $d = 4$ cm.

10. A resonant cavity operating in TM_{101} mode has the following dimensions: $a = 3.5$ cm, $b = 1.25$ cm, and $d = 10$ cm. Calculate the resonant frequency.

11. A resonant cavity operating at a resonant frequency of 4.55 GHz has a –3-dB bandwidth of 100 MHz. Calculate the Q.

12. A cavity with an unloaded Q of 200 exhibits critical coupling. Calculate the loaded Q.

13. Calculate the loaded Q for problem 12 for

    (a) overcritical coupling ($K = 1.75$), (b) undercritical coupling ($K = 0.80$).

14. In a _____ cavity, the walls project into the cavity interior.

15. List three types of cavity.

16. List three methods for tuning a cavity.

17. A _____ _____ is a transmission line device that couples together two circuits in one direction, while providing a high degree of isolation in the opposite direction.

18. In a stripline directional coupler, the group velocity wavelength is 3.5 cm. Calculate the minimum spacing (S) between elements.

19. List two forms of isolator.

KEY EQUATIONS

1. General mode equation for resonant cavity:

$$F_r = \frac{c}{2(\mu\varepsilon)} \sqrt{\left(\frac{m}{a}\right)^2 + \left(\frac{n}{b}\right)^2 = \left(\frac{p}{d}\right)^2}$$

2. Mode equation for air dielectric:

$$F_r = \frac{c}{2} \sqrt{\left(\frac{m}{a}\right)^2 + \left(\frac{n}{b}\right)^2 + \left(\frac{p}{d}\right)^2}$$

3. Quality factor Q of a resonant circuit:

$$Q_0 = \frac{2\pi U_s}{U_d}$$

or

$$Q_0 = \frac{F_r}{BW_{3\,dB}}$$

4. Loaded Q versus unloaded Q:

$$Q = \frac{Q_0}{1 + K}$$

CHAPTER 8

Microwave Filters

OBJECTIVES

1. Learn the different types of filter frequency response.
2. Describe the different practical microwave filters.
3. Define filter parameters.

8-1 PREQUIZ

These questions test your prior knowledge of the material in this chapter. Try answering them before you read the chapter. Look for the answers (especially those you answered incorrectly) as you read the text. After you have finished studying the chapter, try answering these questions again and those at the end of the chapter.

1. A microwave bandpass filter is centered on 4.25 GHz. Calculate the shape factor if the −3-dB bandwidth is 125 MHz, and the −60-dB bandwidth is 400 MHz.
2. Find the Q of the filter in question 1.
3. The *insertion loss* of a filter refers to the attenuation in (a) passband, or (b) stopband.
4. A _____ filter consists of a series of quarter-wavelength stripline transmission line segments parallel with and coupled to each other.

8-2 FREQUENCY SELECTIVE FILTERS

A *filter* in our present context is a device or circuit that selectively discriminates against some frequencies, while favoring other frequencies. The favored frequencies are called the *passband*, while the rejected frequencies are called the *stopband*.

The filter operates by providing a large attenuation for stopband frequencies, and a minimum attenuation (ideally zero) for passband frequencies.

There are four general classes of filter that we will consider here: *low pass, high pass, bandpass,* and *bandstop.* The low-pass filter (LPF) characteristic (Fig. 8-1A) shows that the filter passes frequencies from *dc* or near-*dc* to some *cutoff frequency* (F_c). The attenuation increases above the cutoff frequency until the maximum stopband value is reached. The filter *skirt* is the transition region between the passband and full stopband. The steepness of the skirt slope defines the filter quality.

The skirt slope is usually specified in terms of *decibels of attenuation per octave* (2:1 frequency change) or *per decade* (10:1 frequency change), that is, dB/octave or dB/decade. If, for example, a low-pass filter is specified to exhibit a 10-dB/octave slope for a 2.5-GHz cutoff frequency, the attenuation at 5 GHz is 10 dB greater than the attenuation at 2.5 GHz, and at 10 GHz it is 20 dB greater than the 2.5-GHz value.

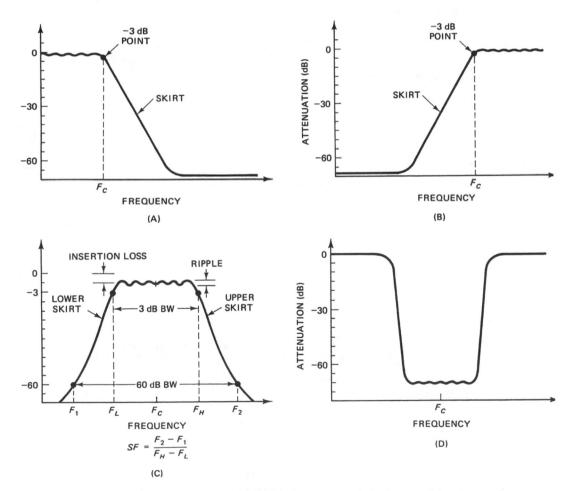

Figure 8-1 Filter passband characteristics: A) low pass; B) high pass; C) bandpass; D) narrow bandpass.

The cutoff frequency is defined as the frequency at which the response falls off –3 dB from its passband response. Because the passband response is not smooth, however, the average attenuation value is used, and the cutoff frequency is found at the point where the attenuation figure increases 3 dB (see the 3-dB point in Fig. 8-1A).

The high-pass filter (HPF) has a response curve that is the inverse mirror image of the LPF response (see Fig. 8-1B). The attenuation is very high below the cutoff frequency and minimum above the cutoff frequency. As was true in the LPF case, the HPF has a skirt or transition region between the stopband and the passband.

The bandpass filter (BPF) is a combination of the LPF and HPF responses in which the respective cutoff frequencies are different (see Fig. 8-1C). In the BPF there is a high-attenuation stopband above and below the minimum-attenuation passband region.

The passband bandwidth is defined as the frequency difference between the upper cutoff frequency (F_H) and lower cutoff frequency (F_L) on the response curve, $F_H - F_L$ in Fig. 8-1C. This expression of bandwidth is usually called the 3-dB bandwidth and is often abbreviated in data sheets and specifications as $BW_{3\,dB}$.

The upper and lower skirts define the sharpness of the cutoff characteristic between the passband and the two stopbands. This parameter is defined by the *shape factor*, which is the ratio of the 60-dB bandwidth to the 3-dB bandwidth. In terms of Fig. 8-1C, the shape factor (SF) is

$$SF = \frac{F2 - F1}{F_H - F_L} \tag{8-1}$$

or

$$SF = \frac{BW_{60\,dB}}{BW_{3\,dB}} \tag{8-2}$$

EXAMPLE 8-1

A microwave bandpass filter is centered on 3.5 GHz. Calculate the shape factor if the 3-dB bandwidth is 120 MHz and the 60-dB bandwidth is 375 MHz.

Solution

$$SF = \frac{BW_{60\,dB}}{BW_{3\,dB}}$$

$$= \frac{375 \text{ MHz}}{120 \text{ MHz}} = 3.1$$

The *figure of merit* or *quality factor* (Q) of a bandpass filter is similar to the Q of cavities discussed earlier. The Q is defined as the ratio of center frequency to the 3-dB bandwidth:

$$Q = \frac{F_c}{BW_{3dB}} \qquad (8\text{-}3)$$

EXAMPLE 8-2

Find the Q of the bandpass filter in Example 8-1.

    Solution

$$Q = \frac{F_c}{BW_{3dB}}$$

$$= \frac{3500 \text{ MHz}}{120 \text{ MHz}} = 29$$

The Q and the shape factor must be considered in designing filtered microwave circuits. The most obvious factor is the relative position of other-frequency signals compared with the center frequency of the filter. Also, the bandwidth must be sufficient to properly pass the spectrum of the expected signals without also being so wide that other signals and excess noise signals are also admitted. For fast rise-time signals (such as pulses or digital signals), a filter that is too narrow (too high Q) will "ring" in the same manner as in *LC* resonant tank circuits.

The passband of an ideal filter is perfectly flat (that is, has constant attenuation) for all frequencies between the cutoff frequencies. But in real filters this ideal condition is never met, so a certain *ripple factor* (see Fig. 8-1C) exists within the passband. In high-quality filters, the passband ripple will be on the order of 0.1 dB to 0.5 dB, although in some cases a larger ripple factor will be acceptable.

The *insertion loss* of a filter is the attenuation of signals inside the passband. Ideally, the insertion loss is zero, but that is not achievable. In most designs, the better the shape factor is, or the higher the Q, the worse the insertion loss. This phenomenon is due to the fact that such filters usually have more elements or poles than lesser types and therefore show greater in-band loss. Many circuit designers opt for a pre- or postfilter amplifier to make up for insertion loss.

The *bandstop filter* (BSP) is the inverse of the bandpass filter. The attenuation is greatest between the cutoff frequencies (see Fig. 8-1D). At frequencies above and below the stopband, signals are passed with minimal insertion loss attenuation. The purpose of a bandstop filter is to remove offending signals. An example is in communications systems where transmitters and receivers on two different frequencies are colocated at the same site. A receiver on frequency F1 will have a front-end bandstop filter on frequency F2, that is, on the frequency of the colocated transmitter.

8-2.1 Typical Microwave Filters

At frequencies lower than microwave bands, filters are often designed using lumped inductance and capacitance (L and C) components. In the microwave bands, implementation of filters is through stripline or (in the low microwaves) chip components. Figure 8-2 shows a microwave stripline implementation of a low-pass filter (HPF, BPF, and BSP designs use the same methods but with different layouts).

As microwave frequencies increase, the dimensions of the stripline components get smaller, eventually becoming too small to either carry required load currents or be easily built using ordinary printed circuit techniques. But stripline width is a function of system impedance, as well as frequency. As a result, microwave filter designers often design a filter for a lower input and/or output impedance than is required by the system and then provide impedance-matching networks to renormalize the circuit. For example, in a 50-Ω system the filter may be designed for 15-Ω impedance, with a 50:15-Ω impedance transformation provided at the input and output terminals. The resultant filter will have wider (more easily built) stripline tracks.

Two forms of resonant stripline bandpass filter are shown in Fig. 8-3. The half-wavelength version is shown in Fig. 8-3A, and the quarter-wavelength version is shown in Fig. 8-3B. This filter is basically a transmission line filter (see Chapter 4 for the frequency-sensitive aspects of transmission lines). These stripline filters are used in the lower portion of the microwave spectrum.

Another form of stripline filter is the interdigital design shown in Fig. 8-4. This type of filter consists of a series of quarter-wavelength stripline transmission line segments. This sort of filter can be used well into the microwave region and is well suited to MMIC and hybrid circuit designs.

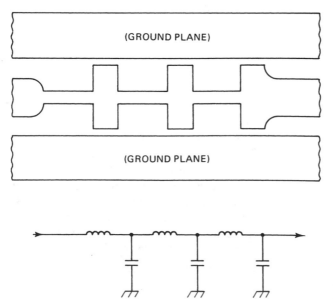

Figure 8-2
Stripline filter and equivalent circuit.

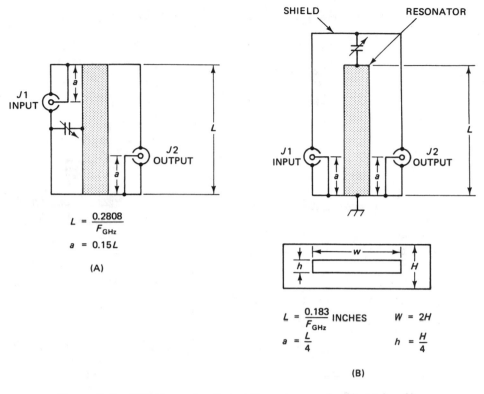

$$L = \frac{0.2808}{F_{GHz}}$$

$$a = 0.15L$$

(A)

$$L = \frac{0.183}{F_{GHz}} \text{ INCHES} \qquad W = 2H$$

$$a = \frac{L}{4} \qquad\qquad h = \frac{H}{4}$$

(B)

Figure 8-3 A) Half-wavelength and B) quarter-wavelength stripline filters.

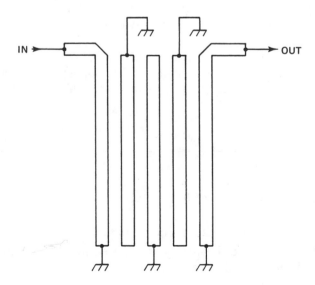

Figure 8-4
Interdigital stripline filter.

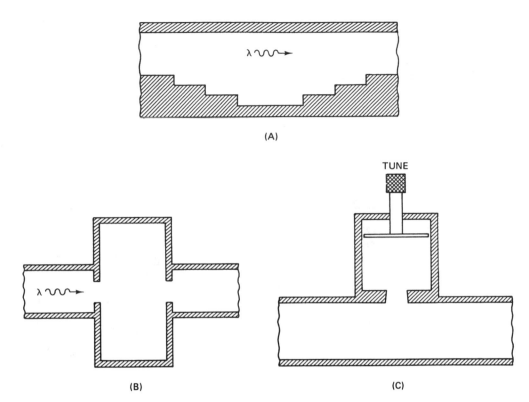

Figure 8-5 Waveguide filters: A) stepped; B) fixed cavity; C) tunable cavity.

Figure 8-5 shows several forms of waveguide frequency-selective filters. Recall from Chapter 5 that the cutoff frequency of waveguides is a function of cross-sectional dimensions. Similarly, inductive and capacitive circuit action is found through the use of restrictive irises in a segment of waveguide. In Fig. 8-5A, we see the *stepped* or *staircase* bandpass filter. In this design, critically dimensioned steps are machined into the internal surfaces of a section of waveguide.

A cavity-type, series-resonant bandpass filter is shown in Fig. 8-3B. Using a reentrant resonant cavity, this filter allows passage of signals with a frequency around the resonant frequency. A parallel-resonant cavity bandpass filter is shown in Fig. 8-3C. This particular version is tunable by virtue of the volume-changing tuning disk inside the cavity.

8-3 SUMMARY

1. Four classes of frequency-selective filter are typically found: *high pass, low pass, bandpass,* and *bandstop.*

2. The steepness of the filter skirts define the characteristics.

3. The Q of the filter, also called the *quality figure,* is defined by the ratio of center frequency to bandwidth of the filter.

4. The shape factor (SF) of the bandpass filter is defined by the bandwidth ratio between the 3- and 60-dB points.

8-4 RECAPITULATION

Now return to the objectives and prequiz questions at the beginning of the chapter and see how well you can answer them. If you cannot answer certain questions, place a check mark by each and review the appropriate parts of the text. Next, try to answer the following questions and work the problems using the same procedure.

QUESTIONS AND PROBLEMS

1. A _____ filter attenuates frequencies above the resonant frequency.
2. A _____ filter attenuates frequencies below the resonant frequency.
3. A _____ filter passes frequencies above the resonant frequency.
4. A _____ filter passes signals above and below two different cutoff frequencies.
5. A _____ filter passes signals between two different cutoff frequencies.
6. The _____ slope is specified in terms of decibels of attenuation per octave.
7. The 3-db bandwidth of a filter is 225 MHz, while the 60-dB bandwidth is 100 MHz. What is the shape factor?
8. The resonant frequency of a filter is 4.5 GHz, while the 3-dB bandwidth is 100 MHz. Calculate the Q of this filter.
9. In high-quality filters, the _____ factor is typically 0.1 to 0.5 dB.
10. The _____ _____ is the attenuation of a filter inside the passband.

KEY EQUATIONS

1. Shape factor:

$$SF = \frac{F2 - F1}{F_H - F_L}$$

or

$$SF = \frac{BW_{60\ dB}}{BW_{3\ dB}}$$

2. Q of a filter:

$$Q = \frac{F_c}{BW_{3\ dB}}$$

CHAPTER 9

Microwave Antennas

OBJECTIVES

1. Review the basics of antenna theory.
2. Learn the types of antennas used in microwave communications.
3. Learn the properties of microwave antennas and how they affect operation.
4. Learn the limitations and advantages of the various microwave antennas.

9-1   PREQUIZ

These questions test your prior knowledge of the material in this chapter. Try answering them before you read the chapter. Look for the answers (especially those you answered incorrectly) as you read the text. After you have finished studying the chapter, try answering these questions again and those at the end of the chapter.

1. A parabolic dish is 15 m in diameter and 2 m deep. It operates on 3.345 GHz. What is its gain?
2. A dipole antenna used as a feed for a parabolic antenna. A splash plate is placed _____ wavelengths behind the antenna.
3. Calculate the effective aperture of a 3.5-m-diameter parabolic reflector antenna at a frequency of 10.25 GHz if the aperture effectiveness is 0.80.
4. What is the directivity of an isotropic radiator? $D =$ _____.

9-2   INTRODUCTION

Antennas are used in communications and radar systems at frequencies from the very lowest to the very highest. In both theory and practice, antennas are used until frequencies reach infrared and visible light, at which point optics become more important. Microwaves are a transition region between ordinary radio waves and optical waves, so, as might be expected, microwave technology makes use of techniques from both worlds. For example, both dipoles and parabolic reflectors are used in microwave systems.

The purpose of an antenna is to act as a *transducer* between either electrical oscillations or propagated guided waves (that is, in transmission lines or wave-guides) and a propagating electromagnetic wave in free space. A principal function of the antenna is to act as an *impedance matcher* between the waveguide or transmission line impedance and the impedance of free space.

Antennas can be used equally well for both receiving and transmitting signals because they obey the *law of reciprocity*. That is, the same antenna can be used for both receive and transmit modes with equal success. Although there may be practical or mechanical reasons to prefer specific antennas for one or the other mode, electrically they are the same.

In the transmit mode, the antenna must radiate electromagnetic energy. For this job, the important property is *gain*, G (Section 9-7). In the receive mode, the job of the antenna is to gather energy from impinging electromagnetic waves in free space. The important property for receiver antennas is the *effective aperture*, A_e, which is a function of the antenna's physical area. Because of reciprocity, a large gain usually implies a large effective aperture, and vice versa.

9.3   ISOTROPIC ANTENNA

Antenna definitions and specifications become useless unless a means is provided for putting everything on a common footing. Although a variety of systems exist for describing antenna behavior, the most common system compares a specific antenna with a theoretical construct called the *isotropic radiator*.

An isotropic radiator is a spherical point source that radiates equally well in all directions. By definition, the directivity of the isotropic antenna is unity (1), and all antenna gains are measured against this standard. Because the geometry of the sphere and the physics of radiation are well known, we can calculate field strength and power density at any point. These figures can then be compared with the actual values from an antenna being tested. From spherical geometry, we can calculate isotropic power density at any distance R from the point source:

$$P_d = \frac{P}{4\pi R^2} \tag{9-1}$$

where

P_d = power density in watts per square meter (W/m^2)
P = power in watts input to the isotropic radiator
R = radius in meters at which point power density is measured

EXAMPLE 9-1

Calculate the power density in W/m^2 at a distance of 1 km (1000 m) from a 1000-W isotropic source.

Solution

$$P_d = \frac{P}{4\pi R^2}$$

$$= \frac{1000 \text{ W}}{4\pi(1000 \text{ m})^2}$$

$$= 7.95 \times 10^{-5} \text{ W/m}^2$$

In the rest of this chapter, antenna gains and directivities will be discussed relative to isotropic radiators.

9-4 NEAR FIELD AND FAR FIELD

Antennas are defined in terms of *gain* and *directivity*, both of which are measured by looking at the radiated field of the antenna. There are two fields to consider: *near field* and *far field*. The patterns published for an antenna tend to reflect far-field performance. The far field for most antennas falls off according to the *inverse square law*. That is, the intensity falls off according to the square of the distance $(1/R^2)$, as in Eq. (9-1).

The near field of the antenna contains more energy than the far field because of the electric and magnetic fields close to the antenna radiator element. The near field tends to diminish rapidly according to a $1/R^4$ function. The minimum distance to the edge of the near field is a function of both the wavelength of the radiated signals and the antenna dimensions:

$$r_{min} = \frac{2d^2}{\lambda} \tag{9-2}$$

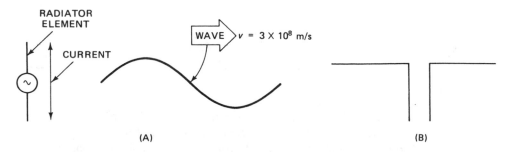

Figure 9-1 A) Signal radiating away from a dipole; B) basic form of a dipole antenna.

where

r_{min} = near-field distance
d = largest antenna dimension
λ = wavelength of the radiated signal

All factors are in the same units.

EXAMPLE 9-2

An antenna with a length of 6 cm radiates a 12-cm wavelength signal. Calculate the near-field distance.

Solution

$$r_{min} = \frac{2d^2}{\lambda}$$

$$= \frac{(2)(6 \text{ cm})^2}{12 \text{ cm}} = \frac{72}{12} \text{ cm} = 6 \text{ cm}$$

9-5 ANTENNA IMPEDANCE

Impedance represents the total opposition to the flow of alternating current (for example, RF), and includes both resistive and reactive components. The reactive components can be either capacitive or inductive or a combination of both. Impedance can be expressed in either of two notations:

$$Z = \sqrt{R^2 + (X_L - X_c)^2} \tag{9-3}$$

or

$$Z = R \pm jX \tag{9-4}$$

Of these, Eq. (9-4) is perhaps the most commonly used in RF applications. The reactive part of antenna impedance results from the magnetic and electrical fields close to the radiator returning energy to the antenna radiator during every cycle. The resistive part of impedance consists of two elements: *ohmic losses* (R_o) and *radiation resistance* (R_r). The ohmic losses are due to heating of the antenna conductor elements by RF current passing through, as when current passes through any conductor.

The radiation resistance results from the radiated energy. An efficiency factor (k) compares the loss and radiation resistances:

$$k = \frac{R_r}{R_r + R_o} \qquad (9\text{-}5)$$

The goal of the antenna designer is to reduce R_o to a minimum. The value of R_r is set by the antenna design and installation and is defined as the quotient of the voltage over the current at the feedpoint, less losses.

9-6 DIPOLE ANTENNA ELEMENTS

The *dipole* is a two-pole antenna (Fig. 9-1) that can be modeled either as a single radiator fed at the center (Fig. 9-1A) or a pair of radiators fed back to back (Fig. 9-2B). RF current from the source oscillates back and forth in the radiator element, causing an electromagnetic wave to propagate in a direction *perpendicular* to the radiator element. The polarity of any electromagnetic field is the direction of the electrical field vector (see Chapter 2). In the dipole, the polarization is parallel to the radiator element: a horizontal element produces a horizontally polarized signal, while a vertical element produces a vertically polarized signal.

Figure 9-2 shows the radiator patterns for the dipole viewed from two perspectives. Figure 9-2A shows the pattern of a horizontal half-wavelength dipole as viewed from above. This plot shows the *directivity* of the dipole: maximum radiation is found in two lobes perpendicular to the radiator length. The plot in Fig. 9-2B shows the end-on pattern of the dipole. This omnidirectional pattern serves for a vertically polarized dipole viewed from above. The end-on pattern of a horizontal dipole is similar, except that it is distorted by ground effects unless the antenna is a very large number of wavelengths above the ground.

A microwave dipole is shown in Fig. 9-3. The antenna radiator element consists of a short conductor at the end of a section of waveguide. Although most low-frequency dipoles are half-wavelength, microwave dipoles might be half-wavelength, less than half-wavelength, or greater than half-wavelength, depend-

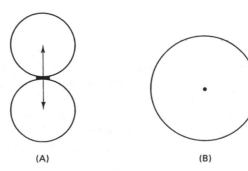

(A)   (B)

Figure 9-2
A) Radiation pattern broadside from a dipole; B) radiation pattern as seen from the ends.

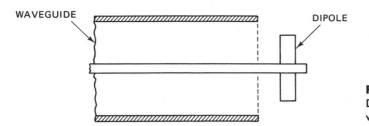

WAVEGUIDE
DIPOLE

Figure 9-3
Dipole element connected to a
waveguide.

ing on the application. For example, because most microwave dipoles are used to illuminate a reflector antenna of some sort, the length of the dipole depends on the exact illumination function required for proper operation of the reflector. Most, however, will be half-wavelength.

9-7   ANTENNA DIRECTIVITY AND GAIN

The dipole discussed in Section 9-6 illustrates two fundamental properties of the type of antennas generally used at microwave frequencies: *directivity* and *gain*. These two concepts are different but so interrelated that they are usually discussed at the same time. Because of the directivity, the antenna focuses energy in only two directions, which means that all the energy is found in those directions (Fig. 9-2A), rather than being distributed over a spherical surface. Thus, the dipole has a gain approximately 2.1 dB greater than isotropic. In other words, the measured power density at any point will be 2.1 dB higher than the calculated isotropic power density for the same RF input power to the antenna.

Directivity.   The directivity of an antenna is a measure of its ability to direct RF energy in a limited direction, rather than in all (spherical) directions equally. As shown previously, the horizontal directivity of the dipole forms a bidirectional figure-8 pattern. Two methods for showing unidirectional antenna patterns are shown in Fig. 9-4. The method of Fig. 9-4A is a polar plot viewed from above. The main lobe is centered on 0°. The plot of Fig. 9-4B is a rectangular method for displaying the same information. This pattern follows a sin X/X function [or (sin x/x^2) for power].

Directivity (D) is a measure of relative power densities:

$$D = \frac{P_{\max}}{P_{\mathrm{av}}} \qquad (9\text{-}6)$$

or, referenced to isotropic,

$$D = \frac{4\pi}{\Phi} \qquad (9\text{-}7)$$

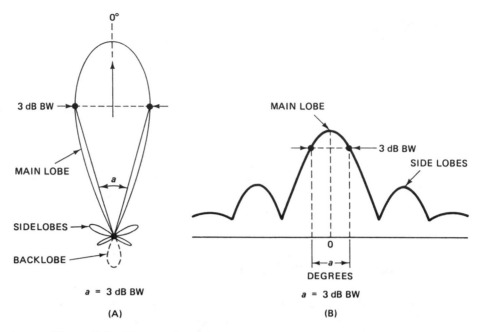

Figure 9-4   Directional antenna pattern: A) polar view; B) graphical view.

where

D	= directivity
P_{max}	= maximum power
P_{av}	= average power
ϕ	= solid angle subtended by the main lobe

The term ϕ is a solid angle, which emphasizes the fact that antenna patterns must be examined in at least two extents: horizontal and vertical.

A common method for specifying antenna directivity is *beamwidth* (BW). The definition of BW is the angular displacement between points on the main lobe (see Figs. 9-4A and 9-4B), where the power density drops to one-half (-3 dB) of its maximum main-lobe power density. This angle is shown in Fig. 9-4A as *a*.

In an ideal antenna system, 100% of the radiated power is in the main lobe, and there are no other lobes. But in real antennas, certain design and installation anomalies cause additional minor lobes, such as the *sidelobes* and *backlobe* shown in Fig. 9-4A. Several problems derive from the minor lobes. First is the loss of usable power. For a given power density required at a distant receiver site, the transmitter must supply whatever additional power is needed to make up for the minor lobe losses.

The second problem is intersystem interference. A major application of directional antennas is the prevention of mutual interference between nearby cochannel stations. In radar systems, high sidelobes translate to errors in detected targets. If, for example, a sidelobe is large enough to detect a target, then the radar display will show this off-axis target as if it were in the main lobe of the antenna.

The result is an azimuth error that could be important in terms of marine and aeronautical navigation.

Gain.   Antenna gain derives from the fact that energy is squeezed into a limited space instead of being distributed over a spherical surface. The term *gain* implies that the antenna creates a higher power, where in fact it merely concentrates into a single direction the power that would otherwise be spread out over a larger area. Even so, it is possible to speak of an apparent increase in power. Antenna-transmitter systems are often rated in terms of *effective radiated power* (ERP). The ERP is the product of the transmitter power and the antenna gain. For example, if an antenna has a gain of +3 dB, the ERP will be twice the transmitter output power. In other words, a 100-W output transmitter connected to a +3-dB antenna will produce a power density at a distant receiver equal to a 200-W transmitter feeding an isotropic radiator. There are two interrelated gains to be considered: *directivity gain* (G_d) and *power gain* (G_p).

The directivity gain is defined as the quotient of the maximum radiation intensity over the average radiation intensity (note the similarity to the directivity definition). This measure of gain is based on the shape of the antenna radiation pattern and may be calculated with respect to an isotropic radiator ($D = 1$) from

$$G_d = \frac{4\pi P_a}{P_r} \tag{9-8}$$

where

   G_d = directivity gain
   P_a = maximum power radiated per unit of solid angle
   P_r = total power radiated by the antenna

The power gain is similar in nature, but slightly different from directivity gain; it includes dissipative losses in the antenna. Not included in the power gain are losses caused by cross-polarization or impedance mismatch between the waveguide or transmission line and the antenna. There are two commonly used means for determining power gain:

$$G_p = \frac{4\pi P_a}{P_n} \tag{9-9}$$

and

$$G_p = \frac{P_{ai}}{P_i} \tag{9-10}$$

where

P_a = maximum power radiated per unit solid angle
P_n = net power accepted by the antenna (that is, less mismatch losses)
P_{ai} = average intensity at a distant point
P_i = intensity at the same point from an isotropic radiator fed the same RF power level as the antenna

The equation assumes equal power to the antenna and comparison isotropic source.

Provided that ohmic losses are kept negligible, the relationship between directivity gain and power gain is given by

$$G_p = \frac{P_r G_d}{P_n}$$

(9-11)

All terms are as previously defined.

9-7.1 Relationship of Gain and Aperture

Antennas obey the law of reciprocity, which means that any given antenna will work as well on receive as on transmit. The function of the receive antenna is to gather energy from the electromagnetic field radiated by the transmitter antenna. The aperture is related to, and often closely approximates, the physical area of the antenna. But in some designs the effective aperture (A_e) is less than the physical area (A), so there is an effectiveness factor (n) that must be applied. In general, however, a high-gain transmitter antenna also exhibits a high receive aperture, and the relationship can be expressed as

$$G = \frac{4\pi A_e\, n}{\lambda^2}$$

(9-12)

where

A_e = effective aperture
n = aperture effectiveness ($n = 1$ for a perfect, lossless antenna)
λ = wavelength of the signal

9-8 HORN ANTENNA RADIATORS

The horn radiator is a tapered termination of a length of waveguide (see Figs. 9-5A through C) that provides the impedance transformation between the waveguide impedance and free-space impedance. Horn radiators are used both as antennas in

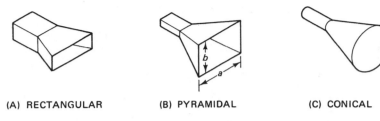

(A) RECTANGULAR   (B) PYRAMIDAL   (C) CONICAL

Figure 9-5   Horn radiators: A) rectangular; B) pyramidal; C) conical.

their own right and as illuminators for reflector antennas (see Sections 9-9 and 9-10). Horn antennas are not a perfect match to the waveguide, although standing-wave ratios of 1.5:1 or less are achievable.

The gain of a horn radiator is proportional to the area (A) of the flared open flange ($A = ab$ in Fig. 9-5B) and inversely proportional to the square of the wavelength:

$$G = \frac{10\,A}{\lambda^2} \qquad\qquad (9\text{-}13)$$

where

A = flange area
λ = wavelength (both in the same units)

The –3-dB beamwidth for vertical and horizontal extents can be approximated from the following:

Vertical:

$$\Phi_v = \frac{51\lambda}{b} \text{ degrees} \qquad\qquad (9\text{-}14)$$

Horizontal:

$$\Phi_h = \frac{70\lambda}{a} \text{ degrees} \qquad\qquad (9\text{-}15)$$

where

ϕ_v = vertical beamwidth in degrees
ϕ_h = horizontal beamwidth in degrees
a, b = dimensions of the flared flange
λ = wavelength

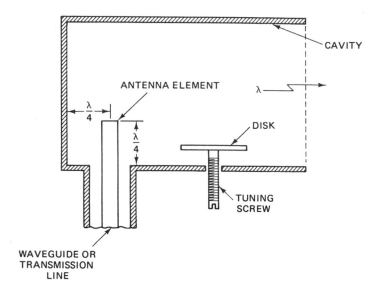

Figure 9-6
Cavity antenna.

A form of antenna related to the horn is the cavity antenna of Fig. 9-6. In this type of antenna, a quarter-wavelength radiating element extends from the waveguide or transmission line connector into a resonant cavity. The radiator element is placed one quarter-wavelength into a resonant cavity and is spaced a quarter-wavelength from the rear wall of the cavity. A tuning disk is used to alter cavity dimensions to provide a limited tuning range for the antenna. Gains to about 6 dB are possible with this arrangement.

9-9   REFLECTOR ANTENNAS

At microwave frequencies, it becomes possible to use *reflector antennas* because of the short wavelengths involved. Reflectors are theoretically possible at lower frequencies, but because of the longer wavelengths, the antennas would be so large that they would be impractical. Several forms of reflector are used (Figs. 9-7 and 9-8). In Fig. 9-7 we see the *corner reflector* antenna, which is used primarily in the high-UHF and low-microwave region. A dipole element is placed at the *focal point* of the corner reflector and so receives in phase the reflected wavefronts from the surface. Either solid metallic reflector surfaces or wire mesh may be used. When mesh is used, however, the holes in the mesh must be 1/12 wavelength or smaller.

Figure 9-8 shows several other forms of reflector surface shapes, most of which are used in assorted radar applications. In Section 9-10, we will discuss the *parabolic reflector* in detail.

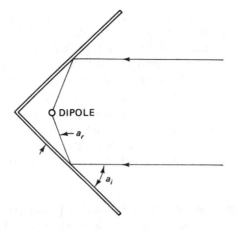

Figure 9-7
Corner reflector antenna.

9-10 PARABOLIC DISH ANTENNAS

The parabolic reflector antenna is one of the most widespread of all microwave antennas and is the type that normally comes to mind when thinking of microwave systems. This type of antenna derives its operation from physics similar to optics, and is possible because microwaves are in a transition region between ordinary radio waves and infrared/visible light.

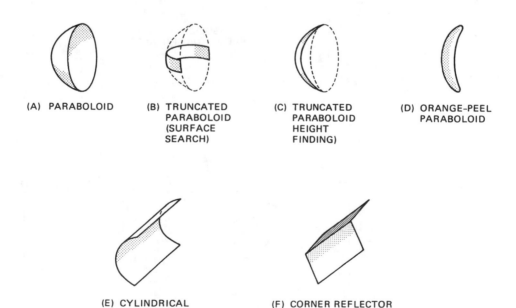

(A) PARABOLOID (B) TRUNCATED PARABOLOID (SURFACE SEARCH) (C) TRUNCATED PARABOLOID HEIGHT FINDING) (D) ORANGE-PEEL PARABOLOID

(E) CYLINDRICAL PARABOLOID (F) CORNER REFLECTOR

Figure 9-8   Assorted parabolic reflectors.

The dish antenna has a paraboloid shape as defined by Fig. 9-9. In this figure, the dish surface is positioned such that the center is at the origin (0,0) of an X-Y coordinate system. For purposes of defining the surface, we place a second vertical axis called the *directrix* (Y') a distance behind the surface equal to the focal length (u). The paraboloid surface follows the function $Y^2 = 4uX$ and has the property that a line from the focal point (F) to any point on the surface is the same length as a line from that same point to the directrix (in other words, $Mn = MF$).

If a radiator element is placed at the focal point (F), it will illuminate the reflector surface, causing wavefronts to be propagated away from the surface in phase. Similarly, wavefronts intercepted by the reflector surface are reflected to the focal point.

Gain.
The gain of a parabolic antenna is a function of several factors: dish diameter, feed illumination, and surface accuracy. The dish diameter (D) should be large compared with its depth. Surface accuracy refers to the degree of surface irregularities. For commercial antennas, 1/8-wavelength surface accuracy is usually sufficient, although on certain radar antennas the surface accuracy specification must be tighter.

The feed illumination refers to how evenly the feed element radiates to the reflector surface. For circular parabolic dishes, a circular waveguide feed produces optimum illumination, while rectangular waveguides are less than optimum. The TE_{11} mode is desired. For best performance, the illumination should drop off evenly from the center to the edge, with the edge being −10 dB down from the center. The diameter, length, and beamwidth of the radiator element or horn must be

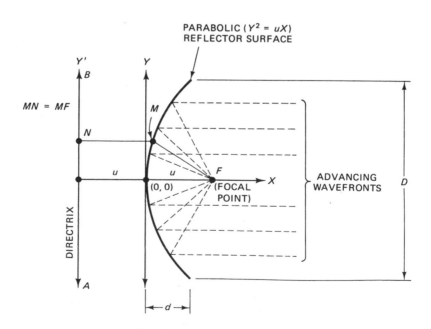

Figure 9-9   Full parabolic dish antenna.

optimized for the specific *F/d* ratio of the dish. The cutoff frequency is approximated from

$$f_{\text{cutoff}} = \frac{175,698}{d_{mm}} \qquad (9\text{-}16)$$

where

F_{cutoff} = cutoff frequency
d = inside diameter of the circular feedhorn

The gain of the parabolic dish antenna is found from

$$G = \frac{k(\pi D)^2}{\lambda^2} \qquad (9\text{-}17)$$

where

G = gain over isotropic
D = diameter
λ = wavelength (same units as D)
k = reflection efficiency (0.4 to 0.7, with 0.55 being most common)

The –3-dB beamwidth of the parabolic dish antenna is approximated by

$$\text{BW} = \frac{70\,A}{D} \qquad (9\text{-}18)$$

and the focal length by

$$f = \frac{D^2}{16d} \qquad (9\text{-}19)$$

For receive applications, the effective aperture is the relevant specification and is found from

$$A_e = k\pi\left(\frac{D}{2}\right)^2 \qquad (9\text{-}20)$$

The antenna pattern radiated by the antenna is similar to Fig. 9-4B. With horn illumination, the sidelobes tend to be 23 to 28 dB below main lobe or 10 to 15 dB below isotropic. It is found that 50% of the energy radiated by the parabolic dish is within the −3-dB beamwidth, and 90% is between the first nulls on either side of the main lobe.

If a dipole element is used for the feed device, then a *splash plate* is placed one quarter-wavelength behind the dipole in order to improve illumination. The splash plate must be several wavelengths in diameter and is used to reflect the backlobe back toward the reflector surface. When added to the half-wave phase reversal inherent in the reflection process, the two-way quarter-wavelength adds another half-wavelength and thereby permits the backwave to move out in phase with the front lobe wave.

9-10.1 Parabolic Dish Feed Geometries

Figure 9-10 shows two methods for feeding parabolic dish antennas, regardless of which form of radiator (horn, dipole, and so on) is used. In Fig. 9-10A, we see the method in which the radiator element is placed at the focal point, and either waveguide or transmission line is routed to it. This method is used in low-cost installations such as home satellite TV receive-only (TVRO) antennas.

In Fig. 9-10B, we see the *Cassegrain feed* system. This system is modeled after the Cassegrain optical telescope. The radiator element is placed at an opening at the center of the dish. A hyperbolic subreflector is placed at the focal point and is used to reflect the wavefronts to the radiator element. The Cassegrain system results in lower-noise operation because of several factors: less transmission line length, lower sidelobes, and the fact that the open horn sees sky instead of earth

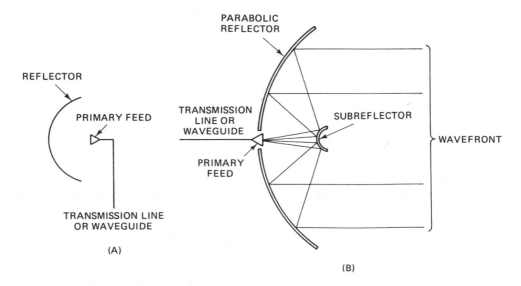

Figure 9-10   A) Standard feed; B) Cassegrain feed.

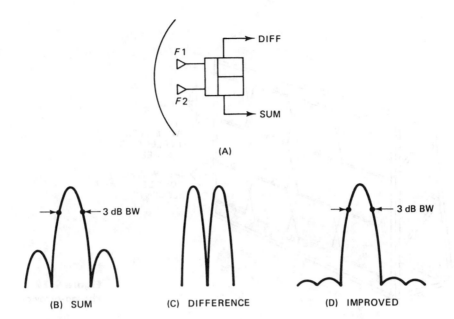

Figure 9-11   A) Monopulse feed; B) sum pattern; C) difference pattern; D) improved pattern due to adding sum and difference.

(which has a lower temperature); on the negative side, galactic and solar noise may be slightly higher on a Cassegrain dish.

Figure 9-11 shows the *monopulse feed* geometry. In this system, a pair of radiator elements is placed at the focal point and fed to a power splitter network that outputs *sum* (Fig. 9-11B) and *difference* (Fig. 9-11C) signals. When these are combined, the resultant beam shape has an improved –3-dB beamwidth due to the algebraic summation of the two.

9-11 ARRAY ANTENNAS

When antenna radiators are arranged in a precision array, an increase in gain occurs. An array might be a series of dipole elements, as in the broadside array of Fig. 9-12 (which is used in the UHF region), or a series of slots, horns, or other radiators. The overall gain of an array antenna is proportional to the number of elements, as well as the details of their spacing. In this and other antennas, a method of *phase shifting* is needed. In Fig. 9-12, the phase shifting is caused by crossed feeding of the elements, while in more modern arrays other forms of phase shifter are used (see Section 9-11.1).

Two methods of feeding an array are shown in Fig. 9-13. The *corporate feed* method connects all elements and their phase shifters in parallel with the source. The *branch feed* method breaks the waveguide network into two or more separate paths; both branch and corporate feed are shown in Fig. 9-13.

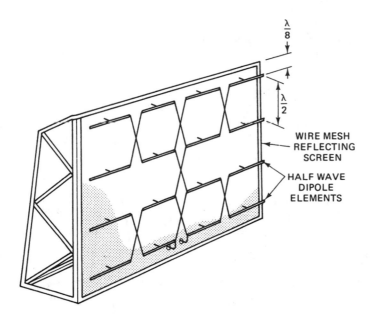

Figure 9-12
Phased-array antenna.

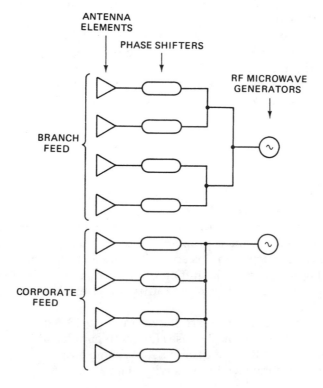

Figure 9-13
Branch and corporate feed for antenna.

9-11.1 Solid-State Array Antennas

Some modern radar sets use solid-state array antennas consisting of a large number of elements, each of which is capable of shifting the phase of a microwave input signal. Two forms are known: *passive* (Fig. 9-14A) and *active* (Fig. 9-14B). In the passive type of element, a ferrite or PIN diode phase shifter is placed in the transmission path between the RF input and the radiator element (usually a slot). By changing the phase of the RF signal selectively, it is possible to form and steer the beam at will. A 3-bit (that is, three discrete states) phase shifter allows phase to shift in 45° increments, while a 4-bit phase shifter allows 22.5° increments of phase shift.

The active element contains a phase shifter in addition to a transmit power amplifier (1 or 2 W) and a low-noise amplifier for receive. A pair of transmit/receive (T/R) switches selects the path to which the RF signal is directed. The total output power of this antenna is the sum of all output powers from all elements in the array. For example, an array of 1000 two-watt elements makes a 2000-W system.

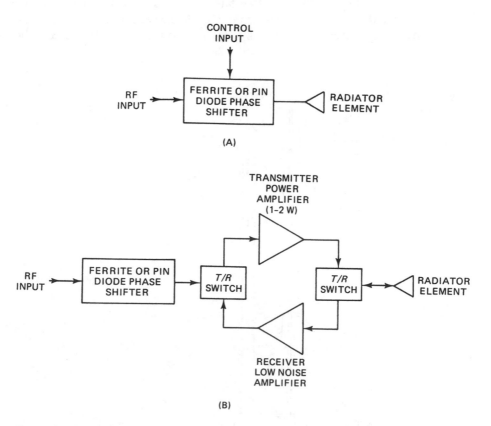

Figure 9-14 A) Phase shifter element; B) block diagram for a T/R element in phased-array antenna.

9-12 SLOT ARRAY ANTENNAS

A resonant slot may be cut into the wall of a section of waveguide in a manner analogous (if not identical) to a dipole. By cutting several slots into the waveguide, we obtain the advantages of an array antenna in which the elements are several slot radiators. Slot array antennas are used for marine navigation radars and telemetry systems and for the reception of microwave television signals in the Multipoint Distribution Service (MDS) on 2.145 GHz.

Figure 9-15 shows a simple slot antenna used in telemetry applications. A slotted section of rectangular waveguide is mounted to a right-angle waveguide flange. An internal wedge (not shown) is placed at the top of the waveguide and serves as a matching impedance termination to prevent internal reflected waves. Directivity is enhanced by attaching flanges to the slotted section of waveguide parallel to the direction of propagation (see end view in Fig. 9-15).

Figure 9-16 shows two forms of *flatplate array* antennas constructed from slotted waveguide radiator elements (shown as insets). In Fig. 9-16A, we see the rectangular array, and in Fig. 9-16B, the circular array. These flatplate arrays are used extensively in microwave communications and radar applications.

The feed structure for a flatplate array is shown in Fig. 9-16C. The antenna element is the same sort as shown in Figs. 9-16A and B. A distribution waveguide is physically mated with the element, and a coupling slot is provided between the

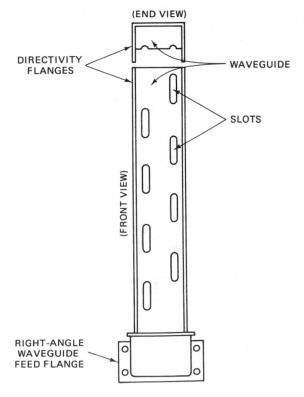

Figure 9-15
Slot antenna.

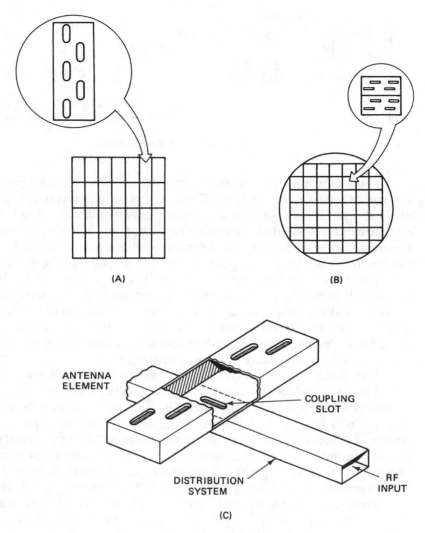

Figure 9-16 Slot array antenna: A) vertically polarized; B) horizontally polarized; C) coupling antenna elements to waveguide.

two waveguides. Energy propagating in the distribution system waveguide is coupled into the antenna radiator element through this slot. In some cases, metallic or dielectric phase-shifting stubs are also used to fine-tune the antenna radiation pattern.

9-13 MISCELLANEOUS ANTENNAS

In addition to the antennas discussed so far, there are several designs that find application but are not widespread enough to be discussed in detail here: *helical, polycone,* and the *ring* or *disk Yagi.*

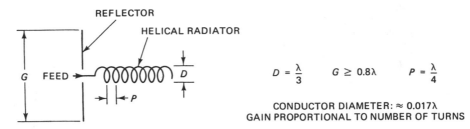

$$D = \frac{\lambda}{3} \qquad G \geq 0.8\lambda \qquad P = \frac{\lambda}{4}$$

CONDUCTOR DIAMETER: $\approx 0.017\lambda$
GAIN PROPORTIONAL TO NUMBER OF TURNS

Figure 9-17   Helical antenna.

The helical antenna is a circularly polarized antenna used primarily in the low end of the microwave region. The design of the antenna (Fig. 9-17) consists of a helical coil of N turns situated in front of a reflector plane (G). The helical antenna is broadband compared with the other types, but only at the expense of some gain. The direction of the helix spiral determines whether the polarity is clockwise or counterclockwise circular. Key design factors are shown in Fig. 9-17.

The polycone antenna consists of a cavity antenna (such as Fig. 9-6) in which a cone of dielectric material is inserted in the open end (that is, the output iris). The dielectric material shapes the beam into a narrower beamwidth than is otherwise obtained. Polycone antennas are used in various low-power applications such as microwave signaling, control system sensors (including Doppler traffic control systems), and in some police CW speed radar guns.

The classical Yagi-Uda array (commonly called Yagi) is used extensively in HF, VHF, UHF, and low-microwave bands; modified versions are used in the mid-microwave region (several gigahertz). The Yagi antenna consists of a half-wavelength center-fed dipole driven element and one or more parasitic elements (microwave versions tend to use ten or more elements). The parasitic elements are placed parallel to the driven element about 0.1 to 0.3 wavelength away from it or each other. Two sorts of parasitic elements are used: *reflectors* and *directors*. The reflectors are slightly longer (approximately 4%) than, and are placed behind, the driven element; directors are slightly longer and placed in front of the driven element. The main lobe is in the direction of the director elements. Yagis are commonly used to frequencies of 800 or 900 MHz. Modified versions of this antenna use rings or disks for the elements and are useful to frequencies in the 3- or 4-GHz region.

9-14 MICROWAVE ANTENNA SAFETY NOTE

Microwave RF energy is dangerous to your health. Anything that can cook a roast beef can also cook you! The U.S. government sets a safety limit for microwave exposure of 10 mW/cm^2 averaged over 6 minutes; some other countries use a level one-tenth of the U.S. standard. The principal problem is tissue heating, and eyes seem especially sensitive to microwave energy. Some authorities believe that cataracts form from prolonged exposure. Some authorities also believe that genetic damage to offspring is possible, as well as other long-term effects due to cumulative exposure.

Because of their relatively high gain, microwave antennas can produce hazardous field strengths in close proximity *even at relatively low RF input power levels*. At least one technician in a TV satellite Earth station suffered abdominal adhesions, solid matter in the urine, and genital disfunction after servicing a 45-m-diameter 3.5-GHz dish antenna with RF power applied.

Be very careful around microwave antennas. Do not service a radiating antenna. When servicing nonradiating antennas, be sure to stow them in a position that prevents inadvertent exposure to humans should power accidentally be applied. A RADIATION HAZARD sign should be prominently displayed on the antenna. Good design practice requires an interlock system that prevents radiation in such situations. Hot transmitter service should be performed with a shielded dummy load replacing the antenna.

9-15 SUMMARY

1. The antenna acts as a transducer between the transmission line or waveguide and free space; as such, it must match the line or guide impedance to the impedance of free space.

2. Antenna measurements and specifications are usually measured relative to an *isotropic radiator*; that is, a spherical point source that radiates equally well in all directions.

3. Two fields are important on an antenna: near field and far field. The near field includes energy from the electrical and magnetic fields close to the antenna and falls off at a rate of $1/R^4$. The far field falls off according to $1/R^2$.

4. The antenna feedpoint impedance includes radiation resistance, ohmic losses, and reactive losses. The reactance is due to some energy being returned to the antenna during each cycle from the near field. Most designers try to reduce the reactive component to zero.

5. *Directivity* is the ability of the antenna to focus energy in a limited direction. The directivity of an isotropic radiator is 1, and all other antennas are measured against this standard. Directivity is defined as the ratio of the isotropic sphere over the solid angle subtended by the main lobe of the antenna.

6. Antenna *beamwidth* is defined as the angular displacement between points on the main lobe where the power density has dropped –3 dB from the maximum power density.

7. Sidelobes and backlobes are antenna defects that rob power from the main lobe and make the system more susceptible to cochannel interference.

8. A receive antenna is rated according to its *effective aperture*. Because antennas obey the law of reciprocity, a large gain usually implies a large effective aperture.

9. Two forms of antenna gain are recognized: *directivity gain* and *power gain*. Directivity gain is the maximum power per unit of solid angle divided by the total power radiated by the antenna; power gain is the maximum power per unit of solid angle divided by the total power accepted by the antenna (less mismatch losses).

10.   Horn antennas are a section of tapered waveguide that radiates a signal into free space. A cavity antenna is a resonant cavity in which a dipole, monopole, or loop antenna is situated.

11.   Reflector antennas concentrate the wave onto or from a radiator element by reflection from a surface located a critical distance behind the radiator. Perhaps the most common reflector is the parabolic dish. The gain of a dish antenna is a function of its diameter, the wavelength, and the reflectivity of the surface.

12.   Three principal feed methods are used for parabolic dish antennas: (a) focal point feed, (b) Cassegrain feed, and (c) monopulse feed.

13.   Array antennas use a number of radiator elements at critical spacing to sum up their respective fields and thus provide gain. The elements of a slot array antenna are slotted waveguide sections; each slot acts as a resonant radiator. Solid-state array antennas use either active or passive phase shifting elements.

14.   Additional antennas used at microwave frequencies are helical, polycone, and ring or disk Yagi.

9-16   RECAPITULATION

Now return to the objectives and prequiz questions at the beginning of the chapter and see how well you can answer them. If you cannot answer certain questions, place a check mark by each and review the appropriate parts of the text. Next, try to answer the following questions and work the problems using the same procedure.

QUESTIONS AND PROBLEMS

1.   Microwaves are a transition region between ordinary radio waves and _____ waves.

2.   The function of an antenna is to act as a _____ between the transmission line or waveguide and free space; as such, it must transform impedances.

3.   The law of _____ states that antennas work as well on receive as on transmit.

4.   The important property of a transmit antenna is gain; the important property of a receive antenna is _____ _____.

5.   The _____ radiator is a theoretical construct used to reference antennas and uses a spherical point source of energy as the comparison.

6.   What is the power density of a point source radiator (as in question 5) at 10 km if the input power is 5000 W?

7.   Calculate the power density of a point source radiator at 1 mile if the input power is 250 W.

8.   The far field of an antenna falls off at $1/R^2$, while the near field falls off at $1/$ _____.

9.   An antenna with a 6- $\times$ 10-cm rectangular horn radiator operates at 12.2 GHz. Calculate the minimum extent of the near field.

10. The antenna in question 9 is shifted to 3.97 GHz. Calculate the extent of the near field.

11. Calculate the efficiency factor of an antenna radiator element if the radiation resistance is 125 Ω, the ohmic losses are 2.9 Ω, and the reactive losses are 0 Ω.

12. Calculate the efficiency factor of an antenna with a 75-Ω radiation resistance and 6 Ω of ohmic losses.

13. Is a horizontally polarized half-wavelength dipole *bidirectional* or *unidirectional*?

14. The directivity of an isotropic antenna is defined as ____.

15. The beamwidth of an antenna is the angular displacement between points on the main lobe where the power density drops off _____ dB from the maximum main-lobe density.

16. Name two other forms of antenna pattern lobe which that be minimized in antenna design.

17. An antenna has a power gain of +9 dB. Calculate the effective radiated power (ERP) if the applied transmitter output power is 5 W.

18. Calculate the input power of an ideal antenna that has a gain of +6 dB and an effective radiated power of 450 W.

19. Calculate the gain of an antenna if the input power is 100 W and the ERP is 1500 W.

20. Calculate the directivity gain of an antenna if 100 W is radiated by the antenna, and the maximum power per unit solid angle is 250 W.

21. In question 20, a total of 107 W is accepted by the antenna. Calculate the power gain.

22. Calculate the gain of an ideal antenna that has an effective aperture of 4 m^2 when operated at a wavelength of 2.75 cm.

23. An antenna has an effective aperture of 1200 m^2. What is its gain at a frequency of 6.35 GHz?

24. Calculate the effective aperture of an antenna if the power gain is 20 dB, and the operating frequency is 12.45 GHz.

25. A horn radiator uses a 4- × 9-cm open flange. What is its gain at 10.25 GHz?

26. Calculate the gain of the antenna in question 25 in decibels.

27. A horn radiator has an open flange of 3 by 5 cm. Calculate the vertical and horizontal field –3-dB beamwidth at 9.95 GHz.

28. The radiator in question 27 is operated at 12.75 GHz. Calculate

    a. the wavelength, b. the gain,

    c. the vertical –3-dB BW, d. the horizontal –3-dB BW.

29. List two kinds of reflector antenna.

30. What form of reflector antenna is probably the most commonly recognized microwave antenna?

31. List the three factors that determine the gain of a parabolic dish antenna.

32. The surface accuracy of a commercial dish antenna should be about ____ wavelength.

33. For a waveguide-fed microwave parabolic dish antenna, a _____ horn is preferred, and it should operate in the _____ mode.

34. For perfect dish illumination, the power density at the edges should be _____dB down from the central illumination power density.

35. A microwave parabolic dish antenna has a diameter of 15 m and a reflection efficiency of 55%. What is its gain at 3.65 GHz?

36. Calculate the gain of a parabolic dish at a wavelength of 2.33 cm if the reflection efficiency is 0.56 and the diameter is 12 m.

37. Calculate the effective aperture of the antennas in questions 35 and 36.

38. What is the effective aperture of a perfectly reflective (ideal) microwave parabolic
dish antenna at 7.5 GHz if the diameter is 45 ft?

39. Calculate the –3-dB beamwidth of a parabolic dish antenna at a wavelength of 6 cm if
the diameter is 3.5 m.

40. Calculate the effective aperture of a parabolic dish antenna with a reflective efficiency
of 0.55 and a diameter of 3 m.

41. List three feed geometries used on parabolic dish antennas.

42. List two methods of feeding microwave RF energy to an array antenna.

43. List two forms of phase shifter for solid-state array antennas.

44. A 3-bit phase shifter allows _____-degree phase shift increments.

45. A 4-bit phase shifter allows _____-degree phase shift increments.

46. A slot array uses slot radiator elements cut into a section of _____.

47. A *flatplate* array antenna is made from a collection of _____ array elements.

48. The maximum exposure averaged over 6 minutes allowed for humans by the U.S.
government is _____mW/cm².

KEY EQUATIONS

1. Power density of an isotropic radiator at distance R:

$$P_d = \frac{P}{4\pi R^2}$$

2. Near-field distance of antenna with dimension d:

$$r_{min} = \frac{2d^2}{\pi}$$

3. Alternate notations for impedance in ac circuits:

$$Z = \sqrt{R^2 + (X_L - Xc^2}$$

or

$$Z = R \pm jK$$

4. Efficiency factor of antenna comparing resistances:

$$k = \frac{R_r}{R_r + R_o}$$

5. Antenna directivity as a function of power densities:

$$D = \frac{P_{max}}{P_{av}}$$

6. Antenna directivity referenced to isotropic:

$$D = \frac{4\pi}{\Phi}$$

7. Directivity gain of an antenna:

$$G_d = \frac{4\pi P_a}{P_r}$$

8. Power gain of an antenna:

$$G_p = \frac{4\pi P_a}{P_n}$$

and

$$G_p = \frac{P_{ai}}{P_i}$$

9. Relationship between directivity gain and power gain:

$$G_p = \frac{P_r G_d}{P_n}$$

10. Relationship between transmit gain and receive aperture:

$$G = \frac{4\pi A_e n}{\lambda^2}$$

11. Gain of a horn radiator:

$$G = \frac{10A}{\lambda^2}$$

12. The –3-dB beamwidth of a horn antenna:

$$Vertical: \ \Phi_v = \frac{51\lambda}{b} \ degrees$$

$$Horizontal: \ \Phi_h = \frac{70\lambda}{a} \ degrees$$

13. Cutoff frequency of waveguide feeding dish antenna:

$$f_{\text{cutoff}} = \frac{175,698}{d_{mm}}$$

14. Gain of parabolic dish antenna:

$$G = \frac{k(\pi D)^2}{\lambda^2}$$

15. The –3-dB beamwidth of parabolic dish antenna:

$$BW = \frac{70\lambda}{D}$$

16. Focal length of parabolic dish antenna:

$$f = \frac{D^2}{16d}$$

CHAPTER 10

Microwave Vacuum Tube Devices
Early Microwave RF Power Generators
(Some History)

OBJECTIVES

1. Learn the various methods used for generating RF power in the early days of radio communications.
2. Understand the limitations of the various early devices used for RF power generation.
3. Learn how the limitations of vacuum tube devices were used to generate microwave power.
4. Learn the types of vacuum tube devices used today to generate microwave RF power.

10-1 PREQUIZ

These questions test your prior knowledge of the material in this chapter. Try answering them before you read the chapter. Look for the answers (especially those you answered incorrectly) as you read the text. After you have finished studying the chapter, try answering these questions again and those at the end of the chapter.

1. List three methods used to generate RF power prior to World War II.
2. A fundamental limitation of vacuum tubes at microwave frequencies is electron _____ time.
3. The first successful RF generator using vacuum tubes was the _____-_____ oscillator.
4. The Hull device (1921) used a _____ _____ in place of a bias voltage to keep the electron beam in orbit.

10-2 RF POWER GENERATION: SOME PROBLEMS

One the earliest and most critical problems inhibiting the full use of radio, radar, and other electromagnetic devices was the generation of sufficient amounts of RF power at frequencies where it could do some good. Problems plagued the early pioneers in radio, and for many years generating RF signals at useful power levels was limited to very low frequencies (VLF). Various problems limited the devices then in common use. But throughout the 1920s, 1930s, and 1940s advances were made that used inherent physical limitations of vacuum tube devices to good advantage to generate higher-frequency signals.

The definition of what constitutes high frequencies has varied over the years according to the difficulty of generating power at those frequencies. Until the mid-1920s, when amateur radio operators serendipitously opened the high-frequency "shortwave" spectrum (3 to 30 MHz), commercial communications were carried on VLF with wavelengths longer than 200 m (that is, those frequencies below 1500 kHz). The technologies used to generate RF power worked well at those frequencies, but their effectiveness dropped off rapidly as frequency increased.

But even in the early days of radio experimentation, higher frequencies were at least investigated. Early radio pioneer Heinrich Hertz in 1887–1888 used frequencies between 31.3 and 1250 MHz for his short range (across the laboratory) investigations. Thus, it is fair to say that some of the very earliest experiments took place at microwave frequencies in the region near 1.25 GHz. Guglielmo Marconi used 500-MHz frequencies for short-range experiments, but switched to VLF when it was found that lower frequencies were more effective for distance communications. Besides the fact that RF signals were easier to generate at those frequencies, the detectors used in those days (iron filing devices called Branly coherers) were far more sensitive at those frequencies. In addition, they also ran into some of the realities of electromagnetic propagation (see Chapter 2). On December 12, 1901, Marconi and his co-workers made radio history by achieving the first confirmed transatlantic transmission. A 313-kHz radio signal was sent from Poldhu, England, on a 10-kW transmitter and was received at Marconi's listening post near St. Johns, Newfoundland.

When vacuum tubes became available, it was possible to operate on higher frequencies. Commercial, military, and amateur radio communications moved to the high-frequency shortwave region during the mid-1920s. Difficulties with devices operated at frequencies higher than 25 MHz caused the region above today's Citizen's Band to be called ultrahigh frequencies (UHF). Compare this with today, when UHF designates frequencies in the 300- to 900-MHz region. Advances in vacuum tube technology during World War II allowed practical use of frequencies up to 450 MHz, so the definition of UHF was changed.

Traditionally, there were three methods for generating RF power: *spark gaps, Alexanderson alternators,* and *vacuum tubes.*

10-3 SPARK GAP GENERATORS

An electric arc produces a tremendous amount of electrical energy both at harmonics and at frequencies that are not harmonically related to the principal frequency. You only need note that on any radio receiver a lightning bolt produces amplitude-modulated RF noise at seemingly all frequencies across the electromagnetic spectrum. Similarly, electrical arcs on devices such as motors, electric ignition systems on gasoline vehicle engines, and other sources also produce large amounts of RF energy spread across a wide portion of the spectrum.

Figure 10-1 shows a simple spark gap RF power generator. Until 1938 when they were declared illegal, circuits such as Fig. 10-1 were used as crude radio transmitters. In fact, some early wireless experimenters and amateur operators used purloined Ford Model A ignition coils from the family car to make impromptu radio transmitters. Even today, surgeons use spark gap RF generators as electrocautery machines. The ragged waveform produced by the spark gap generator is believed to be superior for cauterizing bleeding blood vessels to a sine-wave RF power signal. Oddly, some such medical devices are called microwave even though operating at frequencies considerably lower than the microwave region.

The basic power for the spark gap generator is a high-voltage ac power transformer (T1) connected to the regular ac power mains. The potential across the secondary of T1 is high enough to ionize the air in the space between the electrodes of the spark gap, which creates the spark. A series-resonant *LC* tank circuit (C/L1A) picks off the frequency energy required for the application. Unfortunately, the spark gap produces a very wideband signal. For example, an 800-kHz spark gap generator actually produces significant levels of primary power from 10 to 3000 kHz and at least weak harmonics to the microwave region. An output coupling link wound as a transformer secondary on the tank circuit coil provides the output connection.

Figure 10-2 shows a system used in 1930 to generate microwave signals as high as 75 GHz. A spark gap was placed inside a resonant cavity (see Chapter 6), which functions much like a resonant tank circuit using a different mechanism. An output coupling loop is used to pick off the output signal and deliver it to the load. Unfortunately, a spark gap is a very inefficient device. Because its power is spread over a tremendous spectrum, only a small amount of power is available at any one frequency. In addition, as the frequency gets higher, the available power drops dramatically. At 75 GHz, efficiency is considerably lower than 1% because the major-

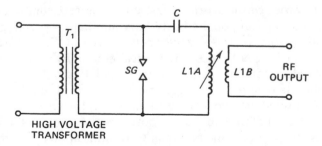

HIGH VOLTAGE
TRANSFORMER

Figure 10-1
Spark gap transmitter.

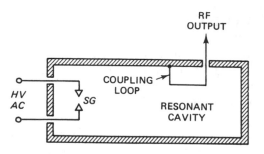

Figure 10-2
Microwave spark gap used many decades ago.

ity of available RF power is distributed to lower frequencies outside the microwave bands.

Spark gap transmitters were useful for on-off radiotelegraphy transmitters during the era when only a few transmitters were on the air. But as the spectrum became more crowded, the usefulness of the spark gap was at an end. The Federal Communications Commission outlawed the use of spark gap transmitters in 1938 at a point when commercial and amateur use had waned to the vanishing point.

10-4 ALEXANDERSON ALTERNATORS

The two main problems with the spark gap were limited efficiency at any given frequency of operation and spectral purity of the output signal. The Alexanderson alternator was an attempt to provide high power at low frequencies in a manner that overcame these difficulties. The alternator was functionally the same (except for rectifiers) as the electrical system charging alternator on modern automobiles. In these machines, a magnet is rotated inside the windings of an electrical coil. The frequency of the ac output of the stationary (stator) coil is a function of the number of poles on the magnet, the number of coil pairs in the stator, and the speed of rotation. For example, if a single two-pole magnet is spun at a rate of one rotation per second inside a single two-pole stator coil, a 1-Hz signal is generated. By increasing the number of magnet pole pairs, the number of stator pole pairs, and the speed of rotation, frequencies up to 1000 kHz could be generated (although most alternators produced signals in the 30- to 200-kHz region).

The alternators used in communications used an electromagnetic coil to generate RF power. On-off telegraphy signals could be created by interrupting the coil current with a telegraph key. In 1916, engineers from the Naval Research Laboratory in Washington, D.C., used the U.S. Navy communications station (NAA) at Arlington, Virginia to produce the world's first voice transmission over radio. NAA, also known as Radio Arlington, had a 100-kW alternator operating on 113 kHz and was a dominant voice on the airwaves prior to World War I. Navy engineers amplitude-modulated the NAA Alexanderson alternator by varying the field current of the rotating electromagnet using a voice signal. Radio Arlington occupied a hillside near the site where the Pentagon was eventually built, but its days as a radio station were limited. Its two 400-ft and one 600-ft towers were an obstacle to air traffic when Washington National Airport was built, so the site was decommissioned. Today it is used for the National Communications System by the

Defense Communications Agency. Because of its low frequency of operation, the Alexanderson alternator was of limited use at microwave frequencies.

10-5 VACUUM TUBE GENERATORS

In 1905, an Englishman, Sir John Fleming, used a phenomenon noted by Thomas Edison 20 years before to invent the first practical vacuum tube. Edison noted that a positively charged electrode inside of an evacuated glass light bulb drew a current. Fleming used that fact to invent a two-element diode vacuum tube that rectified radio signals. The Fleming valve (as the English called it) consisted of a hot cathode that emitted electrons and an anode that collected the electrons. In 1907, American Lee DeForest inserted a third element into Fleming's vacuum tube to form a triode. The third element was a grid that controlled the flow of electrons. Although this book is about microwave devices, we must also take a brief look at these devices for purposes of understanding limitations and problems with vacuum tube devices. In addition, some of the principles will be seen again in Chapter 11 when we discuss M- and O-type microwave tubes. The student should understand, however, that there is a considerable body of knowledge about vacuum tubes that is beyond the scope of this book. Only a few schools continue vacuum tube technology in their curriculum because solid-state devices have replaced tubes in all but the highest-power RF applications. For example, a 5000-W broadcast transmitter will have solid-state devices in all stages, including the high-level exciter, but will have vacuum tube devices in the final amplifier.

Figure 10-3 shows the elementary triode vacuum tube circuit. There are three elements in the triode vacuum tube: *cathode*, *anode*, and *grid*. In some models, the cathode is a filament, while in others the cathode is a hollow tube with an internal heater (as in Fig. 10-3). Whether directly or indirectly heated, the intent is to raise the temperature of the cathode until electrons boil off the surface into the area around the cathode. This process is called *thermionic emission* of electrons. The cloud of electrons produced by thermionic emission is called a *space charge*. When a positively charged anode (also sometimes called a *plate*) is placed inside the vac-

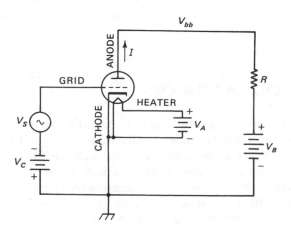

Figure 10-3
Vacuum tube circuit.

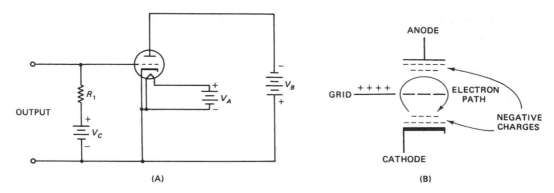

Figure 10-4   Barkhausen-Kurz oscillator.

uum tube, it attracts the negatively charged electrons of the space charge. These electrons form the anode current.

The grid is a porous element placed between the cathode and anode. If it is biased negative (V_c) with respect to the anode, it can control the flow of electrons. If the negative bias is sufficiently high, it can completely shut off the flow of anode current. When the signal voltage (V_s) is superimposed on the bias voltage, the total voltage applied to the grid is the algebraic sum of V_s and V_c. When the signal voltage is negative, the total bias is increased, so the flow of anode current reduces; when the signal is positive, the total bias is less, so the flow of anode current increases. Thus, a varying cathode-anode current is created in response to the signal. This varying current is translated into a varying voltage by anode load resistor R.

Early vacuum tube devices were severely limited in operating frequency, with few devices operating above 15 MHz. The primary problems were *lead inductance, interelectrode capacitance, gain-bandwidth product* (GBP), and *transit time effects*. Interelectrode capacitance is a function of electrode area and the interelectrode spacing. Making the electrodes smaller helped with lower capacitance, but severely limited operating power; this approach was deemed almost useless. Moving the elements farther apart also helped the capacitance, but increased transit angle problems. Transit time, or transit angle, problems exist when the time required for the electrons to make the passage from the cathode to the anode approximates the period of the applied signal.

10-6 BARKHAUSEN-KURZ OSCILLATORS

The solution to the frequency problem in vacuum tube devices was found in working with the inherent transit time limitation and turning it to good advantage. One of the earliest attempts was the *Barkhausen-Kurz oscillator* (BKO).

The BKO circuit (Fig. 10-4A) used the triode device, but reversed bias polarities. In the BKO circuit, the anode receives a negative bias and the grid receives a positive bias. Figure 10-4B shows how the circuit works. Electrons from the cathode are accelerated by the positive charge on the grid. The majority of electrons pass through the grid, however, and head toward the anode. But the negative potential on the anode slows down the electrons and turns them back. Because the cathode is also negatively charged, a similar action takes place as electrons attempt to regain the cathode. The result is that electrons travel in a circular path around the grid structure. The operating frequency is set by the rate of rotation.

Output power from the BKO was taken from the grid. This is a principal limit to the BKO circuit. The grid is small, so it is only capable of outputting a small amount of RF power. In most BKO circuits, the grid runs white hot because of the grid limitations.

10-7 OTHER APPROACHES

Later concepts used magnetic fields to control the electron flow instead of the electrical field of the BKO. These devices include the *magnetron*, an M-type crossed-field device invented by A. W. Hull in 1921, and the parallel-field O-type devices of A. Heil and O. Heil (1935) and Russell Varian and Sigurd Varian (1939). We will discuss these devices in Chapter 11.

10-8 SUMMARY

1.   Three methods were used for RF power generation prior to World War I: spark gaps, Alexanderson alternators, and vacuum tubes. Of these, only the vacuum tube showed promise for generating microwave RF power.

2.   Vacuum tubes were limited to low operating frequencies because of interelectrode capacitance, lead inductance, gain-bandwidth product, and transit time effects (that is, effects found when electron transit time approaches the period of the applied signal).

3.   The Barkhausen-Kurz oscillator (BKO) used the transit time problem to good effect to generate microwave signals. The BKO uses reverse polarity potentials (relative to normal vacuum tube potentials) to keep the electrons emitted from the cathode in a circular orbit around the grid. Output is taken between grid and cathode.

4.   M-type and O-type devices use a magnetic field to keep electrons in orbit or to produce other effects (see Chapter 11) that allow generation of microwave signals.

10-9 RECAPITULATION

Now return to the objectives and prequiz questions at the beginning of the chapter and see how well you can answer them. If you cannot answer certain questions, place a check mark by each and review the appropriate parts of the text. Next, try to answer the following questions and work the problems using the same procedure.

QUESTIONS AND PROBLEMS

1. Heinrich Hertz used frequencies from _____ to _____ MHz for his early radio experiments (1887–1888).

2. Most early radio communications were carried on using (VLF/VHF/HF/UHF) frequencies.

3. Guglielmo Marconi used a 10-kW transmitter on _____ kHz to achieve his history-making transatlantic transmission.

4. A spark gap transmitter was used to generate 75-GHz signals in 1930 by inserting the spark inside a resonant _____.

5. On-off radiotelegraphy and amplitude-modulated signals can be produced on an Alexanderson alternator by varying the _____ current.

6. In a vacuum tube the cloud of electrons is produced by _____ emission.

7. List the three elements of a triode tube.

8. What are the four factors limiting vacuum tube performance at higher frequencies?

9. The anode in a Barkhausen-Kurz oscillator is _____ with respect to the cathode.

CHAPTER 11

Microwave Tubes

OBJECTIVES

1. Learn the theory of operation of M-type vacuum tube devices.
2. Learn the theory of operation of O-type vacuum tube devices.
3. Understand the methods that use inherent device physics limitations to good advantage.
4. Learn the range of applications, strengths, and limitations of common microwave vacuum tube generators.

11-1 PREQUIZ

These questions test your prior knowledge of the material in this chapter. Try answering them before you read the chapter. Look for the answers (especially those you answered incorrectly) as you read the text. After you have finished studying the chapter, try answering these questions again and those at the end of the chapter.

1. A _____-type vacuum tube device uses crossed electrical and magnetic fields.
2. A traveling-wave tube is a _____-type device.
3. The _____ tube uses velocity modulation of an electron beam to generate microwave signals.
4. A magnetron is an _____-type device.

11-2 INTRODUCTION TO MICROWAVE VACUUM TUBE DEVICES

In Chapter 10, you were introduced to classical vacuum tube devices. In that presentation you learned that several factors limited the upper operating frequency of vacuum tubes: interelectrode capacitance, lead inductance, gain-bandwidth product, and transit time effects. In their normal modes of operation, vacuum tubes cannot generate microwave signals because of these limitations, so means had to be found to either overcome those limitations or use them to good advantage.

The first approach was the Barkhausen-Kurz oscillator (BKO), which used an ordinary triode vacuum tube with a cylindrical anode structure. By applying a negative high voltage to the anode, it is possible to keep electrons of the space charge traveling in a near-circular elliptical orbit about the control grid. That rotation occurs at microwave frequencies. Unfortunately, the physical size of the grid in a typical vacuum tube limits the available output power of the BKO circuit.

Other solutions were soon found. In 1921, A.W. Hull discovered that the grid could be eliminated altogether, and the BKO electrical field could be replaced with crossed electric and magnetic fields that interact with the electrons to keep them moving in a circular path. The *magnetrons* of Hull were designated M-type devices. The Hull device was improved by other workers, and by World War II it was the primary means for generating large amounts of microwave RF power.

Work on other tubes continued into the 1930s, 1940s, and 1950s. In the mid-1930s, several researchers had similar ideas for velocity modulating an electron beam to produce a bunching effect that generated a microwave signal. W. W. Hansen of Stanford University, and A. Heil and O. Heil are credited with the discovery of the velocity modulation principle. In 1937, Russell Varian and Sigurd Varian used Hansen's calculations to invent the first *klystron* device (an O-type tube).

During World War II, an Austrian refugee working for the British Navy designed the first *traveling-wave tube* (TWT). This development was extended during 1945–1950 by Bell Telephone Laboratories, and work has continued by various other organizations, including BTL, until the present time. The TWT can be used as either an oscillator or a power amplifier. In the rest of this chapter, we will look at the various M- and O-type vacuum tubes that are commonly used to generate RF power signals at microwave frequencies.

11-3 M-TYPE CROSSED-FIELD DEVICES

The crossed-field magnetron is a microwave generator device that uses electrical and magnetic fields crossed at right angles to each other in an *interaction space* between cathode and anode. The basis for operation of the device is the *magnetron principle* shown in Fig. 11-1. When an electron is injected into an electrical field, it will accelerate from the cathode to the anode in a straight line. But if a perpendicular magnetic field is also present, then the electron moves in a curved cycloidal path. By properly arranging the cathode and anode structure, it is possible to keep the electron cloud of the space charge moving in a curved path. This case is called the *planar magnetron* and is used here for illustration only.

Figure 11-2 shows a simplified *coaxial magnetron* consisting of a circular cathode inside a circular anode. The electrical field is such that the negative side is connected to the cathode and the positive side is connected to the anode. A magnetic field from an external permanent magnet has its lines of force perpendicular to the electrical field, as indicated by the arrow (B) into the page at Fig. 11-2.

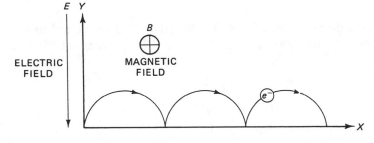

Figure 11-1
Magnetron principle is derived from path of electron in crossed magnetic and electrical fields.

As was true in conventional gridded vacuum tubes, the cathode produces an electron space charge by thermionic emission. The ultimate history of those electrons depends on the applied electric and magnetic fields. Assuming a constant magnetic field (B), there is a critical voltage called the *Hull potential* (V_h) that governs magnetron operation. The Hull potential is given by

$$V_h = \frac{eb^2 B^2}{8m}\left(1 - \frac{a^2}{b^2}\right)^2 \qquad (11\text{-}1)$$

where

V_h = Hull potential in volts
a = cathode radius (Fig. 11-2) in meters
b = anode radius (Fig. 11-2) in meters
e = electronic charge (1.6×10^{-19} coulombs)
m = mass of an electron (9.11×10^{-31} kilograms)
B = magnetic field density in webers per square meter (Wb/m²)

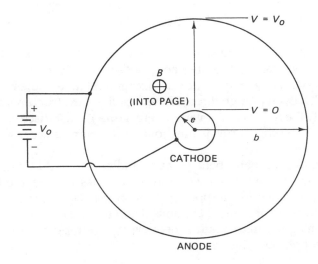

Figure 11-2
Coaxial magnetron.

EXAMPLE 11-1

A magnetron has a cathode radius of 2.5 mm (0.0025 m) and an anode radius of 5 mm (0.005 m). What is the Hull cutoff potential if a 0.27-Wb/m² magnetic field is applied?

Solution

$$V_h = \frac{eb^2 B^2}{8m}\left(1 - \frac{a^2}{b^2}\right)^2$$

$$= \frac{(1.6 \times 10^{-19})(0.005)^2 (0.27\, Wb/m^2)}{(8)(9.11 \times 10^{-31}\, kg)} \times \left(1 - \frac{0.0025\, m^2}{0.005\, m^2}\right)^2$$

$$= \frac{2.9 \times 10^{-25}}{7.3 \times 10^{-30}} \times (1 - 0.25)^2$$

$$= (3.97 \times 104)(0.75)^2 = 22.2\ kV$$

Figure 11-3 shows three separate conditions for Hull potentials less than, equal to, or greater than the critical value. Figure 11-3A shows the electron history for sub-critical Hull potentials (that is, less than V_h). In this case, electrons are collected by the anode and form an anode current (Fig. 11-3D). At the Hull critical potential (Fig. 11-3B), electrons just barely graze the anode before curving back toward the cathode. At potentials greater than the Hull potential (Fig. 11-3C), the rate of curvature becomes greater and the electrons never touch the anode. At this point, anode current drops to zero (again see Fig. 11-3D).

11-3.1 Magnetron Anode Structure

The anodes of the previous figures are only partially useful in describing magnetron operation. Continuous anodes have a nearly infinite number of cyclotron frequencies, so are not terribly valuable. In actual microwave magnetrons, a series of cavity resonators is used, each of which behaves like a resonant LC tank circuit (see Fig. 11-4).

Several different forms of anode structure are commonly used, and these are shown in Fig. 11-5. In each case, the cavities and anode assembly are machined out of a single block of metal. The magnetic field is arranged relative to the resonators, as shown in Fig. 11-5D. Even though three different cavity-slot structures are used, we will deal with only the hole and slot variety (Fig. 11-5A) in the discussion to follow. The operation of the other forms is functionally similar and need not be covered separately.

Each cavity and slot resonator behaves as a parallel-resonant LC tank circuit. In an unstrapped anode, these tanks are effectively in series with each other (Fig. 11-6A). But if *alternate* cavities are strapped (that is, shorted) together (Fig. 11-6B), the circuit becomes an array of parallel LC tanks (Fig. 11-6C). This mode is normal for most microwave magnetron oscillators. Because of the strapping, adjacent resonators are 180° out of phase.

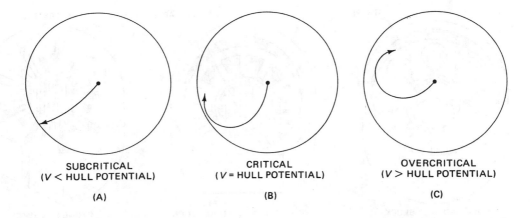

SUBCRITICAL
(V < HULL POTENTIAL)

(A)

CRITICAL
(V = HULL POTENTIAL)

(B)

OVERCRITICAL
(V > HULL POTENTIAL)

(C)

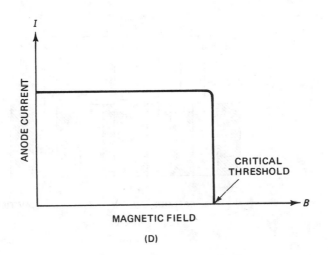

(D)

Figure 11-3 Action of electrons at various potentials: A) less than Hull potential; B) at Hull potential; C) greater than Hull potential; D) magnetic field threshold.

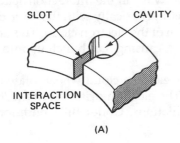

(A)

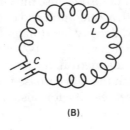

(B)

Figure 11-4
A) Resonant cavity; B) equivalent circuit.

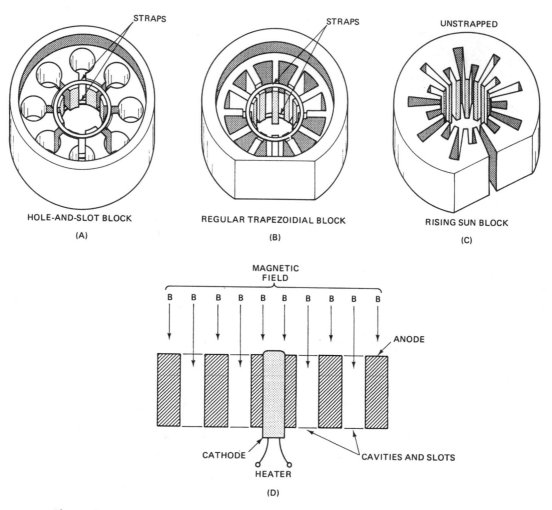

Figure 11-5 Magnetron anodes: A) hole-and-slot block; B) regular trapezoidal block; C) rising sun block; D) arrangement of magnetic fields relative to resonators.

11-3.2 Pi (π) Mode Operation

When an electron cloud sweeps past a resonator, its shock excites the resonator into self-oscillation. This effect is the same as an impulse shock exciting an LC tank circuit. The oscillations combine to set up an electric wave in the interaction space between cathode and anode. Because the wave adds algebraically with the static dc potential, it causes both acceleration and deceleration of the electron cloud. This action, both velocity and density, modulates the cloud, causing it to bunch up into a spoked wheel formation, as in Fig. 11-7.

In an oscillating magnetron, the spokes of the electron cloud wheel rotate in close synchronism with the wave phase velocity. The electron drift velocity (E/B) must match the phase velocity for sustained oscillations. Under this condition,

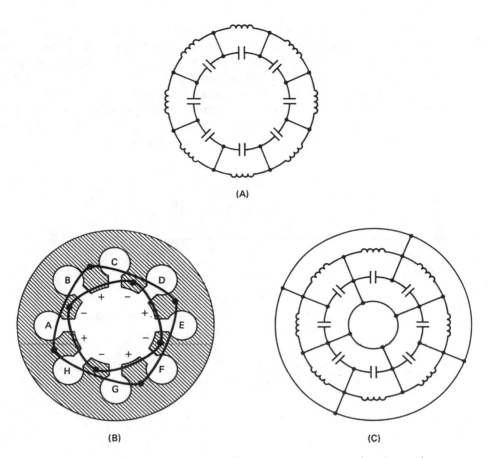

Figure 11-6 A) Resonator magnetron equivalent circuit; B) strapped cavity anode;
C) equivalent circuit.

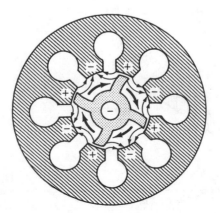

Figure 11-7
Electron cloud motion and
electrical fields in resonant
magnetron.

spokes of the wheel continue to revolve, continually reringing the cavity resonators and creating a sustained wave at a microwave frequency. Because this process causes electrons to lose energy to the wave by interaction, the electrons tend to lose velocity and fall into the cathode, where they are collected. Thus, in the oscillation mode there is no anode current (see again Fig. 11-3D).

Oscillation in the pi mode can occur at voltages between a critical level called the *Hartree potential* and the Hull potential (see Fig. 11-8). The Hull potential was described previously; the Hartree potential is found from

$$V_{Ht} = \frac{2\pi FB}{N}(b^2 - a^2) \tag{11-2}$$

where

V_{Ht} = Hartree potential in volts
B = magnetic field flux density in webers per square meter (Wb/m²)
F = oscillating frequency in hertz (Hz)
N = number of cavity resonators
b = anode radius in meters (m)
a = cathode radius in meters (m)

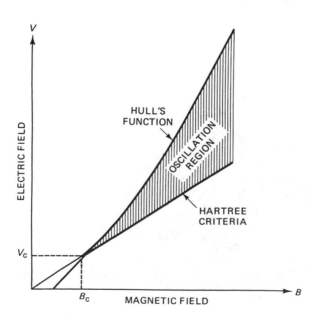

Figure 11-8
Relationship between Hull potential and Hartree potential.

EXAMPLE 11-2

Reconsider the magnetron of Example 11-1 and find the Hartree potential if the magnetron has eight resonators, each tuned to 2.95 GHz (2.95×10^9 Hz).

Solution

$$V_{Ht} = \frac{2\pi FB}{N}(b^2 - a^2)$$

$$= \frac{(2)(3.14)(2.95 \times 10^9 \text{ Hz})(0.27 \text{ Wb/m}^2)}{8} \times [(0.005)^2 - (0.0025)^2]$$

$$= \frac{(5.01 \times 10^9)(1.88 \times 10^{-5})}{8}$$

$$= \frac{9.39 \times 10^4}{8} = 11.7 \text{ kV}$$

11-3.3 Tuning Magnetrons

A severe limitation of magnetrons is their inflexible operating frequency. The frequency of the "maggie" is set by the physical dimensions of the resonant cavities cut into the anode wall. The only practical way to adjust the oscillating frequency is to vary the dimensions of all internal cavities. Even then, only a 10% to 15% change is possible. Figure 11-9 shows two methods commonly used for frequency shifting inside the magnetron. Figure 11-9A shows the inductive method, in which a "crown of thorns" assembly is raised and lowered from the cavity. In Fig. 11-9B, we see the "cookie cutter" method of capacitive tuning. Both methods are cumbersome and result in only a limited amount of frequency change.

11-3.4 Magnetron Startup

New magnetrons, and also those that have been in service but have spent a long time inoperative, may have a small partial pressure of gas inside the anode structure where we expect to find a vacuum. This gas is due to (1) incomplete evacuation at original manufacture, (2) leaks in the seals around electrodes, and (3) outgassing of materials used in construction of the device. Because of this internal gas, it is necessary to bake out (that is, voltage-age) magnetrons prior to initial full-power startup.

Aging a magnetron is done in steps, starting with a low voltage and working upward. Raise the anode potential until internal arcing occurs. This potential is then maintained for a period of time until arcing ceases, and is then raised further until arcing begins again. The new potential is maintained until arcing ceases and is then raised until arcing starts again. This potential is repeated until the magnetron is at its full specified operating potential with no arcing.

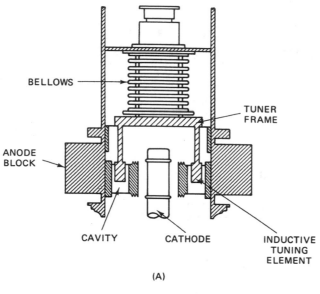

BELLOWS

TUNER
FRAME

ANODE
BLOCK

CAVITY   CATHODE   INDUCTIVE
TUNING
ELEMENT

(A)

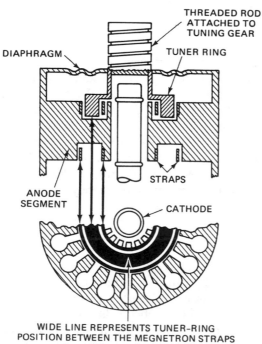

THREADED ROD
ATTACHED TO
TUNING GEAR

DIAPHRAGM   TUNER RING

ANODE
SEGMENT

STRAPS

CATHODE

WIDE LINE REPRESENTS TUNER–RING
POSITION BETWEEN THE MEGNETRON STRAPS

(B)

Figure 11-9
Tunable cavity magnetrons.

11-4 FORWARD WAVE CROSSED-FIELD AMPLIFIER DEVICES

The M-type *crossed-field amplifier* (CFA) shown in Fig. 11-10 is a microwave amplifier device that depends on magnetronlike principles for operation. An electron
stream is emitted from the cathode and accelerated by the positively charged anode.
Because this device uses a magnetic field that is crossed with the electric field, how-

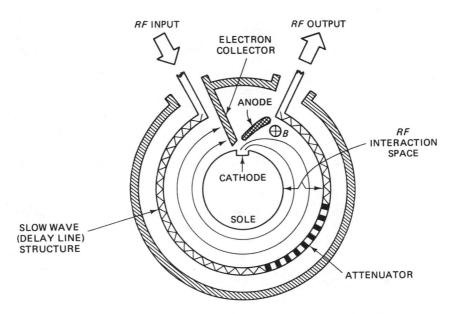

Figure 11-10   Crossed-field amplifier.

ever, the electrons are forced into a curved path away from the anode. A negatively charged *sole plate* electrode repels the electrons, forcing them into the RF interaction space between the sole plate and the *slow wave* or *delay line* structure. When the electrons reach the end of their path, they are absorbed by the *electron collector* electrode. The RF signal is injected into the interaction space through the slow wave structure (SWS). The purpose of the SWS is to reduce the RF wave velocity from the speed of light to nearly the velocity of the electron beam. This velocity synchronism allows an interaction between the RF wave and electrons in which the wave absorbs energy from the beam. The more energetic (that is, amplified) wave is extracted at the RF output port. This phenomenon is discussed more fully when we talk about traveling-wave tubes in Section 11-8.

11-5 M-CARCINOTRON BACKWARD WAVE DEVICES

The M-type *carcinotron* device in Fig. 11-11 is a linear electron beam device that serves as an oscillator. Like the CFA (Section 11-4), it uses crossed electrical and magnetic fields in the RF interaction space. In the carcinotron, however, the electrons are curved away from the anode, but thereafter follow a straight-line path to the collector electrode. In the carcinotron, oscillations build up exponentially along the slow wave structure and are output through a port. The other end of the slow wave structure is terminated in a resistive impedance. Because the RF wave (λ) propagates from the termination end to the output port against the flow of electrons, the carcinotron is called a *backward wave oscillator* (BWO).

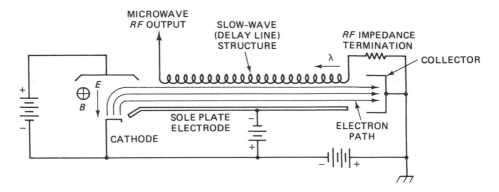

Figure 11-11 M-type carcinotron tube backward wave oscillator.

11-6 O-TYPE MICROWAVE TUBES

The magnetron studied earlier in the chapter operated on the basis of electrical and magnetic fields crossed at right angles in the electron interaction space. A principal feature of such tubes is that electrons travel in a curved path. Those tubes were designated M-type. The O-type tubes differ from M-type in that electrons travel in a *straight line* under the influence of parallel electric and magnetic fields.

Tubes in the O-type category are sometimes called *linear* or *rectilinear beam tubes* in recognition of the straight path taken by the electron beam. In this class of devices, both *velocity* and *density modulation* take place, creating the bunching effect. The electron bundles thus created have a period in the microwave region. Examples of O-type tubes include *klystrons* (Section 11-7) and *traveling-wave tubes* (TWT) (Section 11-8).

11-7 KLYSTRON TUBES

The *klystron* is an O-type, parallel-field microwave tube that can be used as either an oscillator or a power amplifier. Klystrons are capable of power gain between 3 and 90 dB, with peak power outputs from 100 mW to 10 MW and average power levels to 100 kW. In addition, the klystron is capable of high efficiency (35% to 50% in most cases and 70% to 75% in at least one mode) and stable, low-noise operation. Most klystrons are narrow-bandwidth devices, but can be broadbanded at some cost to efficiency. We will examine three basic forms of klystron: *two-cavity*, *multicavity*, and *reflex*.

Figure 11-12A shows the basic structure of the two-cavity klystron. A thermionic cathode emits a stream of electrons that is formed into a narrow beam by the anode. A control grid is used to set the strength of the beam and also to switch the beam on and off. Taken together, the cathode, anode, and control grid form the *electron gun*. The electron beam accelerates through the entire electric field potential before leaving the electron gun.

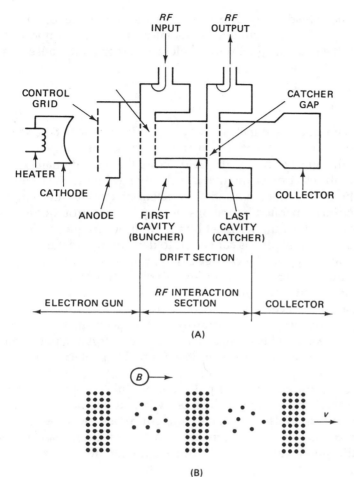

Figure 11-12
A) Klystron tube; B) electron bunching effect.

Following the electron gun are four structures: *buncher resonant cavity*, *drift section*, *catcher resonant cavity*, and *collector*. The job of the collector is to gather electrons after they pass through the *RF interaction region* consisting of the drift section and two cavities. Approximately 50% to 80% of the heat power dissipated by a klystron must be handled by the collector. Thus, we find most of the tube's cooling directed toward the collector.

Beam Focusing.   Klystron action depends on a long, thin electron stream through the interaction region. Because electrons carry negative electrical charges, however, they tend to repel each other, causing the beam to spread. A magnetic field parallel to the electron path is used to overcome beam spreading. In low-power klystrons a series of small permanent magnets, each forming a magnetic lens, is placed along the electron path. In higher-power klystrons, an electromagnet is used. For these cases, a dc solenoid coil is wrapped around the interaction space electron path. Klystron magnets are usually called *focusing elements*.

Loss of the magnetic field can be disastrous. In cases where the focusing element is an electromagnet, sensing circuitry is usually provided to rapidly shut down the klystron in the event magnet power is lost, or other faults cause loss of the magnetic field.

11-7.1 Klystron Operation

When an RF signal is input to the buncher cavity, it sets up an oscillating field inside the cavity. Because of these oscillations, an alternating field is set up across the gap of the cavity. At the buncher cavity, the field across the gap alternately accelerates and decelerates the electron beam (that is, velocity modulates the beam). As the electrons enter the drift section, fast and slow electrons tend to bunch up and form bundles of electrons separated by regions of very low electron density (see Fig. 11-12B). The period of the bundles is set by the RF input frequency. Thus, in the drift section, velocity modulation is converted to density modulation.

When electron bundles pass the catcher cavity, they give up much of their energy by shock exciting the cavity into self-oscillation. RF power is extracted from the catcher cavity by either a coupling loop or a waveguide flange. In cases where a waveguide is used, a ceramic window is required to maintain the klystron internal vacuum; such windows are transparent to electromagnetic waves. Waveguide arcing can destroy a klystron, so an arc detector (often a photocell) is built into the window flange to provide a shutdown signal to protect the klystron.

Modulation.   A klystron can be either pulsed or modulated by several methods. First, the anode accelerating potential can be switched on and off or varied. This method requires the modulator to handle the entire power of the klystron. Alternatively, the control grid is used to turn on and off the electron beam. Finally, a low-level modulating anode can be placed close to the cathode.

11-7.2 Multicavity Klystrons

Additional gain and other benefits are available from inserting one or more *intermediate cavities* between the first buncher cavity and the catcher cavity (Fig. 11-13). The intermediate cavities are additional bunchers and heighten the bunching phenomenon. Each additional cavity provides a 15- to 20-dB increase in power gain, although a larger number of cathode electrons is required to support the increased capability. As we will see next, additional cavities open the way for increasing the bandwidth of a klystron tube.

11-7.3 Klystron Bandwidth

The klystron is a narrow-bandwidth device because of the need for cavity resonators. If all cavities are tuned to the same frequency, a condition called *synchronous tuning*, the bandwidth will be on the order of 0.25% to 0.50%. For example, a 5-GHz klystron that is synchronously tuned and has a 0.25% bandwidth requires input signals within a ±6.25-MHz spread of 5000 MHz.

Stagger tuned klystrons have each cavity in a multicavity device tuned to a slightly different frequency. The overall achievable gain is slightly reduced, and

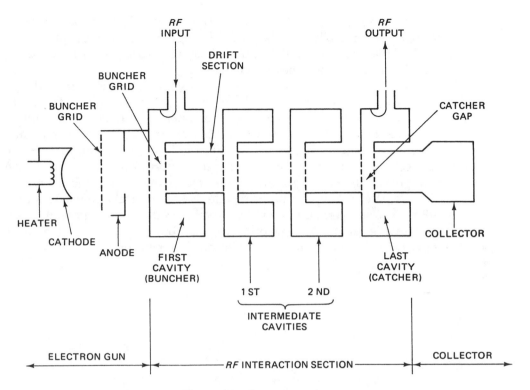

Figure 11-13 Multicavity Klyston.

the bandwidth is increased to the 2.5% to 3% range. In this case, gain is a trade-off for increased bandwidth. It is possible to mistune the catcher cavity to a frequency slightly higher than the buncher cavities in order to increase bandwidth. The gain reduces about 10 dB, but the bandwidth increases to the 15% to 25% range. The loss of gain may be a valid trade-off in many situations because of the increased frequency flexibility.

Finally, it is possible to build a klystron with variable frequency tuning. All tuning methods involve varying the cavity dimensions. It is, for example, possible to design a flexible cavity wall on a synchronously tuned klystron to provide a 2% to 3% tuning range. Unfortunately, flexible-wall cavities have a limited service life. Another method is to insert a capacitive paddle into each cavity. A 10% to 20% tuning range is achieved, but at a loss of efficiency on the low-frequency end of the range. Finally, a sliding contact cavity wall will provide 10% to 15% tuning range at the cost of mechanical complexity.

11-7.3 Klystron Oscillators

Two- and multi-cavity klystrons are power amplifiers, not oscillators. There are, however, two methods for obtaining oscillation: *external feedback to an amplifier klystron* and use of *reflex klystrons*.

In any feedback oscillator, a portion of the output signal is fed back to the input port in phase. On a klystron, a second output coupling loop can be provided in the catcher to provide a feedback port. The input and feedback ports are connected together externally to the klystron through either a waveguide or a transmission line that provides sufficient time delay to ensure in-phase feedback. The amplifier klystron then becomes an oscillator.

A *reflex klystron* (Fig. 11-14) requires only a single resonant cavity and replaces the collector with a *repeller electrode*. This electrode is biased with a negative potential to turn back electrons. The bunching phenomenon occurs in the vicinity of the repeller electrode. Bunched electrons returning to the cavity form an internal positive feedback that shock excites the cavity into self-oscillation. Output power is removed from the cavity by a loop or waveguide port in the usual manner. After electrons give up their energy, they are collected by the walls of the cavity.

The dc potential applied to the repeller is both critical and useful for tuning the reflex klystron operating frequency. A value of the repeller potential must be found that varies the transit time of the repelled bundles so that they arrive back at the cavity in phase with the gap oscillations. A condition for oscillation is

$$t = \left(N + \frac{3}{4}\right)T \tag{11-3}$$

where

t = transit time
T = period of the arriving bundles
N = integer (1, 2, 3 . . .)

Transit time t is adjusted such that electron bundles pass back through the cavity gap during the half-cycle of the RF sine wave when the gap field opposes bundle motion. This criterion must be met in order for the electrons to give up their energy to the cavity.

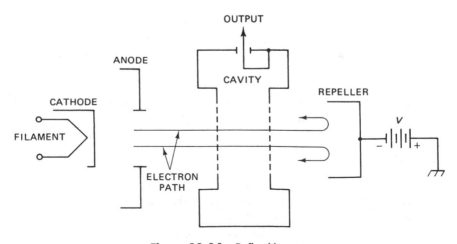

Figure 11-14   Reflex klystron.

Over a limited range of t, the period T will adjust itself to satisfy the equation. Thus, repeller voltage can be "tweaked" to fine-tune the oscillation frequency of the reflex klystron.

11-8 TRAVELING-WAVE TUBES

The traveling-wave tube (TWT) is an O-type, parallel-field, linear beam device, but it differs from the klystron in that the RF field and the electron beam interact with each other over the entire length of the active region, instead of only at the cavity gaps. Although TWTs exist that use resonant cavities, most TWTs are nonresonant devices and hence have wider bandwidths than klystrons. TWTs have been built to operate at frequencies from 0.3 to 50 GHz. Typical low-power TWT amplifier devices have bandwidths from octave (2:1) to 5:1. High-power TWTs used in transmitters typically have 10% to 20% bandwidth.

Figure 11-15 shows the basic structure of the traveling-wave tube. The electron gun is the same as in the klystron, but the RF interaction region differs considerably; its principal feature is a *slow wave structure*. The goal in this design is to slow down the RF wave, which propagates at the speed of light (c), to a phase velocity close to the velocity of the electron beam. Under this condition, direct interaction occurs between the RF wave and the electron beam. Synchronized velocities allow both velocity and density modulation of the electron beam.

The RF wave propagates at the speed of light ($c = 3 \times 10^8$ m/s), while electron beams propagate at much slower velocities. For example, in a 1500-V electric field, electron velocity is about $c/13$. To synchronize with such a beam, an RF wave must be slowed down to one-thirteenth of the normal velocity. The mechanism that reduces RF wave phase *velocity* in a TWT is the *slow wave structure*, also called a *periodic delay line* (Fig. 11-15).

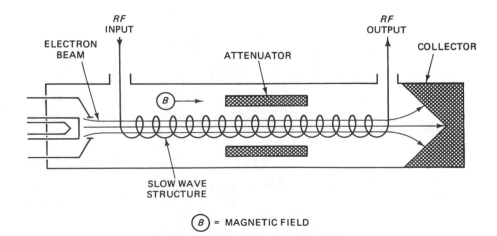

Figure 11-15 Traveling-wave tube.

As the RF wave propagates along the slow wave structure, it oscillates positive and negative. At points where the RF wave is positive, the electrons in the beam are accelerated, and at points where the RF is negative, electrons are decelerated. As a result of this velocity modulation, accelerated and decelerated electrons tend to bunch up, causing the density modulation effect.

11-8.1 Slow Wave Structures

Several different forms of slow wave structure are commonly used in TWTs: *single helix, folded* or *double helix, ring bar,* and *coupled resonant cavity.*

The helix form of slow wave structure uses a conductor wound into a helical shape (Fig. 11-16). In most devices the slow wave helix is wound from flat tungsten or molybdenum, but in a few devices hollow tubing is used. The latter design uses the hollow section of the tubing for cooling fluid. The pitch (*P*) of the helix (Fig. 11-15) is scaled to reduce the RF wave phase velocity to the electron beam velocity. The phase velocity (*V_p*) of an RF signal traveling along a slow wave helix is given by

$$V_p = \frac{cP}{\sqrt{P^2 + (\pi d)^2}} \tag{11-4}$$

where

V_p = phase velocity in meters per second (m/s)
c = 3×10^8 meters per second (m/s)
P = helix pitch in meters (see Fig. 11-15)
d = helix diameter in meters

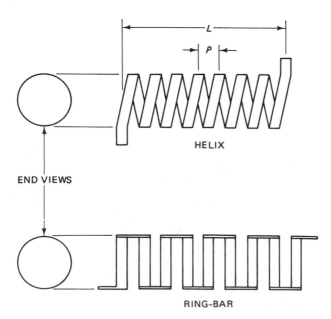

HELIX

END VIEWS

RING-BAR

Figure 11-16
Slow wave structures.

EXAMPLE 11-3

A 2-cm (0.02-m) diameter helix has a pitch of 1 cm (0.01m). Find (a) the phase velocity, V_p, and (b) the percentage of the speed of light represented by the slower phase velocity.

Solution

(a) Phase velocity:

$$V_p = \frac{cP}{\sqrt{P^2 + (\pi d)^2}}$$

$$V_p = \frac{cP}{[P^2 + (\pi d)^2]^{1/2}}$$

$$= \frac{0.01\,\text{cm}\,(3 \times 10^8 \text{ m/s})}{\{(0.01)^2 + [(3.14)\,(0.02)]^2\}^{1/2}}$$

$$= \frac{3 \times 10^6 \text{ m/s}}{(10^{-4} + 3.9 \times 10^{-3})^{1/2}}$$

$$= \frac{3 \times 10^6 \text{ m/s}}{(4.04 \times 10^{-3})^{1/2}}$$

$$= \frac{3 \times 10^6 \text{ m/s}}{0.064} = 4.7 \times 10^7$$

(b) Percent of c:

$$\%c = \frac{V_p}{c}$$

$$= \frac{4.7 \times 10^7 \text{ m/s}}{3 \times 10^8 \text{ m/s}} = 0.156$$

$$= 15.6\%$$

In this example, the velocity of the input RF wave is reduced to 15.6% of its free space velocity.

The ring bar structure is also shown in Fig. 11-16. This type of structure, as well as variants such as the cloverleaf, are sometimes used in high-power TWTs.

In the coupled-cavity form of slow wave structure, a series of resonant cavities aligned adjacent to each other alongside the RF interaction space is used. An inductive or capacitive coupling slot is provided to pass the RF wave from cavity to cavity.

The interaction between RF wave and electron beam along the traveling-wave tube causes both the electron density function and RF voltage to grow exponentially along the length of the slow wave structure. When the larger (that is, amplified) wave reaches the output port, its energy is extracted and fed to an external load.

11-8.2 Gain in TWTs

The gain of a traveling-wave tube is proportional to the length of the slow wave structure and is found from

$$\text{Gain (dB)} = \left(\frac{47.3FL}{2\pi v_c} \sqrt[3]{\frac{IK}{4V}} \right) - 9.54 \qquad (11\text{-}5)$$

where

Gain(dB) = gain in decibels
F = RF frequency in hertz
v_o = electron velocity, $0.593 \times 10^6 (V)^{1/2}$
K = helix impedance in ohms
V = applied dc voltage
I = dc current

The 9.54 factor is to account for losses in the TWT input section.

11-8.3 Oscillations in TWTs

The slow wave structure is bidirectional, so signals can propagate in both directions. When a signal is reflected from the output coupler, it propagates backward along the slow wave structure to become a feedback signal capable of causing oscillations. Some pulsed TWTs exhibit this problem by brief oscillation bursts at turn on or turn off, but do not oscillate during the on time. This type of oscillation is called "rabbit ears" because of its appearance on an oscilloscope display of the pulse. Oscillation due to reflected backwave phenomena can be reduced by inserting an attenuator into the middle third region of the slow wave structure. Another alternative, which is less lossy to the forward wave signal, is the use of internal impedance terminations called *severs*. In most cases, one *sever* is used for each 15 to 20 dB of TWT gain.

11-9 TWYSTRON TUBES

The *Twystron*® tube is a hybrid device that uses elements of both klystron and traveling-wave tubes. The input section is a multicavity klystron, while the output section is a traveling-wave tube slow wave structure and output coupler. The result is a tube with a constant gain over a wide frequency range. Twystrons have been built with up to 5 MW of peak power.

11-10 SUMMARY

1.   Early researchers discovered that a magnetic field can cause electrons to orbit in a circular path and used that effect to create the M-type class of microwave oscillators.

2. In an M-type *magnetron*, the electron beam travels in a circular path, shock exciting a series of resonant cavities. The RF wave set up across the entrances to the cavities interacts with the electron beam to force the electron cloud into a characteristic rotating spoked-wheel shape. As the spokes of this wheel sweep past the resonant cavities, they further shock excite the cavities into self-oscillation, causing the oscillation to be sustained. A coupling loop or waveguide slot is used to extract RF power from the magnetron.

3. The crossed-field amplifier (CFA) is a magnetronlike device that uses crossed electric and magnetic fields and a negatively charged *sole plate* electrode to force the electron beam into a curved path that takes it past a *slow wave* or *delay line* structure. The slow wave structure slows down the RF wave to the same speed as the electron beam, causing an interaction between them. The RF wave extracts energy from the electron beam, causing it to be amplified.

4. An M-type *carcinotron* is similar in operation to the CFA, but forces the electron beam into a straight-line path after a magnetic field bends it away from the anode. Oscillations build up on a slow wave structure due to interaction with the electron beam.

5. O-type microwave devices differ from M-type in that the magnetic field is parallel to the electric field, rather than at right angles to it. Typical O-type devices include the *klystron* and *traveling-wave tube*.

6. In a klystron amplifier, RF signal is injected into an input resonant cavity, where it sets up an oscillatory field. The cavity is arranged such that the oscillation interacts with the electron beam. The interaction causes *velocity modulation* of the beam as the RF wave voltage alternately accelerates and decelerates the electrons. The velocity modulation is converted into *density modulation* in the drift space between input and output cavities. Density modulation leads to bunching of electrons, and these bunches tend to shock excite the output resonant cavity. RF power is extracted from the output cavity. The input cavity is called the *buncher* cavity, while the output cavity is called the *catcher* cavity. Some multicavity klystrons have additional buncher cavities between input and output in an effort to obtain additional gain.

7. A klystron is normally a power amplifier, but can be made to oscillate if an external feedback network is supplied. For most cases, the external feedback network is a length of transmission line or waveguide that provides the necessary phase shift for in-phase feedback.

8. A *reflex klystron* is a single-cavity device in which a negatively charged *repeller* electrode is used in place of the collector. The repeller turns back the electron beam, causing velocity and density modulation. This action causes electron bunching, and the bunches shock excite the resonant cavity into self-oscillation. An output coupling loop or waveguide window is used to extract RF power.

9. The traveling-wave tube (TWT) is an O-type microwave amplifier that uses the parallel magnetic field to focus the electron beam into a thin stream. A slow wave structure serves as a delay line to reduce the RF wave velocity to the electron beam velocity. This synchronism permits interaction between wave and beam, causing velocity and density modulation of the beam. The electron density and RF wave voltage build up exponentially along the length of the RF interaction region, causing amplification.

11-11 RECAPITULATION

Now return to the objectives and prequiz questions at the beginning of the chapter and see how well you can answer them. If you cannot answer certain questions, place a check mark by each and review the appropriate parts of the text. Next, try to answer the following questions and work the problems using the same procedure.

QUESTIONS AND PROBLEMS

1. List the two basic types of microwave tube:
   a. Type _____, b. type _____.
2. Crossed-field oscillator devices are an example of type _____.
3. A magnetron is a _____ _____ oscillator device by virtue of perpendicular electrical and magnetic fields.
4. In a _____, the electron beam follows a curved path and eventually builds up to a spoked-wheel formation.
5. Calculate the Hull potential for a magnetron that has a 2-mm cathode radius, a 4.5-mm anode radius, and a 0.33-Wb/m² magnetic field applied.
6. Calculate the Hull potential for a magnetron with a 3-mm cathode radius, a 5.5-mm anode radius, and a 0.30-Wb/m² magnetic field.
7. At the _____ Hull potential, electrons graze the anode, but are not collected by it.
8. At a potential greater than the potential in question 7, electrons (are/are not) collected by the anode.
9. A practical magnetron _____ has several resonant cavities cut into it.
10. Resonant cavities in a magnetron act like parallel *LC* tank circuits if (alternate/adjacent) cavities are strapped together.
11. In _____-mode oscillation, the electron cloud forms into a spoked wheel.
12. The Hartree potential is the minimum level at which _____ is sustained in the magnetron.
13. Calculate the Hartree potential for the magnetron in question 5 if the device has eight resonators tuned to 3.33 GHz.
14. Calculate the Hartree potential for the magnetron in question 6 above if the device has ten resonators tuned to 2.25 GHz.
15. A bake-in procedure is necessary on a new magnetron or one that has been inoperative for a long time, in order to prevent internal _____ at the operating potential.
16. A crossed-field amplifier uses an internal _____ wave structure.
17. A _____ is similar to a crossed-field amplifier except that the electron path is straight after the beam leaves the region of the cathode and anode.
18. Type- _____ microwave tubes use parallel magnetic and electric fields instead of crossed fields.
19. List two examples of a parallel field microwave tube.
20. A _____ tube uses buncher and catcher resonant cavities.
21. The tube in question 20 is a type- _____ device.

22. The klystron depends on _____ modulation of the electron beam, which in turn sets up _____ modulation that results in electron bunching.

23. A two-cavity or multicavity klystron is an amplifier, but can oscillate if an external _____ network is used.

24. The oscillation network in question 23 can be either _____ or _____, but must have sufficient delay to cause the phase of the signal to be in phase with the input signal.

25. Are the intermediate cavities in a multicavity klystron buncher or catcher cavities?

26. Does a synchronously tuned klystron have a wide or a narrow bandwidth?

27. Does a stagger tuned klystron have a wide or a narrow bandwidth?

28. A _____ klystron is an oscillator and uses only one resonant cavity.

29. A _____ klystron uses a repeller electrode in place of the collector.

30. Calculate the transit time in a reflex klystron in mode 1 ($N = 1$) if the electron bunches have a frequency of 4.5 GHz.

31. Calculate the transit time of a mode 2 ($N = 2$) reflex klystron if the electron bundles have a period of 0.15 ns.

32. The oscillating frequency of a reflex klystron is tweaked by adjusting the _____ voltage to alter transit time (t).

33. List three forms of slow wave structure other than the simple helix found in traveling-wave tubes.

34. A simple helix in a TWT has a 1.25-cm pitch and a diameter of 2.25 cm. Calculate the phase velocity of a 2.75-GHz signal applied to this slow wave structure.

35. A simple helix has a 1-cm pitch and a 1.9-cm diameter. Find the velocity of the RF signal in this slow wave structure.

36. In problem 35, what is the percentage of c represented by the slowed RF wave velocity (expressed as a decimal)?

37. Characteristic "rabbit ears" are _____ that occur when a TWT is pulsed on or off.

38. "Rabbit ears" can be prevented by using either _____ or _____ in the slow wave structure.

39. The _____ tube is a hybrid that uses a klystron input section and a TWT output section.

KEY EQUATIONS

1. Hull cutoff potential in magnetrons:

$$V_h = \frac{eb^2B^2}{8m}\left(1 - \frac{a^2}{b^2}\right)^2$$

2. Hartree potential (criteria for oscillation):

$$V_{Ht} = \frac{2\pi FB}{N}(b^2 - a^2)$$

3. Phase velocity of RF wave in TWT slow wave structure:

$$V_p = \frac{cP}{[P^2 + (\pi d)^2]^{1/2}}$$

4. Gain (in dB) of TWT:

$$\text{Gain (dB)} = \frac{47.3FL}{2\pi v_o} \sqrt[3]{\frac{IK}{4V}} - 9.54$$

CHAPTER 12

Microwave Transistors

OBJECTIVES

1. Understand the limitations of ordinary silicon transistors at microwave frequencies.
2. Learn the types of transistor construction useful at microwave frequencies.
3. Understand the factors that determine the noise figure in microwave transistors.
4. Learn the selection criteria for microwave transistors.

12-1 PREQUIZ

These questions test your prior knowledge of the material in this chapter. Try answering them before you read the chapter. Look for the answers (especially those you answered incorrectly) as you read the text. After you have finished studying the chapter, try answering these questions again and those at the end of the chapter.

1. In microwave power transistor circuits, a high _____ adversely affects device reliability.
2. At the 1-dB _____ point, a transistor amplifier will exhibit a 9-dB output level change in response to a 10-dB input signal level change.
3. _____ gain is measured by dropping a transistor amplifier into a circuit with a 50-Ω input/output impedance and comparing signal levels.
4. The gain in question 3 is used in the _____ case.

12-2 INTRODUCTION

Transistors were developed right after World War II and by 1955 were being used in consumer products. Those early devices were limited to audio and low RF frequencies, however. As a result, only audio products and AM-band radios were widely available. Development continued, and by 1963 solid-state FM broadcast and VHF communications receivers were available. Microwave applications, however, remained elusive.

Early transistors were severely frequency limited by a number of factors, including *electron saturation velocity*, *base structure thickness* (which affects transit time), *base resistance*, and *device capacitances*. In the latter category are junction capacitances and stray capacitances resulting from packaging. When combined with stray circuit inductances and bulk material ohmic resistances, the device capacitance significantly rolled off upper operating frequencies.

The solution to operating frequency limitations was in developing new semiconductor *materials* (for example, gallium arsenide), different device internal *geometries*, and new device *construction* and *packaging* methods. Today, transistor devices operate well into the microwave region, and 40-GHz devices are commercially obtainable. Transistors have replaced other microwave amplifiers in many applications, perhaps especially in low-noise amplifiers.

12-3 SEMICONDUCTOR OVERVIEW

A basic assumption used in preparing this text was that the reader has prior exposure to solid-state electronics and thus needs no instruction in elementary transistor theory. Nonetheless, to set the context, a brief discussion of semiconductor materials seems in order. This discussion is not meant to remedy basic deficiencies, but rather to provide definitions for the discussions to follow.

Semiconductors fall into a gray area between good conductors and poor conductors (that is, insulators). The conductivity of these materials can be modulated by doping with impurities, changes in temperature, and by light (as in the case of phototransistors). Prior to the development of microwave transistors, most devices were made of Group IVA materials, such as silicon (Si) and germanium (Ge). The term Group IVA refers to the group occupied by these materials on the periodic table of the elements. Some modern microwave transistors are made of Group IIIA semiconductors, such as gallium and indium.

Transistor manufacturers alter semiconductor conductivity by adding *dopants* or *impurities* to the basic material. Silicon and germanium are *tetravalent* materials because each atom has four valence electrons in the outer shell. When *pentavalent* (five outer-shell electrons) impurities are added to the semiconductor, it forms covalent bonds with the semiconductor atoms. Combining tetravalent and pentavalent materials leaves an extra electron for each semiconductor atom. These loose electrons become negative charge carriers under the influence of an electric field, and the material is called *N-type*.

P-type material is formed by doping the semiconductor with *trivalent* (three valence electrons) impurities. When such impurity atoms form covalent bonds with the semiconductor atoms, a situation is created where some pairs of atoms are starved for one electron. This situation creates a hole in the material structure. A *hole* is nothing more than *a place where an electron should be, but is not*. Holes are said to be positive charge carriers. When an electron breaks loose from a bond, it soon finds a hole at another atom pair and fills it. Although the actual physical reality is that an electron moved, the appearance is that a hole moved in the direction opposite to the electron movement. The hole can be treated mathematically as if it were a positively charged particle with the mass of an electron; in other words, a positive charge carrier.

Figure 12-1 shows an energy-level diagram for semiconductor materials. There are two permissible bands representing states that are allowed to exist: the *conduction band* and the *valence band*. The region between these permitted bands is a *forbidden band*. This band represents energy states that are not allowed to exist. The width of the forbidden band is the difference between the conduction band energy (E_c) and valence band energy (E_v). Called the *bandgap energy* (E_{BG}), this parameter is unique for each type of material. For silicon at 25°C, the bandgap energy is 1.12 electron volts (eV), while for germanium it is 0.803 eV. A Group IIIA material called *gallium arsenide* (GaAs) is used in microwave transistors and has a bandgap energy of 1.43 eV.

Charge carrier mobility is a gauge of semiconductor material activity and is measured in units of cm²/V-s. Electron mobility is typically more vigorous than hole mobility, as the following table demonstrates. Note particularly the spread for Group IIIA semiconductors.

MOBILITY (CM²/V-S)

Material	Group	Electron	Hole
Ge	IV-A	3900	1900
Si	IA-A	1350	600
GaAs	III-A	8500	400
InP	III-A	4600	150

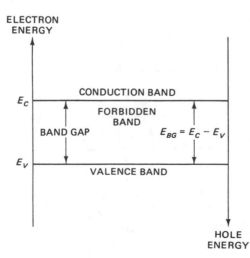

Figure 12-1
Solid-state materials energy diagram.

In the sections to follow, we will examine both bipolar junction transistors and field-effect transistors as applied to microwave situations.

12-4 BIPOLAR TRANSISTORS

A *bipolar* transistor uses both types of charge carriers in its operation. In other words, both electrons and holes are used for conduction, meaning that both N-type and P-type materials are needed. Figure 12-2 shows the basic structure of a bipolar transistor. In this case, the device uses two N-type regions and a single P-type region, thereby forming an *NPN* device. A *PNP* device would have exactly the opposite arrangement. Because *NPN* devices predominate in the microwave world, however, we will consider only that type of device. Silicon bipolar *NPN* devices have been used in the microwave region up to about 4 GHz. At higher frequencies, designers typically use field-effect transistors (see Fig. 12-6).

The first transistors were of the *point contact* type of construction in which "cat's whisker" metallic electrodes were placed in rather delicate contact with the semiconductor material. A marked improvement soon became available in the form of the *diffused junction* device. In these devices, appropriate N- and P-type impurities are diffused into the raw semiconductor material. Metallization methods are used to deposit electrode contact pads onto the surface of the device. *Mesa* devices use a growth technique to build up a substrate of semiconductor material into a tablelike structure.

Modern silicon bipolar transistors tend to be *NPN* devices of either *planar* (Fig. 12-3A) or *epitaxial* (Fig. 12-3B) design. The planar transistor is a diffusion device on which surface passivation is provided by a protective layer of silicon dioxide (SiO_2). The layer provides protection of surfaces and edges against contaminants migrating into the structure.

The epitaxial transistor (Fig. 12-3B) uses a thin, low-conductivity epitaxial layer for part of the collector region (the remainder of the collector region is high-conductivity material). The epitaxial layer is laid down as a condensed film on a substrate made of the same material. Figure 12-3C shows the doping profile versus length for an epitaxial device. Note that the concentration of impurities, and hence region conductivity, is extremely low for the epitaxial region.

The sidewall contact structure (SICOS) transistor is shown in Fig. 12-3D. Regular *NPN* transistors exhibit a high junction capacitance and a relatively large

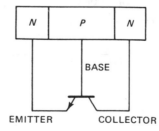

Figure 12-2
NPN transistor.

EMITTER   COLLECTOR

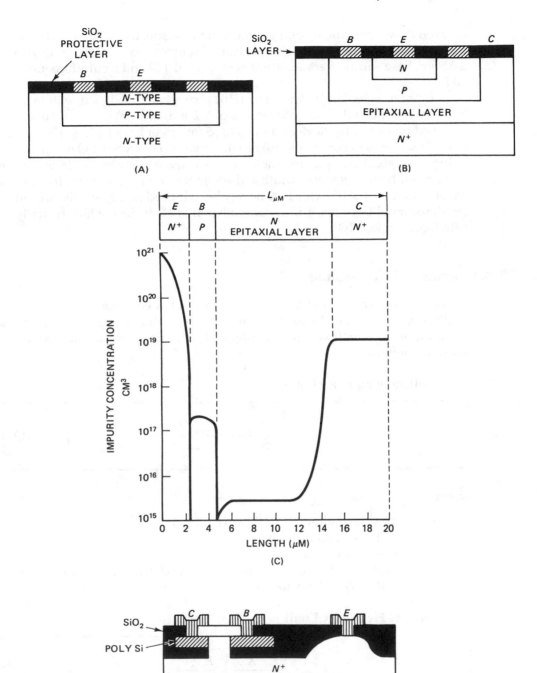

Figure 12-3   A) Planar transistor; B) epitaxial construction; C) ion-doping profile in epitaxial transistor; D) sidewall contact structure.

electron flow across the subemitter junction to the substrate. These factors reduce the maximum cutoff frequency and available current gain. A SICOS device with a 0.5-μm base width offers a current gain (*Hfe*) of 100 and a cutoff frequency of 3 GHz.

Heterojunction bipolar transistors (HBT) have been designed with maximum frequencies of 67 GHz on base widths of 1.2 μm; current gains of 10 to 20 are achieved on very small devices, and up to 55 for larger base widths. The operation of HBTs depends in part on the use of a thin coating of $NA_2S.9H_2O$ material, which reduces surface charge pair recombination velocity. At very low collector currents, a device with a 0.15-μm base width and 40-μm × 100-μm emitter produced current gains in excess of 3800, with 1500 being also achieved at larger collector currents (on the order of 1 mA/cm²). Compare with non-HBT devices, which typically have gain figures under 100.

12-4.1 Johnson Relationships

Microwave bipolar transistors typically obey a set of equations called the *Johnson relationships*. This set of six equations is useful for determining device limitations. In the following equations, terms are defined when first used and subsequent uses are not redefined.

Voltage-Frequency Limit

$$\frac{V_{max}}{2\pi(l/v)} = \frac{E_{max}V_s}{2\pi} \qquad (12\text{-}1)$$

where

V_{max} = maximum allowable voltage ($E_{max}L_{min}$)
V_s = material saturation velocity
E_{max} = maximum electric field
l/v = average charge carrier time at average charge velocity (v) through the length (l) of the material

Current-Frequency Limit

$$\frac{I_{max}X_{co}}{2\pi(l/v)} = \frac{E_{max}V_s}{2\pi} \qquad (12\text{-}2)$$

where

I_{max} = maximum device current
X_{co} = reactance of the output capacitance, ($1/(2\pi(l/v)C_o)$)

Power-Frequency Limit

$$\sqrt{\frac{P_{max}X_{co}}{2\pi(l/v)}} = \frac{E_{max}V_s}{2\pi} \qquad (12\text{-}3)$$

where

P_{max} = maximum power

Power Gain – Frequency Limit

$$\sqrt{\frac{G_{max}KTV_{max}}{e}} = \frac{E_{max}V_s}{2\pi} \qquad (12\text{-}4)$$

where

G_{max} = maximum power gain
K = Boltzmann's constant (1.38×10^{-23} J/K)
T = temperature in kelvins
e = electronic charge (1.6×10^{-19} coulombs)

Maximum Gain

$$G_{max} = \sqrt{\frac{F_t}{F}} \; \frac{Z_o}{Z_{in}} \qquad (12\text{-}5)$$

where

F = operating frequency
Z_o = real component of output impedance
Z_{in} = real component of the input impedance

Impedance Ratio

$$\frac{Z_o}{Z_{in}} = \frac{C_{in}}{C_o} = \frac{\dfrac{I_{max}T_b}{KT/e}}{\dfrac{I_{max}T_b}{V_{max}}} \qquad (12\text{-}6)$$

where

C_{in} = input capacitance
C_o = output capacitance
T_b = charge transit time

12-4.2 Cutoff Frequency

The cutoff frequency (F_t) is the frequency at which current gain drops to unity. Several factors affect cutoff frequency: (1) the saturation velocity for charge carriers in the semiconductor material, (2) the time required to charge the emitter-base junction capacitance (T_{eb}), (3) the time required to charge the base-collector junction capacitance (T_{cb}), (4) the base region transit time (T_{bt}), and (5) the base-collector depletion zone transit time (T_{bc}). These times add together to give us the emitter-collector transit time (T). The expression for cutoff frequency is

$$F_t = \frac{1}{2\pi T} \tag{12-7}$$

or

$$F_t = \frac{1}{2\pi(T_{eb} + T_{bt} + T_{bc} + T_{cb})} \tag{12-8}$$

where

F_t = frequency in hertz (Hz)

Times are in seconds.

12-4.3 Stability

The term stability in reference to a transistor amplifier refers to its freedom from unwanted oscillations. In some circuits, the transistor will oscillate at some natural resonant frequency regardless of input signal; these circuits are *unstable*. Other circuits will not oscillate at all, regardless of the input signal level or the state of input or output impedances; these circuits are said to be *unconditionally stable*. Still other circuits are stable under some conditions and unstable under other conditions; these circuits are *conditionally stable*.

The conditions that determine stability reflect the state of input and output impedances. If internal capacitances and other parameters add up to meet *Barkhausen's criteria* (that is, in-phase feedback and loop gain greater than 1 at the same frequency), then the amplifier is unstable and it will oscillate. Oscillation may also occur if, at any frequency, the real part of either the input or output impedances is negative (-R).

Conditional stability occurs if the real part (R) of either input or output impedances is greater than zero (+R not –R) for positive real input and output impedances at any specific frequency (+R ± jX). In other words, conditionally stable amplifiers may be frequency dependent.

Unconditional stability exists when the preceding criteria are satisfied for *all* frequencies at which the transistor shows gain.

Interestingly, the transistor designed for microwave service (or even VHF/UHF) may not be usable at lower frequencies. While vacuum tubes were generally usable from dc to their cutoff frequency, transistors are sometimes not usable over the entire supposed range from dc upward. VHF/UHF/microwave transistors tend to oscillate when used at low frequencies (usually under 100 MHz).

Another phenomenon, called *squeeging*, is also seen in some cases. If certain transistors are not properly impedance matched to source and load impedances, then the amplified waveform may be chopped or modulated by a low-frequency (audio or subaudio) spurious oscillation.

12-4.4 Bipolar Transistor Geometry

Two problems that limited the high-frequency performance of early transistors were emitter and base resistances and base transit time. Unfortunately, simply reducing the base width in order to improve transit time and reduce overall resistance also reduced current- and voltage-handling capability. Although some improvements were made, older device internal geometries were clearly limited.

The solutions to these problems lay in the geometry of the base-emitter junction. Figure 12-4 shows three geometries that yielded success in the microwave region: *interdigital* (Fig. 12-4A), *matrix* (Fig. 12-4B), and *overlay* (Fig. 12-4C). These construction geometries yielded the thin, wide-area, low-resistance base regions that were needed to increase the operating frequency.

12-4.5 Low-Noise Bipolar Transistors

Geometries that provide the low resistance needed to increase operating frequency also serve to reduce the noise generated by the device. There are three main contributors to the emitter-base noise figure in a bipolar transistor: thermal noise, shot noise in the emitter-base circuit, and shot noise in the collector circuit. The *thermal noise* is a function of temperature and *base region resistance*. By reducing the resistance (a function of internal geometry), we also reduce the noise figure.

The *shot noise* produced by the *PN* junction is a function of the junction current. There is an *optimum collector current* (I_{co}) at which the noise figure is best. Figure 12-5 shows the noise figure in decibels plotted against the collector current for a particular device. The production of the optimum noise figure at practical collector currents is a function of junction efficiency.

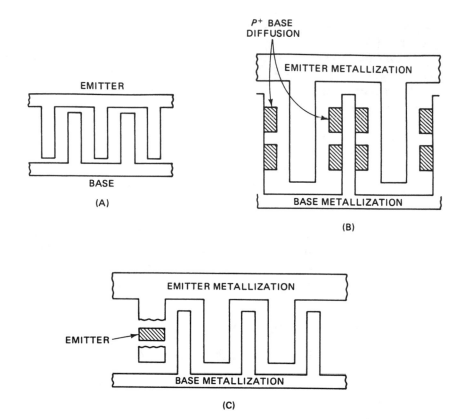

Figure 12-4 Transistor layout structures for increasing frequency.

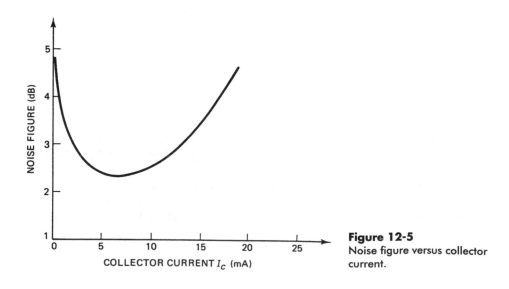

Figure 12-5
Noise figure versus collector current.

12-5 FIELD-EFFECT TRANSISTORS

The field-effect transistor (FET) operates by varying the conductivity of a semi-conductor channel by varying the electric field in the channel. Two elementary types are found: *junction field-effect transistors* (JFET) and *metal oxide semiconductor field-effect transistors* (MOSFET), also sometimes called *insulated gate field-effect transistors* (IGFET). In the microwave world, a number of devices are modifications of these basic types.

Figure 12-6A shows the structure of a generic JFET device that works on the depletion principle. The channel in this case is *N*-type semiconductor material, while the gate is made of *P*-type material. A gate contact metallized electrode is deposited over the gate material. In normal operation the *PN* junction is reverse biased, and the applied electric field extends to the channel material.

The electric field repels charge carriers (in this example electrons) in the channel, creating a depletion zone in the channel material. The wider the depletion zone, the higher the channel resistance. Because the depletion zone is a function of applied gate potential, the gate potential tends to modulate channel resistance. For a constant drain-source voltage, varying the channel resistance also varies the channel current. Because an input voltage varies an output current ($\Delta I_o / \Delta V_{in}$), the FET is called a *transconductance amplifier*.

The MOSFET (Fig. 12-6B) replaces the semiconductor material in the gate with a layer of oxide insulating material. Gate metallization is overlaid on the insulator. The electric field is applied across the insulator in the manner of a capacitor. The operation of the MOSFET is similar to the JFET in broad terms, and in a depletion-mode device the operation is very similar. In an enhancement-mode device, channel resistance drops with increased signal voltage.

12-5.1 Microwave FETs

The microwave field-effect transistor represented a truly giant stride in the performance of semiconductor amplifiers in the microwave region. Using Group IIIA materials, notably gallium arsenide (GaAs), the FET pressed cutoff frequency performance way beyond the 3- or 4-GHz limits achieved by silicon bipolar devices. The *gallium arsenide field-effect transistor* (GaAsFET) offers superior noise perfor-

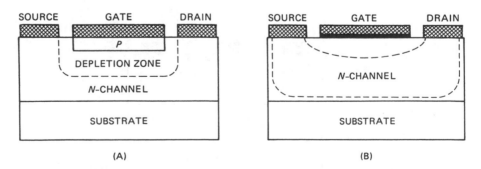

(A)   (B)

Figure 12-6 Junction field-effect (JFET) transistor at different gate potentials.

mance (a noise figure less than 1 dB is achieved!) over silicon bipolar devices, improved temperature stability, and higher power levels.

The GaAsFET can be used as a low-noise amplifier (LNA), class C amplifier, or oscillator. GaAsFETs are also found in monolithic microwave integrated circuits (MIMIC), high-speed analog to digital (A/D) converters, analog/RF applications, and in high-speed logic devices. In addition to GaAs, AlGaAs and InGaAsP are also used.

Figure 12-7 shows the principal player among microwave field-effect transistors: the *metal semiconductor field-effect transistor* (MESFET), also called the *Schottky barrier transistor* (SBT) or *Schottky barrier field-effect transistor* (SBFET). The epitaxial active layer is formed of N-type GaAs doped with either sulphur or tin ions, with a gate electrode formed of evaporated aluminum. The source and drain electrodes are formed of *gold germanium* (AuGe), *gold telluride* (AuTe), or a AuGeTe alloy.

The noise mechanisms in the MESFET are a bit different from those of bipolar transistors, but can be defined by the following equations:

$$NF = 2 + K\left(\frac{E}{E_{sat}}\right)^N \qquad (12\text{-}9)$$

where

NF = noise factor
K = a constant (6 for GaAs)
E = applied electric field in volts per meter
E_{sat} = maximum electric field (300 kV/m for GaAs)
N = a constant (3 for GaAs)

Another form of microwave JFET is the *high electron mobility transistor* (HEMT), shown in Figs. 12-8A and 12-8B. The HEMT is also known as the *two-dimensional electron GaAsFET* (or *TEGFET*) and *heterojunction FET* (HFET). Devices in this category produce power gains up to 11 dB at 60 GHz and 6.2 dB at 90 GHz. Typical noise figures are 1.8 dB at 40 GHz and 2.6 dB at 62 GHz. Power levels at 10 GHz

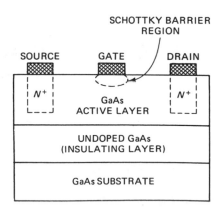

Figure 12-7
The MESFET transistor.

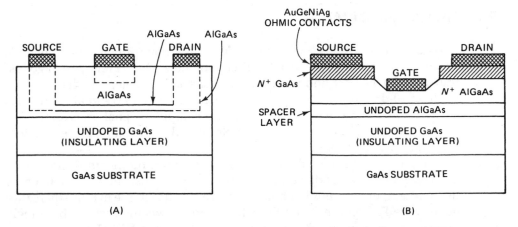

Figure 12-8   High electron mobility transfer (HEMT).

have approached 2 W (CW) per millimeter of emitter periphery dimension ("emitter" is the gate structure junction in JFET-like devices).

HEMT devices are necessarily built with very thin structures in order to reduce transit times (necessary to increasing frequency because of the 1/T relationship). These devices are built using ion implantation, molecular beam epitaxy (MBE), or metal organic chemical vapor deposition (MOCVD).

Power FETs.   The power output capability of field-effect transistors has climbed rapidly over the past few years. Devices in the 10-GHz region have produced 2-W/mm power densities and output levels in the 4- to 5-W region. Figure 12-9 shows a typical power FET internal geometry used to achieve such levels.

An advantage of FETs over bipolar devices is that input impedance tends to be high and relatively frequency stable. It is thus easier to design wideband power amplifiers and either fixed or variable (tunable) frequency power amplifiers.

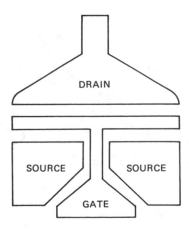

Figure 12-9
Power FET layout.

12-6   NOISE PERFORMANCE

In other sections of this book, we discuss noise in amplifiers from a generic perspective, so we will not repeat the discussion here. Only a few years ago, users of transistors in the UHF and microwave regions had to contend with noise figures in the 8- to 10-dB range. A fundamental truism states that no signal below the noise level can be detected without sophisticated computer signal processing. Thus, reducing the noise floor of an amplifier helps greatly in building more sensitive real-time microwave systems.

Figure 12-10 shows the typical range of noise figures in decibels expected from various classes of device. The silicon bipolar transistor shows a flat noise figure (curve *A*) up to a certain frequency and a sharp increase in noise thereafter. The microwave FET (curve *B*), on the other hand, shows increases in both high- and low-frequency regions. The same is also true for the HEMT (curve *C*), but to a lesser degree. Only the supercooled (-260°C) HEMT performs better.

Most MESFET and HEMT devices show increases in noise figure at low frequencies. Some such devices also tend to become unstable at those frequencies. Thus, microwave and UHF devices may, contrary to what our instincts might suggest, fail to work at frequencies considerably below the optimum design frequency range.

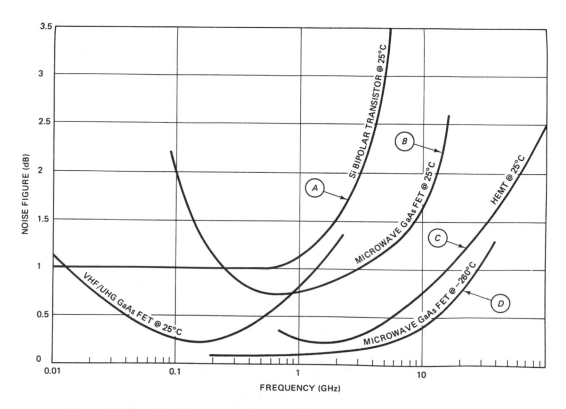

Figure 12-10   Noise figure versus frequency for various forms of microwave transistor.

12-7 SELECTING TRANSISTORS

The criteria for selecting bipolar or field-effect transistors in the microwave range depend a lot on application. The importance of the noise figure, for example, becomes apparent when designing the front end of a microwave satellite receiver system. In other applications, power gain and output may be more important.

Obviously, gain and noise figure are both important, especially in the front ends of receiver systems. For example, Earth communications or receiver terminals typically use a parabolic dish antenna with a low-noise amplifier (LNA) at the feed point. When selecting devices for such applications, however, beware of seemingly self-serving transistor specification data sheets. For example, consider the noise figure specification. As you saw in Fig. 12-10, there is a strong frequency dependence regarding noise figure. Yet device data sheets often list the *maximum frequency* and *minimum noise figure*, even though the two rarely coincide with one another! When selecting a device, consult the *NF* versus *F* curve.

A useful device published by some vendors is the *noise contour chart*. This graph plots the noise figure on a Smith chart. A useful feature of this chart is that stability can be predicted by looking at the noise circles. If any part of a noise circle is off the Smith chart, then for that particular set of conditions the amplifier is unstable.

Device *gain* can also be specified in different ways, so some caution is in order. There are at least three ways to specify gain: maximum allowable gain (G_{max}), gain at the optimum noise figure (G_{NF}), and the insertion gain. The maximum obtainable gain occurs usually at a single frequency where the input and output impedances are conjugately matched (that is, the reactive part of the impedance canceled out and the resistive part transformed for maximum power transfer). The noise figure gain is an impedance-matching situation in which the noise figure is optimized, but not necessarily power gain. Rarely, perhaps never, are the two gains the same.

Another specification to examine is the *1-dB compression point*. This critical point is that at which a 10-dB increase in the input signal level results in a 9-dB increase in the output signal level.

12-8 SUMMARY

1. Early transistors were frequency limited by electron saturation velocity, base structure thickness or width, base resistance, and device capacitance. Solutions to these problems were found in new materials, new internal geometries, new construction methods, and improved packaging concepts.

2. Microwave silicon bipolar transistors operate to about 4 GHz, while newer gallium arsenide (GaAs) devices push the frequency limit much higher. Planar, epitaxial, and heterojunction transistors are used in the microwave region.

3. A set of six equations called the *Johnson relationships* define the operation of microwave bipolar transistors.

4. The cutoff frequency for bipolar transistors is a function of the time required for charge carriers to transit from emitter to collector.

5. The *stability* of transistor amplifiers refers to freedom from spurious oscillations. Three basic conditions are recognized: *unstable* (oscillating), *unconditionally stable*, and *conditionally stable*. The criteria for these conditions depend in part on the nature of the circuit impedances.

6. Three basic geometries contribute to improved microwave performance of bipolar transistors: *interdigited*, *matrix*, and *overlay*.

7. Three main contributors to the overall bipolar device noise figure are *thermal noise, emitter-to-base shot noise*, and *collector-to-base shot noise*. Thermal noise can be reduced both by reducing operating temperature of the device (not always practical) and reducing the internal base resistance; shot noise is reduced by selecting the optimum low collector current.

8. Microwave field-effect transistors (FET), especially GaAsFET devices, are a tremendous improvement in both operating frequency and noise figure. In addition to JFET and MOSFET designs, microwave devices also include *metal semiconductor FETs* (MESFET), or *Schottky barrier transistors* (SBT) as they are also called, and *high electron mobility transistors* (HEMT).

9. The *noise figure* of MESFET and HEMT devices is frequency dependent and rises at frequencies both lower and higher than the optimum frequency. Rarely is the maximum operating frequency of these devices the same as the optimum noise frequency.

10. Microwave transistors often become unstable when operated at frequencies substantially below the microwave region.

12-9 RECAPITULATION

Now return to the objectives and prequiz questions at the beginning of the chapter and see how well you can answer them. If you cannot answer certain questions, place a check mark by each and review the appropriate parts of the text. Next, try to answer the questions and work the problems using the same procedure.

QUESTIONS AND PROBLEMS

1. The 1-dB compression point is the operating point at which a 10-dB input signal level change results in a _____-dB output level change.
2. List four factors that traditionally limited the operating frequency of bipolar transistors.
3. List four factors that helped improve the operating frequency performance of bipolar transistors.
4. Semiconductors are in a gray area between good conductors and _____.
5. The conductivity of semiconductor materials can be modulated by doping with _____, changes in _____, and light.

6. Silicon and germanium semiconductor materials are in a category called Group _____A elements, which refers to their position on the periodic table.

7. Gallium arsenide is an example of a Group ____A material.

8. Silicon is a tetravalent material, so dopants that make the material suitable for transistor manufacture are _____valent and ____valent, resulting in N-type and P-type materials.

9. A ____ is a place where an electron should be, but is not.

10. The region between the conduction band and the valence band on the energy diagram is called the _____ band.

11. Silicon bipolar transistors typically operate to about ____ GHz.

12. Two transistor construction methods used on microwave bipolar devices are planar and _____.

13. In planar transistors, a layer of silicon dioxide (SiO_2) is used for surface _____.

14. The _____ bipolar transistor depends on a thin coating of $NA_2S.9H_2O$ to reduce surface charge recombination activity.

15. In a particular bipolar transistor, emitter-collector transit time is 4×10^{-11} seconds. What is the approximate cutoff frequency?

16. A transistor has a cutoff frequency of 2950 MHz. Calculate the approximate emitter-collector transit time.

17. Transistor amplifier _____ is a measure of its freedom from spurious oscillations.

18. List three conditions describing circuit stability.

19. A transistor amplifier satisfies Barkhausen's criteria at a frequency of 800 MHz. This amplifier will _____ at that frequency.

20. An amplifier has an output impedance that is characterized by $Z = -20 + j2$. Is this amplifier stable or unstable?

21. Microwave transistors tend to _____ when used at frequencies lower than about ____ MHz.

22. The phenomenon called _____ is a condition where bipolar transistor power amplifiers produce a chopped waveform because of impedance mismatch.

23. List three ways of constructing bipolar transistors that tend to increase the operating frequency.

24. List three contributors to transistor noise figure.

25. A _____ uses a Schottky barrier of metal to semiconductor material.

26. The gate electrode on the transistor in question 25 is typically made of evaporated _____.

27. A GaAsFET transistor operating at 5 GHz has an applied electric field of 100 kV/m. Find the noise figure in decibels.

28. True or false: MESFET and HEMT devices have a noise figure that deceases as frequency decreases over the entire region of operation.

29. A noise contour Smith chart shows the noise contour is entirely within the limits of the Smith chart circle. What can you deduce from this fact?

30. At a frequency of 6.89 GHz, the input signal level is increased from –40 to –30 dBm. The output increases from +1.5 to +10.5 dBm. This is the ____ _____ point.

KEY EQUATIONS

1. Voltage-frequency limit:

$$\frac{V_{max}}{2\pi(l/v)} = \frac{E_{max}V_s}{2\pi}$$

2. Current-frequency limit:

$$\frac{I_{max}X_{co}}{2\pi(l/v)} = \frac{E_{max}V_s}{2\pi}$$

3. Power-frequency limit:

$$\frac{\sqrt{P_{max}X_{co}}}{2\pi(l/v)} = \frac{E_{max}V_s}{2\pi}$$

4. Power gain-frequency limit:

$$\sqrt{\frac{G_{max}KTV_{max}}{e}} = \frac{E_{max}V_s}{2\pi}$$

5. Maximum gain:

$$G_{max} = \sqrt{\frac{F_t}{F}}\frac{Z_o}{Z_{in}}$$

6. Impedance ratio:

$$\frac{Z_o}{Z_{in}} = \frac{C_{in}}{C_o} = \frac{(I_{max}T_b)/(KT/e)}{(I_{max}T_o)/V_{max}}$$

CHAPTER 13

Discrete Microwave Amplifiers

OBJECTIVES

1. Understand *noise figure, noise factor,* and *noise temperature* specifications of discrete microwave amplifiers.
2. Learn the parameters needed to specify the noise performance of microwave amplifiers.
3. Understand the means for impedance matching and tuning of microwave amplifiers.
4. Understand the operation of microwave parametric microwave amplifiers.

13-1 PREQUIZ

These questions test your prior knowledge of the material in this chapter. Try answering them before you read the chapter. Look for the answers (especially those you answered incorrectly) as you read the text. After you have finished studying the chapter, try answering these questions again and those at the end of the chapter.

1. Two 10-dB amplifiers are connected in cascade; A1 is the input stage and A2 is the output stage. Calculate the total noise figure if A1 has a noise figure (NF1) of 2.8 dB and A2 has a noise figure (NF2) of 4.5 dB.
2. Calculate the noise figure in question 1 if the amplifiers are reversed such that A1 is the output stage and A2 is the input stage.
3. A noise figure of 5 dB is equivalent to a noise factor of _____.
4. True or false: The overall best noise figure always results when the input and source impedances of the amplifier are matched.

13-2 INTRODUCTION

The general subject of amplifiers is usually covered in earlier courses than one on microwave devices, so it is assumed that the reader has at least a basic understanding of amplifier terminology. In this chapter, we will concentrate on topics of particular importance in microwave amplifiers.

Any study of microwave amplifiers must consider *noise* problems. As you will see shortly, low-noise amplifiers (LNAs) are critical to proper operation of many microwave systems. Be aware that the topic of low-noise amplifiers also pertains to a discussion of microwave receivers, so you are advised to review this chapter when you study receivers in Chapter 18.

The chapter material also covers general input/output tuning and impedance-matching methods. Also discussed is an amplifier that is found only in microwave systems: *parametric amplifier*. But before discussing those topics, we will take a look at the noise problem.

13-3 NOISE, SIGNALS, AND AMPLIFIERS

Although gain, bandwidth, and the shape of the passband are important amplifier characteristics, we must concern ourselves about circuit *noise*. In the spectrum below VHF, artificial and natural atmospheric noise sources are so dominant that receiver noise contribution is trivial. But at VHF and above, receiver and amplifier noise sets the performance of the system.

At any temperature above absolute zero (0 K or –273°C), electrons in any material are in constant random motion. Because of the inherent randomness of that motion, however, there is no detectable current in any direction. In other words, electron drift in any single direction is canceled over short times by equal drift in the opposite direction. There is, however, a continuous series of random current pulses generated in the material, and those pulses are heard by the outside world as a noise signal. This signal is called by several names: *thermal agitation noise, thermal noise*, or *Johnson noise*.

It is important to understand what we mean by noise in this context. In a communications system, the designer may regard all unwanted signals as noise, including electrical spark signals from machinery and adjacent channel communications signals, as well as Johnson noise. In other cases, the harmonic content generated in a linear signal by a nonlinear network could be regarded as noise. But in the context of microwave amplifiers and of Chapters 18 and 19 (receivers), noise refers to thermal agitation (Johnson) noise.

Amplifiers and other linear networks are frequently evaluated using the same methods, even though the two classes appear radically different. In the generic sense, a passive network is merely an amplifier with negative gain or a complex transfer function. We will consider only amplifiers here, but keep in mind that the material herein also applies to other forms of linear (passive) network.

Amplifiers are evaluated on the basis of *signal-to-noise ratio* (S/N ratio or SNR). The goal of the designer is to enhance the SNR as much as possible. Ultimately, the minimum signal detectable at the output of an amplifier is that which appears above the noise level. Therefore, the lower the system noise, the smaller the *minimum detectable signal* (MDS).

Noise resulting from thermal agitation of electrons is measured in terms of *noise power* (P_n), and carries the units of power (watts or its subunits). Noise power is found from

$$P_n = KTB \qquad (13\text{-}1)$$

where

P_n = noise power in watts (W)
K = Boltzmann's constant (1.38×10^{-23} J/K)
B = bandwidth in hertz (Hz)
T = temperature in kelvins (K)

Notice in Eq. (13-1) that there is no center frequency term, only a bandwidth. True thermal noise is *gaussian* or near gaussian in nature, so frequency content, phase, and amplitudes are equally distributed across the entire spectrum; this is called *equipartition distribution* of energy. Thus, in bandwidth-limited systems such as a practical amplifier or network, the total noise power is related only to temperature and bandwidth. We can conclude that a 20-MHz bandwidth centered on 1 GHz produces the same thermal noise level as a 20-MHz bandwidth centered on 4 GHz or some other frequency.

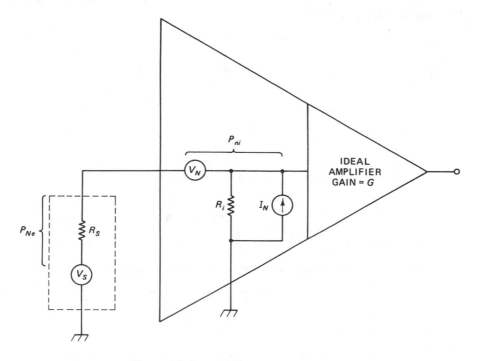

Figure 13-1 Amplifier equivalent circuit.

Noise sources can be categorized as either *internal* or *external*. The internal noise sources are due to thermal currents in the semiconductor material resistances. It is the noise component contributed by the amplifier under consideration. If noise, or SNR, is measured at both the input and the output of an amplifier, the output noise is greater. The internal noise of the device is the difference between output noise level and input noise level.

External noise is the noise produced by the signal source and so is often called *source noise*. This noise signal is due to thermal agitation currents in the signal source, and even a simple zero-signal input termination resistance has some amount of thermal agitation noise.

Both types of noise generator are shown schematically in Fig. 13-1. Here we model a microwave amplifier as an ideal noiseless amplifier with a gain of G and a noise generator at the input. This noise generator produces a noise power signal at the input of the ideal amplifier. Although noise is generated throughout the amplifier device, it is common practice to model all noise generators as a single input-referred source. This source is shown as voltage V_i and current I_i.

13-3.1 Noise Factor, Noise Figure, and Noise Temperature

The noise of a system or network can be defined in three different but related ways: *noise factor* (F_n), *noise figure* (NF), and *equivalent noise temperature* (T_e); these properties are definable as a ratio, decibel, and temperature, respectively.

Noise Factor (F_n). The noise factor is the ratio of output noise power (P_{no}) to input noise power (P_{ni}):

$$F_n = \frac{P_{no}}{P_{ni}} \quad T = 290\,\text{K} \tag{13-2}$$

To make comparisons easier, the noise factor is always measured at the standard temperature (T_o) of 290 K (approximately room temperature).

The input noise power P_{ni} can be defined as the product of the source noise at standard temperature (T_o) and the amplifier gain:

$$P_{ni} = G\,K\,B\,T_o \tag{13-3}$$

It is also possible to define noise factor F_n in terms of output and input SNR:

$$F_n = \frac{SNR_{in}}{SNR_{out}} \tag{13-4}$$

which is also

$$F_n = \frac{P_{no}}{K \, T_o \, B \, G} \tag{13-5}$$

where

SNR$_{in}$ = input signal-to-noise ratio
SNR$_{out}$ = output signal-to-noise ratio
P_{no} = output noise power in watts (W)
K = Boltzmann's constant (1.38×10^{-23} J/K)
T_o = 290 K
B = network bandwidth in hertz (Hz)
G = amplifier gain

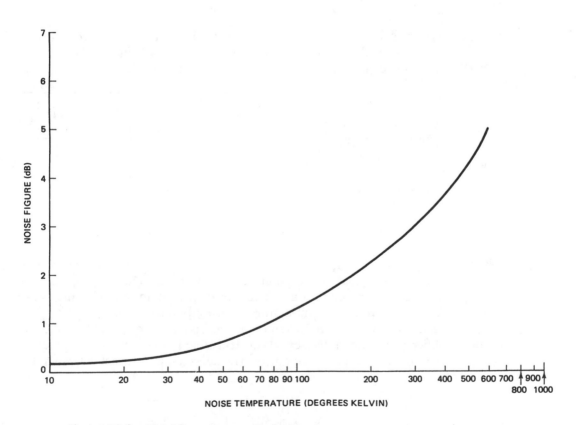

Figure 13-2   Noise figure versus noise temperature.

The noise factor can be evaluated in a model that considers the amplifier ideal and therefore only amplifies through gain G the noise produced by the input noise source:

$$F_n = \frac{K T_o B G + \Delta N}{K T_o B G} \tag{13-6A}$$

or

$$F_n = \frac{\Delta N}{K T_o B G} \tag{13-6B}$$

where

ΔN = noise added by the network or amplifier

All other terms are as defined previously.

Noise Figure (NF).   The noise figure is a frequently used measure of an amplifier's quality, or its departure from idealness. Thus, it is a figure of merit. The noise figure is the noise factor converted to decibel notation:

$$\text{NF} = 10 \log F_n \tag{13-7}$$

where

NF = noise figure in decibels (dB)
F_n = noise factor

Log refers to the system of base 10 logarithms.

Noise Temperature (T_e).   The noise temperature is a means of specifying noise in terms of an equivalent temperature. Evaluating Eq. (13-1) shows that the noise power is directly proportional to temperature in kelvins and also that noise power collapses to zero at the temperature of absolute zero (0 K).

Note that the equivalent noise temperature T_e is *not* the physical temperature of the amplifier, but rather a theoretical construct that is an *equivalent* temperature that produces that amount of noise power. The noise temperature is related to the noise factor by

$$T_e = (F_n - 1) T_o \tag{13-8}$$

and to the noise figure by

$$T_e = \left[\text{antilog} \left(\frac{\text{NF}}{10 - 1} \right) \right] K\, T_o \qquad (13\text{-}9)$$

Now that we have noise temperature T_e, we can also define the noise factor and noise figure in terms of noise temperature:

$$F_n = \left(\frac{T_e}{T_o} \right) + 1 \qquad (13\text{-}10)$$

and

$$\text{NF} = 10 \log \left[\left(\frac{T_e}{T_o} \right) + 1 \right] \qquad (13\text{-}11)$$

The total noise in any amplifier or network is the sum of internally generated and externally generated noise. In terms of noise temperature,

$$P_{n(\text{total})} = G\, K\, B\, (T_o + T_e) \qquad (13\text{-}12)$$

where

$P_{n\ \text{total)}}$ = total noise power

All other terms are as previously defined.

13-3.2 Noise in Cascade Amplifiers

A noise signal is seen by a following amplifier as a valid input signal. Thus, in a cascade amplifier, the final stage sees an input signal that consists of the original signal and noise amplified by each successive stage. Each stage in the cascade chain both amplifies signals and noise from previous stages and also contributes some noise of its own. The overall noise factor for a cascade amplifier can be calculated from *Friis's noise equation*:

$$F_n = F_1 + \frac{F_2 - 1}{G1} + \frac{F_3 - 1}{G1\, G2} + \ldots + \frac{F_{n-1}}{G1G2\ldots G_{n-1}} \qquad (13\text{-}13)$$

where

F_n = overall noise factor of N stages in cascade
F_1 = noise factor of stage 1
F_2 = noise factor of stage 2
F_n = noise factor of the nth stage
$G1$ = gain of stage 1
$G2$ = gain of stage 2
G_{n-1} = gain of stage (n-1)

As you can see from Eq. (13-13), the noise factor of the entire cascade chain is dominated by the noise contribution of the first stage or two. Typically, high-gain amplifiers use a low-noise device for only the first stage or two in the cascade chain.

EXAMPLE 13-1

A three-stage amplifier (Fig. 13-3) has the following gains: $G1 = 10$, $G2 = 10$, and $G3 = 25$. The stages also have the following noise factors: $F1 = 1.4$, $F2 = 2$, and $F3 = 3.6$. Calculate (a) the overall gain of the cascade chain in decibels, (b) the overall noise factor, and (c) the overall noise figure.

Solution

$$(a)\ G = G1 \times G2 \times G3$$
$$= 10 \times 10 \times 25$$
$$= 2500$$
$$G = 10 \log(2500)$$
$$G = (10)(3.4) = 34\ dB$$

$$(b)\ F_n = F1 + \frac{F_2 - 1}{G1} + \frac{F_3 - 1}{G1G2}$$

$$= 1.4 + \frac{2 - 1}{10} + \frac{3.6 - 1}{(10)(10)}$$

$$= 1.4 + \frac{1}{10} + \frac{2.6}{100}$$

$$= 1.4 + 0.1 = 0.026 = 1.53$$

$$(c)\ NF = 10 \log F_n$$
$$= 10(\log 1.53)$$
$$= (10)(0.19) = 1.9\ dB$$

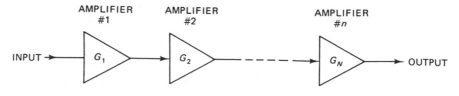

Figure 13-3   Cascade amplifiers.

Note in example 13-1 that the overall noise factor (1.53) is only slightly worse than the noise factor of the input amplifier (1.4) and is better than the noise factors of the following stages (2 and 3.6, respectively). Clearly, the overall noise factor is set by the input stage.

STUDENT EXERCISE

The student may want to perform the following calculations: (a) convert noise factors $F1$, $F2$, and $F3$ to noise temperatures and rework the problem in Example 13-1; (b) reverse the positions of amplifiers 1 and 3 so that new $G1 = 25$, $G2 = 10$, and $G3 = 10$, and $F1 = 3.6$, $F2 = 2$, and $F3 = 1.4$. Compare the results of this calculation with the results obtained in example 13-1.

13-4 DISCRETE MICROWAVE AMPLIFIER CONFIGURATION

As is true with other RF amplifiers, the discrete microwave amplifier can be built with any of several transistor devices in either wideband or bandpass configurations. The type of transistor selected may be based on the noise figure required in the specific application. Figure 13-4 shows the noise figures associated with silicon

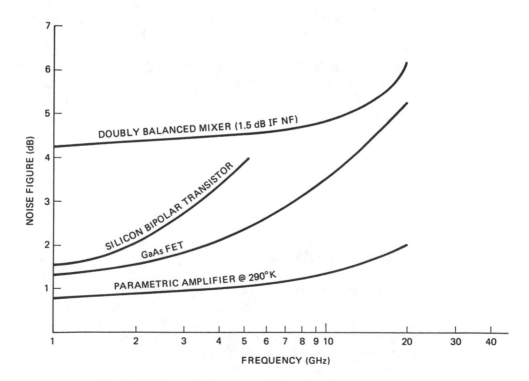

Figure 13-4   Noise figure versus frequency for various amplifiers.

bipolar transistors and gallium arsenide (GaAs) field-effect transistors. Also shown for the sake of comparison is the noise figure of the doubly balanced mixer circuit and an uncooled parametric amplifier (Section 13-5).

Impedance matching is a requirement on most amplifiers because maximum power transfer in any electrical circuit requires that source and load impedances be matched. The source impedance must match the amplifier input impedance, and the load impedance must match the amplifier output impedance. In most RF systems these impedances are standardized at 50 Ω, or 75 Ω in the case of television systems. Rarely do the transistor input and output impedances match the system impedance (for an exception see the MMIC discussion in Chapter 14). In most circuits, therefore, some means of matching the input and output impedances of the amplifier to the system impedance is required.

In lower-frequency amplifiers, coil transformers (sometimes resonant), tapped inductors, special inductor-capacitor (*LC*) networks, or capacitor voltage-divider networks are used to provide the impedance-matching function. These methods can also provide the bandpass tuning required in some (perhaps most) amplifiers. Such bandpass tailoring not only enhances reception of the desired frequencies, but perhaps more importantly in many systems, it also attenuates unwanted out-of-band signals and limits the input noise power [see the bandwidth term in Eq. (13-1)].

Figure 13-5 shows how both impedance matching and input-output tuning can be accomplished in microwave amplifiers. Figure 13-5A shows a method similar to low-frequency amplifiers in which an *LC* resonant tank circuit is used for input and output tuning. The inductor is a straight length of printed circuit track. Impedance matching is provided on the input side of Q1 by tapping inductor *L*1. On the output side, impedance matching is provided by a pi network consisting of *L*2, *C*2, and *C*3.

A different method is shown in Fig. 13-5B. In this circuit the transformation is provided by a quarter-wavelength transmission line *Q*-section transformer. The impedance transformation occurs if the characteristic impedance of the transmission line segment is equal to

$$Z_o = \sqrt{R_i \times R_o} \qquad (13\text{-}14)$$

13-5 PARAMETRIC AMPLIFIERS

The *parametric amplifier* is capable of high-gain and low-noise operation (see Fig. 13-4) in the UHF and microwave regions, but is fundamentally different from conventional forms of amplifier. The parametric amplifier takes its name from the fact that amplification occurs through exciting a circuit parameter. This amplifier is actually misnamed because it is the reactance parameters (X_c and X_L) that are excited. Perhaps a better name is *reactance amplifier*.

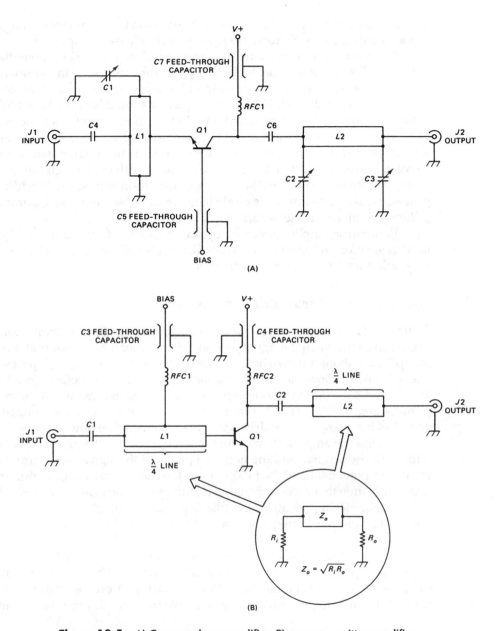

Figure 13-5 A) Common base amplifier; B) common emitter amplifier.

A reactance differs from a resistance in that the latter dissipates power, while reactances store energy and redeliver it to the circuit without any power dissipation. If the reactance can be varied at a rapid rate, then the energy stored and discharged by the reactance can be used to amplify the signal. Although either capacitors or inductors can be used in parametric amplifiers, it is the capacitive reactance that is used in practical circuits. The reason is that suitable voltage variable capacitance diodes (varactors) are readily available (see Chapter 15).

In a varactor the capacitance is a function of the reverse bias potential applied across the *PN* junction of the diode. A typical varactor useful in parametric amplifiers has a breakdown voltage of –4 to –12 V and a zero-bias junction capacitance of 0.2 to 5 picofarads (pF). The cutoff frequency should be 20 GHz or higher. Generally, the noise figure is improved with higher diode cutoff frequencies.

The low-noise figure of the parametric amplifier derives from its use of a reactance as the active element. In an ideal circuit, the noise generated is zero. In real circuits, however, resistive losses are associated with the tank circuit and the varactor, and these give rise to thermal agitation (Johnson) noise. In addition, other processes take place inside the diode to generate noise. As a result, parametric amplifiers exhibit low-noise factors, but not zero.

Parametric amplifiers can be operated in any of three modes: *degenerative*, *nondegenerative*, and *regenerative*. We will consider all modes and provide a tool for evaluating parametric amplifier circuits.

13-5.1 Degenerative Parametric Amplifiers

Figure 13-6A shows the basic parametric amplifier. The varactor diode is connected so as to switch the signal on and off to the load as an external pump signal is applied. Although shown here as a series-connected switch, both series- and parallel-connected diodes are used. The signal and pump waveform (see Fig. 13-6B) are phased such that the diode capacitance is fully charged when the peak of the pump signal arrives. The charge is constant, so by $V = Q/C$ the voltage must increase as the pump voltage drives the diode capacitance down.

Parametric amplification occurs when the peak of the pump signal coincides with both the positive and the negative peaks of the signal waveform. Increasing the pump potential (as at the peak) drives the diode capacitance to minimum, and it is at this time that capacitor charge is dumped to the load. To achieve degenerative parametric amplification, the phasing must be precise, and this requirement means that the pump frequency must be the second harmonic of the signal frequency.

A severe limitation to the degenerative parametric amplifier is the necessity of precisely phasing the pump and signal waveforms. Drift in either signal can reduce the gain or prevent the circuit from operating. A broader bandwidth method is to use the nondegenerative or regenerative parametric amplifier circuits.

13-5.2 Nondegenerative and Regenerative Parametric Amplifiers

The requirement for precise phasing in the degenerative parametric amplifier is relieved somewhat in the nondegenerative case (Fig. 13-7) by the addition of a third frequency. In addition to the signal frequency (f_s) and the pump frequency (f_p) we now also have an *idler frequency* (f_i). In Fig. 13-7, note that the third resonant tank circuit (*L3C3*) is tuned to f_i. The idler frequency is the output frequency of the circuit, which, incidentally, also operates as a frequency translator or converter.

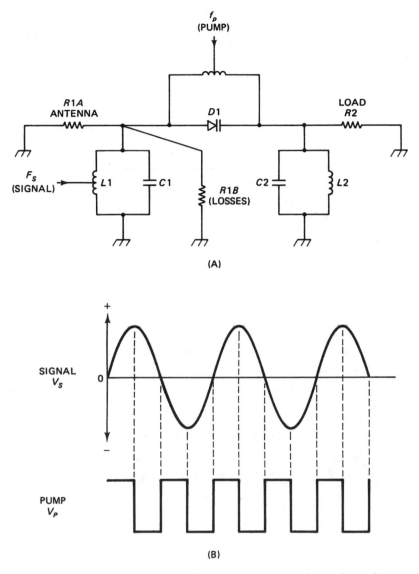

Figure 13-6 Parametric amplifier: A) circuit; B) waveform relationships.

There are two general cases of nondegenerative parametric amplifiers: *up-converters* and *down-converters*. In the up-converter case the idler frequency is the sum of the pump and signal frequencies:

$$f_i = f_s + f_p \qquad\qquad (13\text{-}15)$$

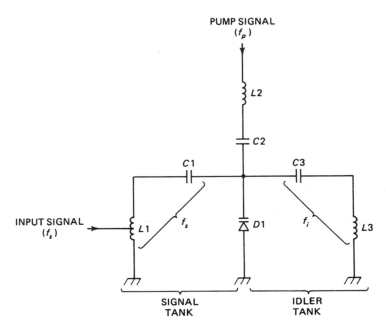

Figure 13-7
Equivalent circuit.

In the down-converter case, the idler frequency is the difference between pump and signal frequencies:

$$f = f_s - f_p \tag{13-16}$$

Power gain is defined as the ratio of the output power to the input power. In the case of a lossless circuit, the gain of the up-converter (f_i greater than f_s) is

$$G = \frac{f_i}{f_s} \tag{13-17}$$

The down-converter case is actually a loss (attenuation), rather than a power gain.

The third category of parametric amplifier is the regenerative circuit and is actually a special case of the nondegenerative amplifier. In the regenerative amplifier, the pump frequency is the sum of the signal and idler frequencies. In this case, power gain is negative, which implies a negative resistance characteristic. As a result, the circuit is regenerative. Implicit in this property, if the circuit can be kept out of oscillation, is very low noise coupled with very high gain.

13-5.3 Noise in Parametric Amplifiers

The low-noise capability of the parametric amplifier is a result of the fact that the amplifier element is a reactor rather than a resistor. In an ideal parametric amplifier, the noise figure is zero, but in practical circuits we have two noise contributors: *circuit losses* and *frequency conversion noise*. These sources combine to create a nonzero noise factor on the order of

$$F_{noise} = \frac{R_a}{R1} + \frac{f_s}{f_i} \qquad (13\text{-}18)$$

where

F_{noise} = noise factor
R_a = antenna impedance resistive component
$R1$ = sum of circuit resistive losses
f_i = idler frequency
f_s = signal frequency

Some authorities recommend a pump frequency seven to ten times the signal frequency for lowest noise operation.

13-5.4 Microwave Configuration for Parametric Amplifiers

The circuit examples presented thus far show inductor-capacitor (*LC*) resonant tank circuits for the various frequencies. These circuits work well in the UHF and lower microwave region. At higher microwave frequencies, however, the *LC* tank circuit fails to work well and is not practical. Therefore, we see parametric amplifiers with resonant cavities (Fig. 13-8) in place of resonant tank circuits. A tuning disk tunes the cavities to resonance.

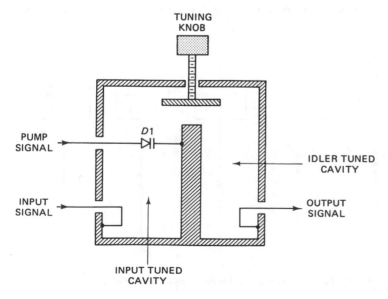

Figure 13-8
Cavity parametric amplifier.

13-5.5 Manley-Rowe Relationships

In 1957, J. M. Manley and H. E. Rowe proposed a means for evaluating parametric amplifier circuits. Consider the equivalent circuit in Fig. 13-9. In this circuit, we have a variable capacitance as the reactor element and two signal sources: the signal frequency (f_s) and the pump frequency (f_p), both of which are shown as generators. In series with both generators are filters that pass the generator frequency and totally reject all other frequencies. There is also a series of loads, each of which is isolated from the others by the same kind of ideal narrowband filter. The frequencies of these filters are ($f_p + f_s$), ($f_p - f_s$), up to ($mf_p \pm nf_s$) (where m and n are integers). The Manley-Rowe relationships are

I.

$$\sum_{m,n} \frac{mP_{m,n}}{mf_p + nf_s} = 0 \qquad (13\text{-}20)$$

II.

$$\sum_{m,n} \frac{nP_{m,n}}{mf_p + nf_s} = 0 \qquad (13\text{-}21)$$

In working with Manley-Rowe equations, we recognize the following algebraic sign conventions regarding power:

1. $+P$ is assigned to power flowing either into the capacitor or from the pump and input signal generators.
2. $-P$ is assigned to power flowing out of the capacitor or into a load resistance.

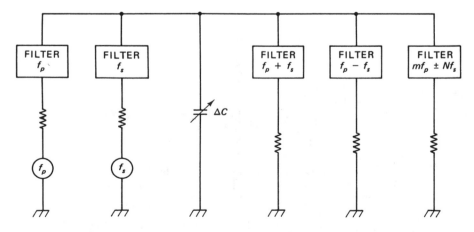

Figure 13-9   Equivalent circuit for Manley-Rowe calculations.

The stability of the parametric amplifier is determined by the sign of the power flowing with respect to the capacitor. If the power from the signal flows into the capacitor, then the stage is stable. Because we deal with integers from 0 through the ith, we can check not only the fundamental frequencies (m and n are 1), but also their respective harmonics (m,n greater than 1). Some of these combinations are stable, while others are unstable.

13-6 SUMMARY

1. Amplifiers add noise to the signal. This noise is due to thermal agitation of electrons flowing in the device. We characterize noise in terms of *noise factor*, *noise figure*, and *noise temperature*.

2. Noise in cascade amplifiers is defined in terms of the Friis noise equation. The overall noise figure of a cascade amplifier is reduced by having an input stage with a very low noise figure. The noise figure of that stage dominates the chain.

3. Microwave amplifiers use stripline tuners, transmission line Q sections, and other methods for limiting bandwidth and matching impedance.

4. Parametric amplifiers operate by switching power into and out of a reactor (capacitive or inductive). Low-noise operation results from the fact that ideal reactors do not dissipate power. Three cases are known: *degenerative*, *nondegenerative*, and *regenerative*. The parametric amplifier can be evaluated using the Manley-Rowe equations.

13-7 RECAPITULATION

Now return to the objectives and prequiz questions at the beginning of the chapter and see how well you can answer them. If you cannot answer certain questions, place a check mark by each and review the appropriate parts of the text. Next, try to answer the following questions and work the problems, using the same procedure.

QUESTIONS AND PROBLEMS

1. The principal noise source in microwave amplifier is Johnson noise, which is due to _____ agitation of electrons.
2. Calculate the noise power of a source at 300 K if the bandwidth is 100 MHz.
3. Calculate the noise power of a 75-MHz bandwidth source at a temperature of +330 K.
4. An amplifier has a noise factor (F) of 3.5; calculate the noise figure (NF).
5. Calculate the noise figure (NF) of an amplifier of a noise factor (F) of 2.75.
6. Calculate the equivalent noise temperature of an amplifier that has a noise factor of 2.75.

7. Calculate the equivalent noise temperature of an amplifier that has a noise factor of 3.5.

8. An amplifier has an equivalent noise temperature of 125 K. Calculate the noise factor.

9. What is the noise factor of a 100-K amplifier?

10. Calculate the noise figure of the amplifiers of problems 8 and 9.

11. Calculate the overall noise factor of a three-stage cascade amplifier if the following specifications apply: amplifier A1: $G1 = 10$, $F1 = 1.7$; amplifier A2: $G2 = 20$, $F2 = 3.5$; and amplifier A3: $G3 = 10$, $F3 = 3.5$.

12. In problem 11, convert the noise factors into equivalent noise temperatures and work the problem again. Convert the result of problem 11 and compare with the result of this problem.

KEY EQUATIONS

1. Noise power:

$$P_n = KTB$$

2. Noise factor definition:

$$F_n = \frac{P_{no}}{P_{ni}}$$

3. Standardized input noise power:

$$P_{ni} = GKBT_o$$

4. Noise factor as a function of SNR:

$$F_n = \frac{SNR_{in}}{SNR_{out}}$$

5. Noise factor as a function of output noise power:

$$F_n = \frac{P_{no}}{KT_0 BG}$$

6. Noise factor model:

a.
$$F_n = \frac{KT_0 BG + \Delta N}{KT_0 BG}$$

or

b.
$$F_n = \frac{\Delta N}{KT_0 BG}$$

7. Noise figure as a function of noise factor:

$$NF = 10 \log F_n$$

8. Noise temperature as a function of noise factor:

$$T_e = (F_n - 1)T_0$$

9. Noise temperature as a function of noise figure:

$$T_e = \left[\text{antilog} \left(\frac{NF}{10} \right) - 1 \right] KT_0$$

10. Noise figure as a function of noise temperature:

$$NF = 10 \log \left(\frac{T_e}{T_0} + 1 \right)$$

11. Total noise in a system:

$$P_{n\,(\text{total})} = GKB(T_o + T_e)$$

12. Noise factor of amplifiers in cascade:

$$F_n = F_1 + \frac{F_2 - 1}{G1} + \frac{F_3 - 1}{G1G2} + \ldots + \frac{F_n - 1}{G1G2 \ldots G_{n-1}}$$

13. Noise temperature of amplifiers in cascade:

$$T_e = T_1 + \frac{T_2 - 1}{G1} + \frac{T_3 - 1}{G1G2} + \ldots + \frac{T_n - 1}{G1G2 \ldots G_{n-1}}$$

14. Characteristic impedance required of a quarter-wave Q-section transformer:

$$Z_o = \sqrt{R_i \times R_o}$$

15. Idler frequency of an up-converter parametric amplifier:

$$f_i = f_s + f_p$$

16. Idler frequency of a down-converter parametric amplifier:

$$f_i = f_s - f_p$$

17. Gain of a parametric amplifier:

$$G = f_i/f_s$$

18. Noise in a parametric amplifier:

$$F_{\text{noise}} = \frac{R_a}{R1} + \frac{f_s}{f_i}$$

19. Manley-Rowe relationship for parametric amplifiers:

$$\text{I.} \quad \sum_{m,n} \frac{mP_{m,n}}{mf_p + nf_s} = 0$$

$$\text{II.} \quad \sum_{m,n} \frac{nP_{m,n}}{mf_p + nf_s} = 0$$

CHAPTER 14

Hybrid and Monolithic Microwave Integrated Circuit Amplifiers

---

OBJECTIVES

1. Learn the identifying properties of microwave integrated circuit (MIC) devices.
2. Learn the key design parameters of MIC devices.
3. Understand the design procedure for MIC amplifier circuits.
4. Learn the strengths and limitations of MIC amplifiers.

14-1 PREQUIZ

These questions test your prior knowledge of the material in this chapter. Try answering them before you read the chapter. Look for the answers (especially those you answered incorrectly) as you read the text. After you have finished studying the chapter, try answering these questions again and those at the end of the chapter.

1. Interconnections in HMIC and MMIC devices are usually in the form of _____ transmission lines.
2. Amplifier gain measurements at microwave frequencies are difficult if the _____ is higher than 1.15.
3. Solid-state devices in MIC amplifiers are usually made of _____ instead of silicon.
4. A _____ power divider is sometimes used at microwave frequencies to parallel connect two or more MIC devices.

14-2 INTRODUCTION

Very wideband amplifiers (those operating from near dc to UHF or into the microwave regions) have traditionally been very difficult to design and build with consistent performance across the entire passband. Many such amplifiers have either gain irregularities, such as "suck-outs," or peaks. Others suffer large variations of input and output impedance over the frequency range. Still others suffer spurious oscillation at certain frequencies within the passband. Barkhausen's criteria for oscillation require (1) loop gain of unity or more, and (2) 360° (in-phase) feedback at the frequency of oscillation. At some frequencies, the second of these criteria may be met by adding the normal 180° phase shift inherent in the amplifier to phase shift due to stray *RLC* components. The result will be oscillation at the frequency where the *RLC* phase shift was an additional 180°. In addition, in the past only a few applications required such amplifiers. Consequently, such amplifiers were either very expensive or did not work nearly as well as claimed. Hybrid microwave integrated circuit (HMIC) and monolithic microwave integrated circuit (MMIC) devices were low-cost solutions to the problem.

14-3 WHAT ARE HMICS AND MMICS?

MMICs are tiny, gain block monolithic integrated circuits that operate from dc or near dc to a frequency in the microwave region. HMICs, on the other hand, are hybrid devices that combine discrete and monolithic technology. One product (Signetics NE-5205) offers up to +20 dB of gain from dc to 0.6 GHz, while another low-cost device (Minicircuits Laboratories, Inc. MAR-x) offers +20 dB of gain over the range from dc to 2 GHz, depending on the model. Other devices from other manufacturers are also offered, and some produce gains to +30 dB and frequencies to 18 GHz. Such devices are unusual in that they present input and output impedances that are a good match to the 50 or 75 Ω normally used as system impedances in RF circuits.

Monolithic integrated circuit devices are formed through photoetching and diffusion processes on a substrate of silicon or some other semiconductor material. Both active devices (such as transistors and diodes) and some passive devices can be formed in this manner. Passive components such as on-chip capacitors and resistors can be formed using various thin- and thick-film technologies. In the MMIC device, interconnections are made on the chip via built-in *planar transmission lines*.

Hybrids are a level closer to regular discrete circuit construction than ICs. Passive components and planar transmission lines are laid down on a glass, ceramic, or other insulating substrate by vacuum deposition or other methods. Transistors and unpackaged monolithic chip dies are cemented to the substrate and then connected to the substrate circuitry by mil-sized gold or aluminum bonding wires.

Because the material in this chapter can sometimes apply to either HMIC or MMIC devices, the convention herein will be to refer to all devices in either subfamily as microwave integrated circuits (MIC), unless otherwise specified.

Three things specifically characterize the MIC device. First is simplicity. As you will see in the circuits discussed, the MIC device usually has only input, output, ground, and power supply connections. Other wideband IC devices often have up to 16 pins, most of which must be either biased or capacitor bypassed. The second feature of the MIC is the very wide frequency range (dc to GHz) of the devices. Third is the constant input and output impedance over several octaves of frequency.

Although not universally the case, MICs tend to be unconditionally stable because of a combination of series and shunt negative feedback internal to the device. The input and output impedances of the typical MIC device are a close match to either 50 or 75 Ω, so it is possible to make a MIC amplifier without any impedance-matching schemes, a factor that makes it easier to broadband than if tuning was used. A typical MIC device generally produces a standing-wave ratio (SWR) of less than 2:1 at all frequencies within the passband, provided that it is connected to the design system impedance (for example, 50 Ω). The MIC is not usually regarded as a low-noise amplifier (LNA), but can produce noise figures (NF) in the 3- to 8-dB range. Some MICs are LNAs, however, and the number available should increase in the near future.

Narrowband and passband amplifiers can be built using wideband MICs. A narrowband amplifier is a special case of a passband amplifier and is typically tuned to a single frequency. An example is the 70-MHz IF amplifier used in microwave receivers. Because of input and/or output tuning, such an amplifier will respond only to signals in the 70-MHz frequency band.

14-4 VERY WIDEBAND AMPLIFIERS

Engineering wideband amplifiers, such as those used in MIC devices, seems simple, but has traditionally caused a lot of difficulty for designers. Figure 14-1A shows the most fundamental form of MIC amplifier; it is a common-emitter *NPN* bipolar transistor amplifier. Because of the high-frequency operation of these devices, the amplifier in MICs is usually made of a material such as *gallium arsenide* (GaAs).

In Fig. 14-1A, the emitter resistor (R_e) is unbypassed, so introduces a small amount of negative feedback into the circuit. Resistor R_e forms *series feedback* for transistor $Q1$. The *parallel feedback* in this circuit is provided by collector-base bias resistor R_f. Typical values for R_f are in the 500-Ω range, and for R_e in the 4- to 6-Ω range. In general, the designer tries to keep the ratio R_f/R_e high in order to obtain higher gain, higher output power compression points, and lower noise figures. The input and output impedances (R_o) are equal, and defined by the patented equation

$$Ro = \sqrt{R_f \times R_e} \qquad (14\text{-}1)$$

where

R_o = output impedance in ohms
R_f = shunt feedback resistance in ohms
R_e = series feedback resistance in ohms

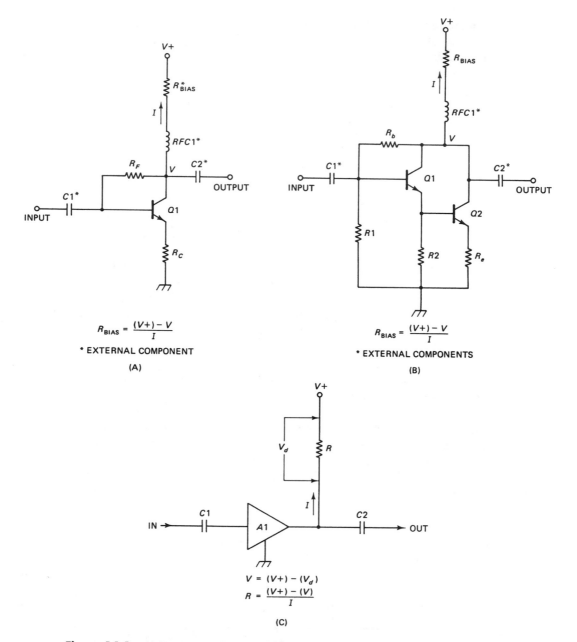

$$R_{BIAS} = \frac{(V+) - V}{I}$$

* EXTERNAL COMPONENT

(A)

$$R_{BIAS} = \frac{(V+) - V}{I}$$

* EXTERNAL COMPONENTS

(B)

$$V = (V+) - (V_d)$$
$$R = \frac{(V+) - (V)}{I}$$

(C)

Figure 14-1 A) Common emitter amplifier; B) Darlington amplifier; C) MMIC amplifier with external components.

EXAMPLE 14-1

An amplifier such as Fig. 14-1A has a series feedback resistance of 5 Ω and a shunt feedback resistance of 500 Ω. Calculate the input and output impedance, R_o.

Solution

$$Ro = \sqrt{R_f \times R_e}$$

$$= \sqrt{(500\Omega)(5\Omega)}$$

One factor that makes this form of amplifier so useful over such a wide band is that R_o remains essentially constant over several octaves of frequency. With the values of resistors shown in the preceding example, the MIC amplifier is a good impedance match for standard 50-Ω RF systems. Other MIC devices, with slightly different values of resistor, are designed to match standard 75-Ω TV/video systems.

The gain of the amplifier in Fig. 14-1A is also a function of the feedback resistance and is found from

$$G_{dB} = 20 \log \left[\frac{R_f - R_e}{R_o + R_e} \right] \tag{14-2}$$

where

G_{dB} = voltage gain in decibels (dB)

R_f, R_e, and R_o are as defined previously.

EXAMPLE 14-2

Calculate the voltage gain of the amplifier in Example 14-1. Recall that R_o = 50 Ω, R_f = 500 Ω, and R_e = 5 Ω.

Solution

$$G_{dB} = 20 \log \left[\frac{R_f - R_e}{R_o + R_e} \right]$$

$$= 20 \log \left[\frac{(500 - 5)\ \Omega}{(50 + 5)\ \Omega} \right]$$

$$= 20 \log (495/55)$$

$$= 20 \log (9)$$

$$= 20 \log (9) = (20)(0.95) = 19 \text{ db}$$

A more common form of MIC amplifier circuit is shown in Fig. 14-1B. Although based on the *Darlington amplifier* circuit, this amplifier still has the same sort of series and shunt feedback resistors (R_e and R_f) as the previous circuit. All resistors except R_{bias} are internal to the MIC device.

A Darlington amplifier, also called a *Darlington pair* or *superbeta transistor*, consists of a pair of bipolar transistors (Q1 and Q2) connected such that Q1 is an emitter follower driving the base of Q2, and with both collectors connected in parallel with each other. The Darlington connection permits both transistors to be treated as if they were a single transistor with higher than normal input impedance and a *beta* gain (β) equal to the product of the individual *beta* gains. For the Darlington amplifier, therefore, the beta (β or H_{fe})

$$\beta_o = \beta_{Q1} \times \beta_{Q2} \tag{14-3}$$

where

β_o = beta gain of the Q1/Q2 pair
β_{Q1} = beta gain of Q1
β_{Q2} = beta gain of Q2

EXAMPLE 14-3

Calculate the beta gain of a Darlington pair if Q1 and Q2 are identical transistors in which $\beta = 100$ each.

Solution

$$\beta_o = \beta_{Q1} \times \beta_{Q2}$$
$$= 100 \times 100$$
$$= 10,000$$

From Example 14-3, you should be able to see two facts. First, the beta gain is very high for a Darlington amplifier that is made with relatively modest transistors. Second, the beta of a Darlington amplifier made with identical transistors is the square of the common beta rating.

External Components.   Figures 14-1A and B show several components that are usually external to the MIC device. The bias resistor (R_{bias}) is sometimes internal, although on most MIC devices it is external. RF choke *RFC1* is in series with the bias resistor and is used to enhance operation at the higher frequencies; *RFC1* is considered optional by some MIC manufacturers. The reactance of the RF choke is in series with the bias resistance and increases with frequency according to the $2\pi FL$ rule. Thus, the transistor sees a higher impedance load at the upper end of the passband than at the lower end. Use of *RFC1* as a peaking coil thus helps overcome the adverse effect of stray circuit capacitance that ordinarily causes a similar decreasing frequency-dependent characteristic. A general rule of thumb is to make the combination of R_{bias} and X_{RFC1} form an impedance of at least 500 Ω at the lowest frequency of operation. The gain of the amplifier may drop about 1 dB if *RFC1* is deleted. This effect is caused by the bias resistance shunting the output impedance of the amplifier.

The capacitors are used to block dc potentials in the circuit. They prevent intracircuit potentials from affecting other circuits, as well as preventing potentials in other circuits from affecting MIC operation. More will be said about these capacitors later, but for now understand that practical capacitors are not ideal; real capacitors are complex *RLC* circuits. While the *L* and *R* components are negligible at low frequencies, they are substantial in the microwave region. In addition, the *LC* characteristic forms a self-resonance that can either "suck out" or enhance gain at specific frequencies. The result is an uneven frequency response characteristic at best and spurious oscillations at worst.

14-5 GENERIC MMIC AMPLIFIER

Figure 14-1C shows a generic circuit representing MIC amplifiers in general. As you will see when we look at an actual product, this circuit is nearly complete. The MIC device usually has only input, output, ground, and power connections; some models do not have separate dc power input. There is no dc biasing, no bypassing (except at the dc power line) and no seemingly useless pins on the package. MICs tend to use either microstrip packages like UHF/microwave small-signal transistor packages or small versions of the miniDIP or metallic IC packages. Some HMICs are packaged in larger transistorlike cases, while others are packaged in special hybrid packages.

The dc bias resistor (R_{bias}) connected to either the power supply terminal (if any) or the output terminal must be set to a value that limits the current to the device and drops the supply voltage to a safe value. MIC devices typically require a low dc voltage (4 to 7 V dc) and a maximum current of about 15 to 25 milliamperes

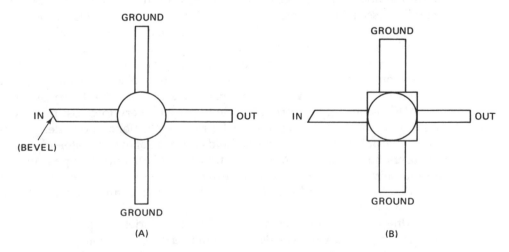

Figure 14-2 MMIC package styles.

(mA) depending on type. There may also be an optimum current of operation for a specific device. For example, one device is advertised to operate over a range of 2 to 22 mA, but the optimum design current is 15 mA. The value of resistor needed for R_{bias} is found from Ohm's law:

$$R_{bias} = \frac{(V+) - V}{I_{bias}} \qquad\qquad (14\text{-}4)$$

where

R_{bias} = resistor bias in Ω
$V+$ = dc power supply potential in volts
V = rated MIC device operating potential in volts
I_{bias} = operating current in amperes

EXAMPLE 14-4

A MIC device requires 15 mA at +5 V dc for optimum operation. Calculate R_{bias} for use with a +12-V dc power supply.

Solution

$$R_{bias} = \frac{(V+) - V}{I_{bias}}$$

$$= \frac{(12 - 5)\,V}{0.015\,A}$$

The construction of amplifiers based on MIC devices must follow microwave practices. This requirement means short, wide, low-inductance leads made of printed circuit foil and stripline construction. Interconnection conductors tend to behave like transmission lines at microwave frequencies and must be treated as such. In addition, capacitors should be capable of passing the frequencies involved, yet have as little inductance as possible. In some cases, the series inductance of common capacitors forms a resonance at some frequency within the passband of the MMIC device. These resonant circuits can sometimes be detuned by placing a small ferrite bead on the capacitor lead. Microwave chip capacitors are used for ordinary bypassing.

MIC technology is currently able to provide very low-cost microwave amplifiers with moderate gain and noise figure specifications, and better performance is available at higher cost; the future holds promise of even greater advances. Manufacturers have extended MIC operation to 18 GHz and dropped noise figures substantially. In addition, it is possible to build the entire front end of a microwave receiver into a single HMIC or MMIC, including the RF amplifier, mixer, and local oscillator stages.

Although MIC devices are available in a variety of package styles, those shown in Fig. 14-3 are typical. Because of the very high frequency operation of these devices, MICs are packaged in *stripline* transistorlike cases. The low-inductance leads for these packages are essential in UHF and microwave applications.

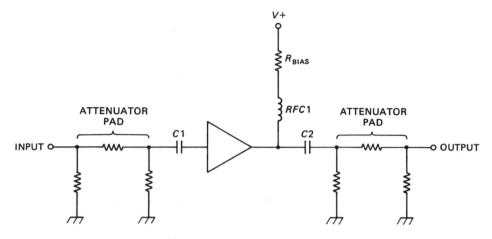

Figure 14-3   Use of input and output attenuators to stabilize impedance excursions.

14-5.1 Attenuators in Amplifier Circuits?

It is common practice to place attenuator pads in series with the input and output signal paths of microwave circuits in order to swamp out impedance variations that adversely affect circuits that ordinarily require either impedance matching or a constant impedance. Especially when dealing with devices such as *LC* filters (low pass, high pass, bandpass), VHF/UHF amplifiers, matching networks, and MIC devices, it is useful to insert 1-, 2-, or 3-dB resistor attenuator pads in the input and output lines. The characteristics of many RF circuits depend on seeing the design impedance at input and output terminals. With the attenuator pad (see Fig. 14-4) in the line, source and load impedance changes do not affect the circuit nearly as much.

The attenuator tactic is also sometimes useful when a designer is confronted with seemingly unstable very wideband amplifiers. Insert a 1-dB pad in series with both the input and the output lines of the unstable amplifier. This tactic will cost about 2 dB of voltage gain, but often cures instabilities that arise out of frequency-dependent load or source impedance changes.

14-6 CASCADE MIC AMPLIFIERS

MIC devices can be connected in cascade (Fig. 14-5) to provide greater gain than is available from only a single device, although a few precautions must be observed. It must be recognized, for example, that MICs possess a substantial amount of gain from frequencies near dc to well into the microwave region. In all cascade amplifiers, attention must be paid to preventing feedback from stage to stage. Two factors must be addressed. First, as always, is component layout. The output and input circuitry external to the MIC must be physically separated to prevent coupling feedback. Second, it is necessary to decouple the dc power supply lines that feed two or more stages. Signals carried on the dc power line can easily couple into one or more stages, resulting in unwanted feedback.

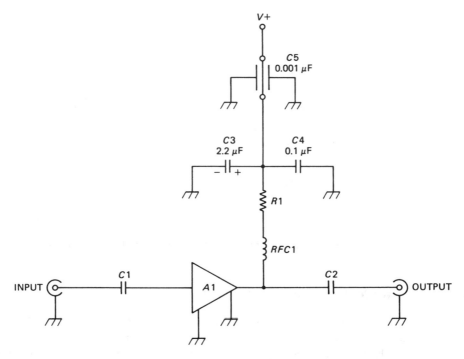

Figure 14-4 Typical MMIC circuit.

Figure 14-5 shows a method for decoupling the dc power line of a two-stage MIC amplifier. In lower-frequency cascade amplifiers, the $V+$ ends of resistors $R1$ and $R2$ would normally be joined together and connected to the dc power supply. Only a single capacitor would be needed at that junction to ensure adequate decoupling between stages. But as the operating frequency increases, the situation becomes more complex, in part because of the nonideal nature of practical components. For example, in an audio amplifier the electrolytic capacitor used in the power supply ripple filter may provide sufficient decoupling. At RF frequencies, however, electrolytic capacitors are essentially useless as capacitors (they act more like resistors at those frequencies).

The decoupling system in Fig. 14-5 consists of RF chokes $RFC3$ and $RFC4$ and capacitors $C4$ through $C9$. The RF chokes help block high-frequency ac signals from traveling along the power line. These chokes are selected for a high reactance at VHF through microwave frequencies, while having a low dc resistance. For example, a 1-microhenry (1-μH) RF choke might have only a few milliohms of dc resistance, but (by $2\mu FL$) has a reactance of more than 3000 Ω at 500 MHz. It is important that $RFC3$ and $RFC4$ be mounted so as to minimize mutual inductance due to interaction of their respective magnetic fields.

Capacitors $C4$ through $C9$ are used for bypassing signals to ground. It should be noted that a wide range of values and several types of capacitor are used in this circuit. Each has its own purpose. Capacitor $C8$, for example, is an electrolytic type and is used to decouple very low-frequency signals (that is, those up to several hundred kilohertz). Because $C8$ is ineffective at higher frequencies, it is shunted by ca-

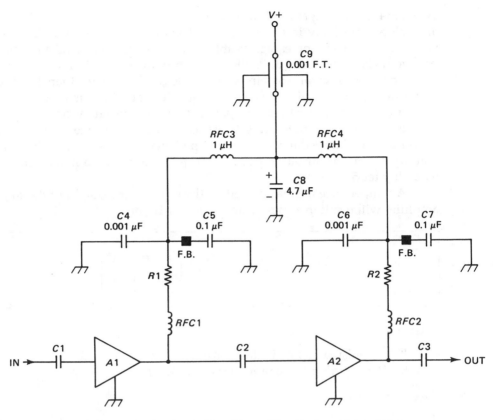

Figure 14-5 Cascade MMIC amplifiers for lower frequencies.

pacitor C9, shown in Fig. 14-5 as a feedthrough capacitor. Such a capacitor is usually mounted on the shielded enclosure housing the amplifier. Capacitors C5 and C7 are used to bypass signals in the HF region. Because these capacitors are likely to exhibit substantial series inductance, they will form undesirable resonances within the amplifier passband. Ferrite beads (FB) are sometimes installed on each capacitor lead to detune capacitor self-resonances. Like C9, capacitors C4 and C6 are used to decouple signals in the VHF-and-up region. These capacitors must be of microwave chip construction, or they may prove ineffective above 200 MHz or so.

14-7 GAIN IN CASCADE AMPLIFIERS

In low-frequency amplifiers, we might reasonably expect the composite gain of a cascade amplifier to be the product of the individual stage gains:

$$G = G1 \times G2 \times G3 \times \ldots \times G_n \qquad (14\text{-}5)$$

While that reasoning is valid for low-frequency voltage amplifiers, it fails for RF amplifiers (especially in the microwave region) where input/output standing-wave ratio (SWR) becomes significant. In fact, the gain of any RF amplifier cannot be accurately measured if the SWR is greater than about 1.15:1.

There are several ways in which SWR can be greater than 1:1, and all involve an impedance mismatch. For example, the amplifier may have an input or output resistance other than the specified value. This situation can arise due to design errors or manufacturing tolerances. Another source of mismatch is the source and load impedances. If these impedances are not exactly the same as the amplifier input or output impedance, respectively, then a mismatch will occur (see Chapter 5).

An impedance mismatch at either the input or the output of the single-stage amplifier will result in a gain mismatch loss (ML) of

$$ML = -10 \log \left[1 - \left(\frac{SWR - 1}{SWR + 1} \right)^2 \right] \qquad (14\text{-}6)$$

where

ML   = mismatch loss in decibels (dB)
SWR = standing wave ratio (dimensionless)

EXAMPLE 14-5

A 13-dB gain amplifier has an input impedance of 100 Ω and is driven from a 50-Ω transmission line. Assuming that the source is matched properly to the line, the SWR is $Z_{in}/Z_o = 100/50 = 2$:1. Find the mismatch loss of this system.

Solution

$$ML = -10 \log \left[1 - \left(\frac{SWR - 1}{SWR + 1} \right)^2 \right] dB$$

$$= -10 \left[1 - \left(\frac{2 - 1}{2 + 1} \right)^2 \right] dB$$

$$= -10 \left[1 - \left(\frac{1}{3} \right)^2 \right] dB$$

$$= -10 \log (1 - 0.11) dB$$

The actual gain of the 13-dB amplifier, therefore, is 13 – 0.51, or 12.49 dB, because of the 2:1 SWR in the system.

In a cascade amplifier, we have the distinct possibility of an impedance mismatch, and hence an SWR, at more than one point in the circuit. An example (Fig. 14-6) might be where neither the output impedance (R_o) of the driving amplifier

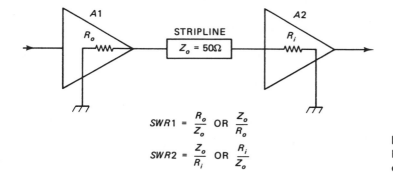

$$SWR1 = \frac{R_o}{Z_o} \text{ OR } \frac{Z_o}{R_o}$$

$$SWR2 = \frac{Z_o}{R_i} \text{ OR } \frac{R_i}{Z_o}$$

Figure 14-6
Higher-frequency cascade amplifiers.

($A1$) nor the input impedance (R_i) of the driven amplifier ($A2$) are matched to the 50-Ω (Z_o) microstrip line that interconnects the two stages. Thus, R_o/Z_o or its inverse forms one SWR, while R_i/Z_o or its inverse forms the other. For a two-stage cascade amplifier, the mismatch loss is

$$ML = 20 \log \left[1 - \left(\frac{SWR1 - 1}{SWR1 + 1} \right) \left(\frac{SWR2 - 1}{SWR2 + 1} \right) \right] \qquad (14\text{-}7)$$

EXAMPLE 14-6

An amplifier ($A1$) with a 25-Ω output resistance drives a 50-Ω stripline transmission line. The other end of the stripline is connected to the input of another amplifier ($A2$) in which $R_i = 100 \ \Omega$: (a) Calculate the maximum and minimum gain loss for this system, and (b) calculate the range of system gain if $G1 = 6$ dB and $G2 = 10$ dB.

Solution

$$SWR1 = Z_o/R_o = 50/25 = 2\text{:}1$$
$$SWR2 = R_i/Z_o = 100/50 = 2\text{:}1$$

$$ML1 = 20 \log \left[1 \pm \left(\frac{SWR1 - 1}{SWR1 + 1} \right) \left(\frac{SWR2 - 1}{SWR2 + 1} \right) \right]$$

$$= 20 \log \left[1 + \left(\frac{2 - 1}{2 + 1} \right) \left(\frac{2 - 1}{2 + 1} \right) \right] dB$$

$$= 20 \log \left[1 + (1/3)(1/3) \right] db$$

$$= 20 \log \left[1 + 0.11 \right] db$$

$$= 20 \log \left[1.11 \right] db$$

$$= (20)(0.045) = 0.91 db$$

$$ML1 = 20 \log (1 - 0.11) \text{ db}$$

$$= 20 \log (0.89) \text{ dB}$$

$$= (20)(-0.051) \text{ db}$$

$$= -1.02 \text{ db}$$

Thus,

$$ML1 = +1.11 \text{ dB}$$

$$ML2 = -1.02 \text{ dB}$$

Without the SWR, gain in decibels is $G = G1 + G2 = (6 \text{ dB}) + (10 \text{ dB}) = 16 \text{ dB}$. With SWR considered we find that $G = G1 + G2 \pm ML$. So

$$G_a = G1 + G2 + ML1$$

$$= (6 \text{ dB}) + (10 \text{ dB}) + (1.11 \text{ dB})$$

$$= 17.11 \text{ dB}$$

and

$$G_b = G1 + G2 + ML2$$

$$= (6 \text{ dB}) + (10 \text{ dB}) + (-1.02 \text{ dB})$$

$$= 14.98 \text{ dB}$$

The mismatch loss can vary from a negative loss resulting in less system gain (G_b) to a "positive loss" (which is actually a gain in its own right) resulting in greater system gain (G_a). The reason for this apparent paradox is that it is possible for a mismatched impedance to be connected to its complex conjugate impedance.

14-8 NOISE FIGURE IN CASCADE AMPLIFIERS

It is common practice in microwave systems, especially in communications and radar receivers, to place a low-noise amplifier (LNA) at the input of the system. The amplifiers that follow the LNA need not be of LNA design and so are less costly. The question is sometimes asked: "Why not use an LNA in each stage?" The answer can be deduced from Friis's equation:

$$NF_{total} = NF1 + \frac{NF2 - 1}{G1} = \frac{NF3 - 1}{G1G2} + \ldots + \frac{NF_n - 1}{G1G2 \ldots G_n} \qquad (14\text{-}8)$$

where

NF_{total} = system noise figure
$NF1$ = noise figure of stage 1

NF2   = noise figure of stage 2
NF_n = noise figure of stage n
G1 is the gain of stage 1
G2 is the gain of stage 2
G_n is the gain of stage n

All quantities are dimensionless ratios rather than decibels.

EXAMPLE 14-7

Two MIC amplifiers, $A1$ and $A2$, are connected in cascade. Amplifier $A1$ has a gain of +6 dB and a noise figure of 4 dB; amplifier $A2$ has a gain of +7 dB and a noise figure of 4.5 dB. Calculate the system noise figure.

Solution

Convert G1, G2, NF1, and NF2 to dimensionless ratios.

$$G1' = 10^{G1/10}$$
$$= 10^{6/10}$$
$$= 3.98$$
$$G2' = 10^{G2/10}$$
$$= 10^{7/10}$$
$$= 5.01$$
$$NF1' = 10^{NF1/10}$$
$$= 10^{4/10}$$
$$= 2.5$$
$$NF2' = 10^{NF2/10}$$
$$= 10^{4.5/10}$$
$$= 2.82$$

Calculate system noise figure using Friis's equation.

$$NF_{total} = NF1' + \frac{NF2' - 1}{G1}$$

$$= 2.5 + \frac{2.82 - 1}{3.98}$$

$$= 2.5 + 0.46 = 2.96$$

Reconvert to decibel notation if required.

$$NF(dB) = 10 \log (2.96)$$
$$= (10)(0.47) = 4.7 \text{ dB}$$

A lesson to be learned from this example is that the noise figure of the first stage in the cascade chain dominates the noise figure of the combination of stages. Note that the overall noise figure increased only a small amount when a second amplifier was used.

14-9 MINI-CIRCUITS MAR-X SERIES DEVICES: A MIC EXAMPLE

The Mini-Circuits Laboratories, Inc. MAR-x series of MIC devices offers gains from +13 dB to +20 dB and top-end frequency response of either 1 or 2 GHz, depending on type. The package used for the MAR-x device (Fig. 14-3A) is similar to the case used for modern UHF and microwave transistors. Pin No. 1 (RF input) is marked by a color dot and a bevel.

The usual circuit for the MAR-x series devices is shown in Fig. 14-4. The MAR-x device requires a voltage of +5 V dc on the output terminal and must derive this potential from a dc supply of greater than +7 V dc.

The RF choke (*RFC1*) is called optional in the engineering literature on the MAR-x, but it is recommended for applications where a substantial portion of the total bandpass capability of the device is used. The choke tends to pre-emphasize the higher frequencies, and thereby overcomes the de-emphasis normally caused by circuit capacitance; in traditional video amplifier terminology, that coil is called a peaking coil because of this action (that is, it peaks up the higher frequencies).

It is necessary to select a resistor for the dc power supply connection. The MAR-x device wants to see +5 V dc, at a current not to exceed 20 mA. In addition, *V*+ must be greater than +7 V. Thus, we need to calculate a dropping resistor (R_d) of

$$R_d = \frac{(V+) - 5\,\text{V dc}}{I} \qquad (14\text{-}9)$$

where R_d is in ohms and I is in amperes.

In an amplifier designed for +12-V dc operation, we might select a trial bias current of 15 mA (0.015 A). The resistor value calculated is 467 Ω (a 470-Ω, 5% film resistor should prove satisfactory). As recommended previously, a 1-dB attenuator pad is inserted in the input and output lines. The V+ is supplied to this chip through the output terminal.

14-10 COMBINING MIC AMPLIFIERS IN PARALLEL

Figures 14-7 through 14-10 show several MIC applications involving parallel combinations of MIC devices. Perhaps the simplest of these is the configuration of Fig. 14-7. Although each input must have its own dc blocking capacitor to protect the MIC device's internal bias network, the outputs of two or more MICs may be connected in parallel and share a common power supply connection and output coupling capacitor.

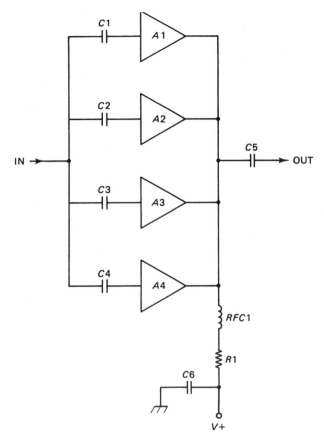

Figure 14-7
Parallel MMIC amplifiers to
increase output power.

Several advantages are realized with the circuit of Fig. 14-7. First, the power output increases even though total system gain (P_o/P_m) remains the same. As a consequence, however, drive power requirements also increase. The output power increases 3 dB when two MICs are connected in parallel and 6 dB when four are connected (as shown). The 1-dB output power compression point also increases in parallel amplifiers in the same manner: 3 dB for two amplifiers and 6 dB for four amplifiers in parallel.

The input impedance of a parallel combination of MIC devices reduces to R_i/N, where N is the number of MIC devices in parallel. In the circuit shown in Fig. 14-7, the input impedance would be $R_i/4$ or 12.5 Ω if the MICs are designed for 50-Ω service. Because 50/12.5 represents a 4:1 SWR, some form of input impedance matching must be used. Such a matching network can be either broadbanded or frequency specific as the need dictates. Figures 14-8 through 14-9D show methods for accomplishing impedance matching.

The method shown in Fig. 14-8 is used at operating frequencies up to about 100 MHz and is based on broadband ferrite toroidal RF transformers. These transformers dominate the frequency response of the system because they are less broadbanded than the usual MIC device. This type of circuit can be used as a gain block in microwave receiver IF amplifiers (which are frequently in the 70-MHz region), or in the exciter section of master oscillator power amplifier (MOPA) transmitters.

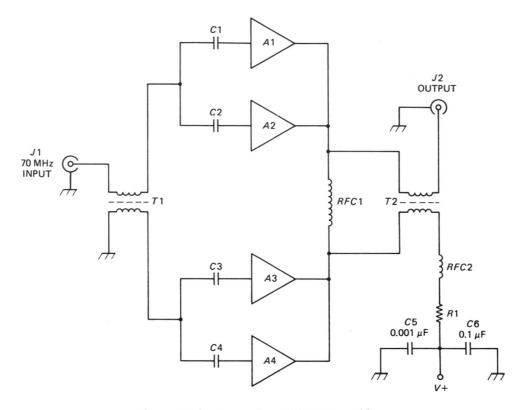

Figure 14-8 Push-pull parallel MMIC amplifiers.

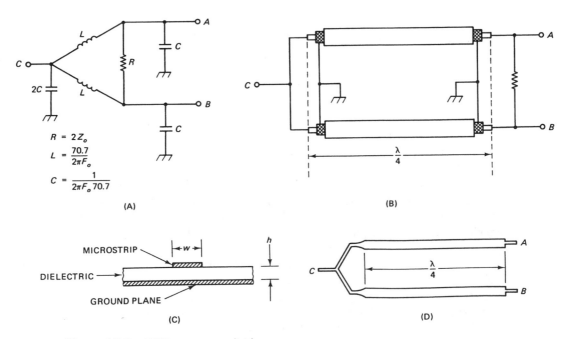

$$R = 2Z_o$$

$$L = \frac{70.7}{2\pi F_o}$$

$$C = \frac{1}{2\pi F_o 70.7}$$

(A)

(B)

(C)

(D)

Figure 14-9 Wilkinson power dividers.

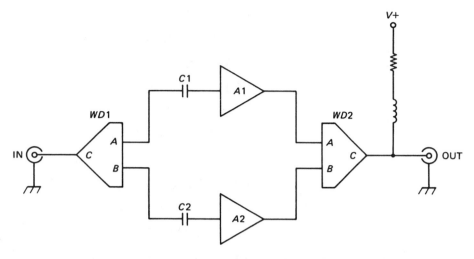

Figure 14-10   Use of Wilkinson dividers to combine MMIC amplifiers.

Another method is more useful in the UHF and microwave regions. In Fig. 14-9, we see several forms of the *Wilkinson power divider* circuit. An *LC* network version is shown here for comparison, although coaxial or stripline transmission line versions are used more often in microwave applications. The *LC* version is used to frequencies of 150 MHz. This circuit is bidirectional and so can be used as either a *power splitter* or a *power divider*. RF power applied to port C is divided equally between port A and port B. Alternatively, power applied to ports A/B are summed together and appear at port C. The component values are found from the following relationships:

$$R = 2Z_o \qquad\qquad (14\text{-}10)$$

$$L = \frac{70.7}{2\pi F_o} \qquad\qquad (14\text{-}11)$$

$$C = \frac{1}{2\pi\,70.7F_o} \qquad\qquad (14\text{-}12)$$

where R is in ohms, L is in henrys (H), C is in farads (F), and F_o is in hertz (Hz).

EXAMPLE 14-8

An IF output amplifier requires a Wilkinson power divider such as Fig. 14-9A. Calculate the resistance, capacitance, and inductance values needed for this circuit to work in a 50-Ω system at 70 MHz.

$$DE = 1. \ R = 2Z_o$$
$$= (2)(50 \ \Omega) = 100 \ \Omega$$
$$2. \ L = 70.7/(2\pi F_o)$$
$$= 70.7/(2\pi(7 \times 10^7))$$
$$= 70.7/(4.4 \times 10^8)$$
$$= 1.61 \times 10^{-7} \ H = 0.161 \ \mu H$$
$$3. \ C = 1/(2\pi \ 70.7 \ F_o)$$
$$= 1/((2)(\pi)(70.7)(7 \times 10^7))$$
$$= 1/(3.11 \times 10^{10})$$
$$= 3.22 \times 10^{-11} \ F = 32.2 \ pf$$

Figure 14-9B shows a coaxial cable version of the Wilkinson divider that can be used at frequencies up to 2 GHz. The lower frequency limit is set by practicality because the transmission line segments become too long to be handled easily. The upper frequency limit is set by the practicality of handling very short lines and by the dielectric losses, which are frequency dependent. The transmission line segments are each quarter-wavelength; their length is found from

$$L = \frac{2952 \ V}{F} \tag{14-13}$$

where

L = physical length of the line in inches (in.)
F = frequency in megahertz (MHz)
V = velocity factor of the transmission line (0 to 1)

EXAMPLE 14-9

Calculate the length of Teflon® transmission line ($V = 0.77$) required to make a Wilkinson divider at 400 MHz.

Solution

$$L = \frac{2952 \ V}{F}$$

$$= \frac{(2952)(0.77)}{400 \ MHz}$$

$$= \frac{2273}{400} = 5.68 \ in.$$

An impedance transformation can take place across a quarter-wavelength transmission line if the line has a different impedance than the source or load impedances being matched. Such an impedance-matching system is often called a Q section. The required characteristic impedance for the transmission line is found from

$$Z'o = \sqrt{Z_L Z_O} \tag{14-14}$$

where

$Z_o' =$ characteristic impedance of the quarter-wavelength section
$Z_L =$ load impedance
$Z_o =$ system impedance (for example, 50 Ω)

EXAMPLE 14-10

Calculate the impedance required of a quarter-wavelength transmission line Q section used to transform 50 Ω to 100 Ω.

Solution

$$Z'_o = \sqrt{Z_L Z_o}$$

$$= \sqrt{(100)(50)}$$

$$= \sqrt{5000} = 70.7\Omega$$

In the case of parallel MIC devices, the nominal impedance at port C of the Wilkinson divider is one-half of the reflected impedance of the two transmission lines. For example, if the two lines are each 50-Ω transmission lines, then the impedance at port C is 50/2 Ω, or 25 Ω. Similarly, if the impedance of the load (the reflected impedance) is transformed to some other value, then port C sees the parallel combination of the two transformed impedances. In the case of a parallel MIC amplifier, we might have two devices with 50-Ω input impedance each. Placing these devices in parallel halves the impedance to 25 Ω, which forms a 2:1 SWR with a 50-Ω system impedance. But if the quarter-wavelength transmission line transforms the 50-Ω input impedance of each device to 100 Ω, then the port C impedance is 100/2, or 50 Ω, which is correct.

At the upper end of the UHF spectrum and in the microwave spectrum, it may be better to use a stripline transmission line instead of coaxial cable. A stripline (see Fig. 14-9C) is formed on a printed circuit board. The board must be double-sided so that one side can be used as a ground plane, while the stripline is etched into the other side. The length of the stripline depends on the frequency of operation; either half-wave or quarter-wave lines are usually used. The impedance of the stripline is a function of three factors: (1) stripline width (w), (2) height of the stripline above the groundplane (h), and (3) dielectric constant (ε) of the printed circuit material:

$$Z_o = 377 \frac{h}{w\sqrt{\varepsilon}} \tag{14-15}$$

where
h = height of the stripline above the ground plane
w = width of the stripline (h and w in same units)
Z_o = characteristic impedance in ohms

EXAMPLE 14-11

Calculate the characteristic impedance of a 0.125-in. stripline on 3/16-in. (0.1875-in.) printed circuit material that has a dielectric constant of 4.75.

Solution

$$Z_o = 377 \frac{h}{w\sqrt{\varepsilon}}$$

$$377 \frac{0.1875 \text{ in.}}{0.125 \text{ in.} \sqrt{4.75}}$$

$$= 377 \frac{0.1875}{(0.125)(2.18)} = 259 \, \Omega$$

The stripline transmission line is etched into the printed circuit board as in Fig. 14-9D. Stripline methods use the printed wiring board to form conductors, tuned circuits, and the like. In general, for microwave operation the conductors must be very wide (relative to their simple dc and RF-power carrying size requirements) and very short in order to reduce lead inductance. Certain elements, the actual striplines, are transmission line segments and follow transmission line rules. The printed circuit material must have a *large permittivity* and a *low loss tangent*. For frequencies up to about 3 GHz, it is permissible to use ordinary glass-epoxy double-sided board ($\varepsilon = 5$), but for higher frequencies a low-loss material such as *Rogers Duroid* ($\varepsilon = 2.17$) must be used.

When soldering connections in these amplifiers, it is important to use as little solder as possible and keep the soldered surface as smooth and flat as possible. Otherwise, the surface wave on the stripline will be interrupted and operation will suffer.

Figure 14-10 shows a general circuit for a multi-MIC amplifier based on one of the previously discussed Wilkinson power dividers. Because the divider can be used as either splitter or combiner, the same type can be used as input (*WD1*) and output (*WD2*) terminations. In the case of the input circuit, port C is connected to the amplifier main input, while ports A/B are connected to the individual inputs of the MIC devices. At the output circuit, another divider (*WD2*) is used to combine power output from the two MIC amplifiers and direct it to a common output connection. Bias is supplied to the MIC amplifiers through a common dc path at port C of the Wilkinson dividers.

14-11 SUMMARY

1.   Microwave integrated circuits (MIC) are monolithic or hybrid gain blocks that have (a) bandwidth from dc or near dc to the microwave range, (b) constant

input and output impedance over at least several octaves of frequency, and (c) only the bare essentials for connections (for example, input, output, ground, and sometimes dc power). Some MIC devices are unconditionally stable over the entire frequency range.

2. MICs have all components built in except for input and output coupling capacitors and an external bias resistor. For some types an external peaking inductor or *RF choke* is recommended. Narrowband or bandpass amplifiers can be built by using tuning devices or frequency selective filtering in the input and/or output circuits.

3. MIC devices can be connected in cascade in order to increase overall system gain. Care must be taken to ensure that the SWR is low; that is, the impedances are matched. Otherwise, matching must be provided or gain errors tolerated.

4. The noise figure in a cascade MIC amplifier is dominated by the noise figure of the input amplifier stage.

5. MICs can be connected in parallel. The output terminals may be directly connected in parallel, but the input terminals of each device must be provided with its own dc blocking capacitor. MICs can also be combined using broadband RF transformers, Wilkinson power dividers/splitters, or other methods.

6. Printed circuit boards used for MIC devices must have a high permittivity and a low-loss tangent. For frequencies up to 3 GHz, glass-epoxy is permissible, but for higher frequencies a low-loss product is required.

14-12 RECAPITULATION

Now return to the objectives and prequiz questions at the beginning of the chapter and see how well you can answer them. If you cannot answer certain questions, place a check mark by each and review the appropriate parts of the text. Next, try to answer the following questions and work the problems using the same procedure.

QUESTIONS AND PROBLEMS

1. What condition may result in a microwave amplifier if the loop gain is unity or greater, and the feedback is inadvertently in phase with the input signal?

2. On-chip interconnections in MIC devices are usually made with _____ transmission lines.

3. What type of semiconductor material is usually used to make the active devices in MIC amplifiers?

4. A MIC amplifier contains a common-emitter Darlington amplifier transistor and internal _____ and _____ negative feedback elements.

5. In a common-emitter MIC amplifier, the _____ ratio must be kept high in order to obtain higher gain, higher output power compression points, and lower noise figure.

6. A common-emitter amplifier inside a MIC device has a collector-base resistor of 450 Ω, and an emitter resistor of 12.5 Ω. Is this amplifier a good impedance match for 75-Ω cable television applications?

7. Calculate the input and output impedances of a MIC amplifier in which the series feedback resistor is 5 Ω and the shunt feedback resistor is 550 Ω.

8. Calculate the gain of the amplifiers in questions 6 and 7.

9. Two transistors are connected in a Darlington amplifier configuration. Q1 has a beta of 100, and Q2 has a beta of 75. What is the beta gain of the Darlington pair?

10. A MIC amplifier must be operated over a frequency range of 0.5 to 2.0 GHz. The dc power circuit consists of an RF choke and a bias resistor in series between $V+$ and the MIC output terminal. If the resistor has a value of 220 Ω, what is the minimum appropriate value of the RF choke required to achieve an impedance of at least 500 Ω?

11. An MCL MAR-1 must operate from a +12-V dc power supply. This device requires a +5-V dc supply and has an optimum current of 17 mA. Calculate the value of the bias resistor.

12. A MIC amplifier has an optimum current of 22 mA and requires +7 V dc at the output terminal. Calculate the bias resistor required to operate the device from

    (a) 12 V dc, (b) 9 V dc.

13. Because they operate into the microwave region, MIC devices are packaged in _____ cases.

14. To overcome problems associated with changes of source and load impedance, it is sometimes the practice to insert resistive _____ in series with the input and output signal paths of a MIC device.

15. The SWR in a microwave amplifier should be less than _____ to make accurate gain measurements.

16. Calculate the mismatch loss in a single-stage MIC amplifier if the output impedance is 50 Ω and the load impedance is 150 Ω.

17. Calculate the mismatch loss in a single-stage MIC amplifier if the input SWR is 1.75:1.

18. Calculate the mismatch loss in a cascade MIC amplifier if the output impedance of the driver amplifier and the input impedance of the final amplifier are both 50 Ω, but they are interconnected by a 90-Ω transmission line. Remember that there are two values.

19. A cascade MIC amplifier has two SWR mismatches: SWR1 = 2:1 and SWR2 = 2.25:1. Calculate both values of mismatch loss.

20. The amplifiers in question 19 have the following gains: $A1 = 16$ dB and $A2 = 12$ dB. Calculate the range of gain that can be expected from the cascade combination $A1 \times A2$, considering the possible mismatch losses.

21. An MIC amplifier has two stages, $A1$ and $A2$. A1 has a gain of 20 dB and a noise figure of 3 dB; $A2$ has a gain of 10 dB and a noise figure of 4.5 dB. Calculate the noise figure of the system (a) when $A1$ is the input amplifier and $A2$ is the output amplifier, and (b) when $A2$ is the input amplifier and $A1$ is the output amplifier. What practical conclusions do you draw from comparing these results?

22. Two MIC amplifiers are connected in parallel. What is the total input impedance if each amplifier has an input impedance of 75 Ω?

23. Four MIC amplifiers are connected in parallel. The output compression point is now (raised/lowered) _____dB.

24. An LC Wilkinson power divider must be used to connect two MIC devices in parallel. The system impedance, Z_o, is 50 Ω, and each MIC amplifier has an impedance of 50 Ω. Calculate the values of R, L, and C required for this circuit.

25. A quarter-wavelength transmission line for 1.296 GHz must be made from Teflon® coaxial cable that has a dielectric constant of 0.77. What is the physical length in inches?

26. A 75-Ω load must be transformed to 125 Ω in a quarter-wave Q section transmission line. Calculate the required characteristic impedance required for this application.

27. A stripline transmission line is being built on a double-sided printed circuit board made with low-loss ($\varepsilon = 2.5$) material. If the board is 0.125 in. thick, what width must be stripline be in order to match 70.7 Ω?

KEY EQUATIONS

1. Input/output impedance of MIC internal amplifier:

$$R_o = \sqrt{R_f \times R_e}$$

2. Gain in MIC internal amplifier:

$$G_{dB} = 20 \log \left[\frac{R_f - R_e}{R_o = R_e} \right]$$

3. Beta gain of a Darlington amplifier pair:

$$\beta_o = \beta_{Q1} \times \beta_{Q2}$$

4. Value of MIC bias resistor:

$$R_{bias} = \frac{(V+) - V}{I_{bias}}$$

5. Gain in a cascade amplifier:

$$G = G1 \times G2 \times G3 \times \ldots \times Gn$$

6. Mismatch loss due to SWR in a single-stage amplifier:

$$ML = -10 \log \left[1 - \left(\frac{SWR - 1}{SWR + 1} \right)^2 \right]$$

7. Mismatch loss due to multiple SWR sources in cascade amplifiers:

$$ML = 20 \log \left[1 \pm \left(\frac{SWR1 - 1}{SWR1 + 1} \right) \left(\frac{SWR2 - 1}{SWR2 + 1} \right) \right]$$

8. Noise figure of a cascade amplifier:

$$NF_{total} = NF1 + \frac{NF2 - 1}{G1} + \frac{NF3 - 1}{G1G2} + \ldots + \frac{NFn - 1}{G1G2Gn}$$

9. Component values in the LC version of the Wilkinson power divider:

$$R = 2Z_o$$

$$L = \frac{70.7}{2\pi F_o}$$

$$C = \frac{1}{2\pi \, 70.7 F_o}$$

10. Physical length of a quarter-wavelength coaxial cable section:

$$L = \frac{2952V}{F}$$

11. Impedance required of a quarter-wavelength Q matching section:

$$Z'_o = \sqrt{Z_L Z_O}$$

12. Characteristic impedance of a stripline section:

$$Z_o = 377 \, \frac{h}{w\sqrt{\varepsilon}}$$

CHAPTER 15

Microwave Diodes

OBJECTIVES

1. Learn the history and evolution of microwave diode development.
2. Understand the limitations of ordinary *PN* junction diodes that prevent their operation at microwave frequencies.
3. Learn the operation of point contact, Schottky, varactor, *PIN*, and tunnel diodes at microwave frequencies.
4. Be introduced to the concept of *negative resistance*.

15-1 PREQUIZ

These questions test your prior knowledge of the material in this chapter. Try answering them before you read the chapter. Look for the answers (especially those you answered incorrectly) as you read the text. After you have finished studying the chapter, try answering these questions again and those at the end of the chapter.

1. The earliest diodes used at microwave frequencies were of the _____ _____ type of construction, such as types 1N21 and 1N23.
2. The _____ diode, also called the Esaki diode, has a negative resistance *I* versus *V* characteristic.
3. A _____ diode uses a lightly doped intrinsic semiconductor sandwiched between *P* and *N* regions.
4. An Esaki diode can be used in the astable mode (as an oscillator) by virtue of its _____ _____ characteristic.

15-2 EVOLUTION OF MICROWAVE DIODES

Di- means two and *–ode* is derived from electrode. Thus, the word *diode* refers to a two-electrode device. Although in conventional low-frequency, solid-state terminology the word diode indicates only *PN* junction diodes, in the microwave world we use both *PN* diodes and others as well. Some diodes, for example, are multiregion devices, while others have but a single semiconductor region. The latter depend on bulk effect phenomena to operate. In this chapter and Chapter 16, we will examine some of the various microwave devices that are available on the market. The student is also referred to Chapter 3 for an elementary review of *PN* junction diode operation.

World War II created a huge leap in microwave capabilities because of the needs of the radar community. Even though solid-state electronics was in its infancy, developers were able to produce crude microwave diodes for use primarily as mixers in superheterodyne radar receivers. Theoretical work done in the late 1930s, 1940s, and early 1950s predicted microwave diodes that would be capable not only of nonlinear mixing in superheterodyne receivers, but of amplification and oscillation as well. These predictions were carried forward by various researchers, and by 1970 a variety of commercial devices were on the market.

15-3 EARLY MICROWAVE DIODES

The development of solid-state microwave devices suffered the same sort of problems as microwave vacuum tube development. Just as electron transit time limits the operating frequency of vacuum tubes, a similar phenomenon affects semiconductor devices. The transit time across a region of semiconductor material is determined not only by the thickness of the region, but also by the *electron saturation velocity*, about 10^7 cm/s, which is the maximum velocity attainable in a given material. In addition, minority carriers are in a storage state that prevents them from being easily moved, causing a delay that limits the switching time of the device. Also, like the vacuum tube, interelectrode capacitance limits the operating frequency of solid-state devices. Although turned to an advantage in variable-capacitance diodes (see Section 15-5), capacitance between the *N* and *P* regions of a *PN* junction diode seriously limits operating frequency and/or switching time.

During World War II, microwave diodes such as the 1N21 and 1N23 devices were developed. The capacitance problem was partially overcome by using the *point-contact* method of construction (shown in Fig. 15-1). The point-contact diode uses a cat's whisker electrode to effect a connection between the conductive electrode and the semiconductor material.

Point-contact construction is a throwback to the very earliest days of radio when a commonly used detector in radio receivers was the naturally occurring *galena crystal* (a lead compound). Those early crystal set radios used a cat's whisker electrode to probe the crystal surface for the spot where rectification of radio signals could occur.

The point-contact diode of World War II was used mainly as a mixer element to produce an intermediate-frequency (IF) signal that was the difference between

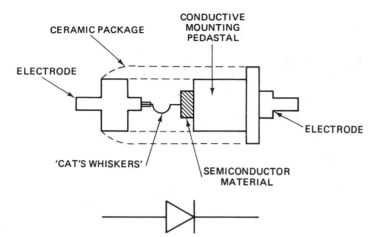

CONDUCTIVE
MOUNTING
PEDASTAL

CERAMIC PACKAGE

ELECTRODE

ELECTRODE

'CAT'S WHISKERS'

SEMICONDUCTOR
MATERIAL

Figure 15-1
Point-contact diode.

a radar signal and a local oscillator (LO) signal. These early superheterodyne radar receivers did not use RF amplifier stages ahead of the mixer, as was common in lower-frequency receivers. Although operating frequencies to 200 GHz have been achieved in point-contact diodes, the devices are fragile and are limited to very low power levels.

15-4 SCHOTTKY BARRIER DIODES

The Schottky diode (also called the hot carrier diode) is similar in mechanical structure to the point-contact diode, but is more rugged and is capable of dissipating more power. The basic structure of the Schottky diode is shown in Fig. 15-2A, while the equivalent circuit is shown in Fig. 15-2B. This form of microwave diode was predicted in 1938 by William Schottky during research into majority carrier rectification phenomena.

A *Schottky barrier* is a quantum mechanical potential barrier set up by virtue of a metal-to-semiconductor interface. The barrier is increased by increasing reverse-bias potential and reduced (or nearly eliminated) by forward-bias potentials. Under reverse-bias conditions, the electrons in the semiconductor material lack sufficient energy to breach the barrier, so no current flow occurs. Under forward-bias conditions, however, the barrier is lowered sufficiently for electrons to breach it and flow from the semiconductor region into the metallic region.

The key to junction diode microwave performance is its switching time, or switching speed as it is sometimes called. In general, the faster the switching speed (the shorter the switching time), the higher the operating frequency. In ordinary *PN* junction diodes, switching time is dominated by the minority carriers resident in the semiconductor materials. The minority carriers are ordinarily in a rest or storage state and so are neither easily nor quickly swept along by changes in the electric field potential. But majority carriers are more mobile and so respond rapidly to potential changes. Thus, Schottky diodes, by depending on majority

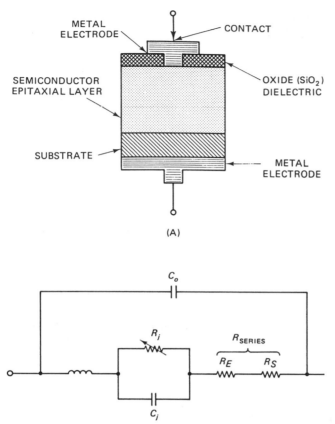

Figure 15-2
Schottky diode: A) structure;
B) equivalent circuit.

carriers for switching operation, operate with very rapid speeds, well into the microwave region for many models.

Operating speed is also a function of the distributed-circuit RLC parameters (see Fig. 15-2B). In Schottky diodes that operate into the microwave region, series resistance tends to be about 3 to 10 Ω, capacitance is around 1 picofarad (1 pF) and series lead inductance tends to be 1 nanohenry (1 nH) or less.

The Schottky diode is basically a rectifying element, so it finds a lot of use in mixer and detector circuits. The double-balanced mixer (DBM) circuit, which is used for a variety of applications, from heterodyning to the generation of double-sideband suppressed carrier signals to phase sensitive-detectors, is usually made from matched sets of Schottky diodes. Many DBM components work into the low end of the microwave spectrum.

15-5 VARACTOR DIODES

Diodes can be used as capacitors in microwave circuits. Special enhanced-capacitance diodes intended for this operation are called by several names, perhaps the most common of which are *varactor* (variable reactor) and *varicap*

Figure 15-3
Varactor diode: A) symbol;
B) typical package.

(A)   (B)

(variable capacitor). Although varactors are specially designed for use as electrically variable capacitors, all *PN* junction diodes exhibit the variable-junction capacitance phenomena to some extent. Even ordinary, low-leakage, 1-A silicon rectifier diodes have been used as low-frequency varactors in laboratory experiments.

Figure 15-3A shows the usual circuit symbol for varactors (although several other symbols are also used). In some cases, the capacitor at the top end of the diode symbol has an arrow through it to denote variable capacitance. Varactors come in several different standard diode packages, including the two-terminal type 182 package shown in Fig. 15-3B. Some variants bevel the edge of the package to indicate the cathode. In other cases, the package style will be like other forms of diode package.

15-5.1 How Do Varactors Work?

Varactors are specially made *PN* junction diodes designed to enhance the control of the *PN* junction capacitance with a reverse-bias voltage. A *PN* junction consists of *P*- and *N*-type semiconductor regions in juxtaposition with each other, as shown in Fig. 15-4. When the diode is forward biased (Fig. 15-4A), the charge carriers (electrons and holes) are forced to the junction interface, where positively charged holes and negatively charged electrons annihilate each other (causing a current to flow). But under reverse-bias situations (for example, those shown in Fig. 15-4B), the charges are drawn away from the junction, thereby forming a high-resistance *depletion zone* at the junction interface. Varactors operate under reverse-bias conditions.

The depletion zone is essentially an insulator between the two charge-carrying *P* and *N* regions, and this situation fulfills the criterion for a capacitor: two conductors separated by an insulator. The depletion zone width is increased by increasing the reverse-bias potential, which is analogous to increasing the separation between plates. Thus, the capacitance decreases. Similarly, reducing the reverse-bias potential reduces the depletion zone width, thereby increasing the capacitance.

The diode capacitance (C_j) as a function of applied reverse-bias potential (neglecting stray capacitances) is found from

$$C_j = \frac{CK}{(V_b - V)^m} \qquad (15\text{-}1)$$

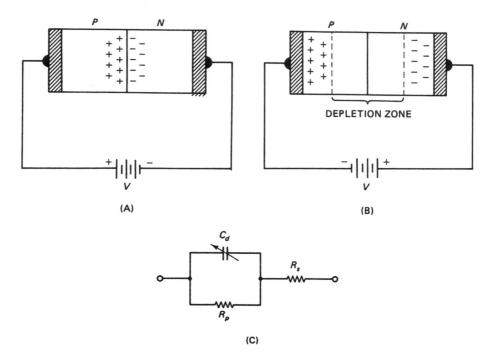

Figure 15-4 A) PN diode at high-capacitance bias; B) low-capacitance bias; C) equivalent circuit.

where

C_j = diode capacitance
C = diode capacitance when the device is unbiased ($V = 0$), expressed in the same units as C_j, and is a constant
V = applied reverse-bias potential
V_b = barrier potential at the junction (0.55 – 0.7 V in silicon)
m = constant that is material dependent (tends to be one-third to one-half)
K = a constant (often 1)

EXAMPLE 15-1

A silicon varactor has a zero-potential ($V = 0$) capacitance of 100 pF. If the m factor is 1/2, what is the capacitance when the applied voltage is –4 V dc? Assume a barrier potential of 0.65 V and $K = 1$.

Solution

$$C_j = \frac{CK}{(V_b - V)^m}$$

$$= \frac{(100 \, \text{pF})\,(1)}{[(0.65 \, \text{V} - (-4 \, \text{V})]^{1/2}}$$

$$= \frac{100 \, \text{pF}}{\sqrt{4.65}} = \frac{100 \, \text{pF}}{2.16} = 46.4 \, \text{pF}$$

A simplified equivalent circuit for a varactor diode is shown in Fig. 15-4C. The series resistance and diode capacitance determine the maximum operating frequency according to the relationship

$$F = \frac{1}{2\pi R_s C_j} \tag{15-2}$$

where

F = cutoff frequency in hertz
R_s = series resistance in ohms
C_j = diode capacitance in farads

EXAMPLE 15-2

A variable-capacitance diode has a capacitance of 5 pF at a certain reverse-bias potential and a series resistance of 12 Ω. Find the cutoff frequency in megahertz (MHz).

Solution

$$F = \frac{1}{2\pi R_s C_j}$$

$$= \frac{1}{(2)(3.14)(12\,\Omega)\left[5\,\text{pF} \times \left(\dfrac{1\,\text{farad}}{10^{12}\,\text{pF}}\right)\right]}$$

$$= \frac{1}{3.77 \times 10^{-10}}\,\text{Hz}$$

$$= 2.65 \times 10^9\,\text{Hz} \times \frac{1\,\text{MHz}}{10^6\,\text{Hz}} = 2,650\,\text{MHz}$$

Because maximum operating frequency is a function of diode capacitance, and diode capacitance is in turn a function of applied reverse-bias potential, we can conclude that the maximum operating frequency is also somewhat dependent on applied voltage.

The *quality factor* (Q) of a varactor diode is a function of the operating frequency and is given as the ratio of cutoff frequency (F) and operating frequency (f):

$$Q = \frac{F}{f} \tag{15-3}$$

EXAMPLE 15-3

A varactor diode has a cutoff frequency of 2450 MHz. What is the operating Q at a frequency of 54 MHz?

Solution

$$Q = F/f$$
$$= 2450 \text{ MHz}/54 \text{ MHz}$$
$$= 45.4$$

It is common practice for manufacturers to specify Q at a given test frequency such as 1 MHz. The actual Q at any given operating frequency can thus be quickly calculated from the preceding expression. It is also possible to guess the maximum operating frequency (F) if the Q and its test frequency are known.

Figure 15-5 shows a typical test circuit for the varactor. This circuit is also representative of typical operating circuits. A variable reverse-bias dc voltage is applied across the diode. A series resistor serves both to limit the current should the voltage exceed the avalanche or zener points (which could destroy the diode) and also to isolate the diode from the rest of the circuitry. Without a high-value resistor (10 kΩ to 1 MΩ is the normal range; 100 kΩ is typical) in series with the dc supply, stray circuit capacitances, the capacitance of C2, and the power supply output capacitance would swamp the typically low value of varactor capacitance. The capacitor at the output (C1) is used both to block dc bias and thereby prevent it from affecting other circuits and to prevent the dc in other circuits from affecting the diode. The value of this capacitor must be very large with respect to the diode capacitance to prevent it from affecting the total capacitance (C_t). The total capacitance is found from the standard series capacitor equation:

$$C_t = \frac{C1 \times C_d}{C1 + C_d} \qquad (15\text{-}4)$$

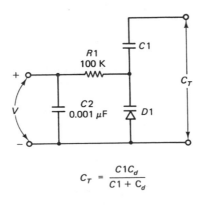

$$C_T = \frac{C1C_d}{C1 + C_d}$$

Figure 15-5
Dc tuning-bias circuit for varactors.

15-5.2 Varactor Applications

Because the varactor is a variable capacitor, it can be used in most applications where an ordinary mechanical capacitor of the same value might be used. For example, varactors are used to tune variable *LC* resonant tank circuits. Because the frequency of such a circuit is a function of the capacitance, it is also a function of the voltage in those circuits.

Various *LC* tank circuit applications are possible. For example, in a receiver *automatic frequency control* (AFC), we can use a varactor diode to pull the local oscillator (LO) signal to the correct point under command of an error signal generated by the detector; AFC circuits are commonplace in FM receivers. In both transmitters and receivers, the operating frequency can be set by a *phase-locked-loop* (PLL) frequency synthesizer (Fig. 15-6).

The main oscillator is a *voltage-controlled oscillator* (VCO) in which a varactor sets the operating frequency. When the frequency-divided VCO output signal strays away from the stable reference oscillator frequency, the *phase-sensitive detector* (PSD) issues an error signal to the varactor. The output of the PSD is integrated in a low-pass filter and sometimes scaled in a dc amplifier (not shown) before being applied to the varactor input of the VCO.

Varactors are also used as frequency modulators. Piezoelectric crystals are generally used in transmitters to set operating frequency. The frequency of oscillation for these crystals is a function of circuit capacitance, so by connecting a varactor either in series or parallel with the crystal, we can cause its frequency to change with changes in applied voltage. Thus, the audio or analog telemetry signal voltage can frequency modulate the crystal oscillator.

When an RF sine wave of a frequency $F = \sin(\omega t)$ is applied to a varactor, the current in the varactor obeys the relationship

$$I = I_1 \cos(\omega t) + I_2 \cos(2\omega t) + \ldots + I_n \cos(n\omega t)$$

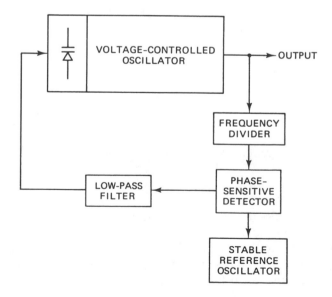

Figure 15-6
Varactor tuning of the voltage-controlled oscillator in a phase-locked-loop synthesizer.

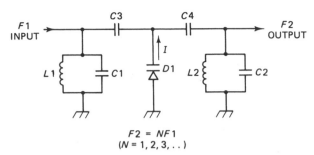

$$F2 = NF1$$
$$(N = 1, 2, 3, \ldots)$$

IF $V = \sin \omega t$
THEN $I = I_1 \cos \omega T + I_2 \cos 2\omega T + \ldots + I_N \cos N\omega T$

Figure 15-7
Varactor frequency multiplier.

From this relationship, you should be able to predict that the varactor diode can serve as a *frequency multiplier* by virtue of the fact that the current components are a function of the fundamental frequency and are integer multiples (2, 3, ..., n) thereof. A typical multiplier circuit is shown in Fig. 15-7. In this circuit, the input tank circuit ($L1C1$) is tuned to the fundamental frequency, while the output tank ($L2C2$) is tuned to an integer multiple ($N = 2, 3, \ldots n$) of the input frequency.

Varactor frequency multipliers are often used in microwave receiver local oscillator circuits. The actual oscillator frequency is a subharmonic of the required LO frequency. For example, consider a 2145-MHz microwave receiver designed to pick up Multipoint Distribution Service television signals (which send movie channel services to apartment buildings and the like). A typical down-converter for this channel translates the 2145-MHz signal to channels 5 or 6. If we assume that channel 6 is used, then the frequency is 82 MHz. The required LO frequency is (2145 MHz to 82 MHz), or 2063 MHz. The down-converter designer chose to use the fourth subharmonic or a frequency of 2063/4, or 515.75 MHz. A cascade chain of two 1:2 frequency multipliers, with an intervening amplifier to overcome circuit losses, provided the required LO signal to the down-converter front end.

Another common microwave application for varactors is as an RF *phase shifter*. Because the capacitance is a function of applied reverse-bias voltage, the varactor serves as a continuously variable analog phase shifter. The varactor will, however, handle only a few milliwatts of RF power.

15-6 PIN DIODES

The *P-I-N* or *PIN* diode is different from the *PN* junction diode (see Fig. 15-8): it has an insulating region between the *P*- and *N*-type materials. It is therefore a multiregion device despite having only two electrodes. The *I* region is not really a true semiconductor insulator, but rather is a very lightly doped *N* type region. It is called an *intrinsic* region because it has very few charge carriers to support the flow of an electrical current. When a forward-bias potential is applied to the *PIN* diode, charge carriers are injected into the *I* region from *N* and *P* regions. But the lightly doped design of the intrinsic region is such that the *N*- and *P*-type charge carriers do not immediately recombine (as in *PN* junction diodes). There is always a *delay*

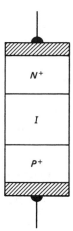

Figure 15-8
PIN diode structure.

period for recombination. Because of this delay phenomenon, there is always a small but finite number of carriers in the *I* region that are uncombined. As a result, the resistivity of the *I* region is very low.

Radio-frequency ac signals can pass through the *PIN* device and, in fact, under some circumstances see it as merely a parallel plate capacitor. We can use *PIN* diodes as electronic switches for RF signals and as an RF delay line or phase shifter, or as an amplitude modulator.

15-6.1 *PIN* Diode Switch Circuits

PIN diodes can be used as switches in either series or parallel modes. Figure 15-9 shows three switch circuits. In the series circuit (Fig. 15-9A), the diode (*D1*) is placed in series with the signal line. When the diode is turned on, the signal path has a low resistance, and when the diode is turned off, it has a very high resistance (thus providing the switching action). When switch *S1* is open, the diode is unbiased, so the circuit is open by virtue of the very high series resistance. But when *S1* is closed, the diode is forward biased and the signal path is now a low resistance. The ratio of off/on resistances provides a measure of the isolation of the circuit.

Figure 15-9B shows the circuit for a shunt switch. In this case, the diode is placed across the signal line. When the diode is turned off, the resistance across the signal path is high, so operation of the circuit is unimpeded. But when the diode is turned on (*S1* closed) a near short circuit is placed across the line. This type of circuit is turned off when the diode is forward biased. This action is in contrast to the series switch, in which a forward-biased diode is used to turn the circuit on.

A combination series-shunt circuit is shown in Fig. 15-9C. In this circuit, *D1* and *D2* are placed in series with the signal line, while *D3* is in parallel with the line. *D1* and *D2* will turn on with a positive potential applied, while *D3* turns on when a negative potential is applied. When switch *S1* is in the on position, a positive potential is applied to the junction of the three diodes. As a result, *D1* and *D2* are forward biased and thus take on a low resistance. At the same time, *D3* is hard reverse biased and so has a very high resistance. Signal is passed from input to output essentially unimpeded (most *PIN* diodes have a very low series resistance).

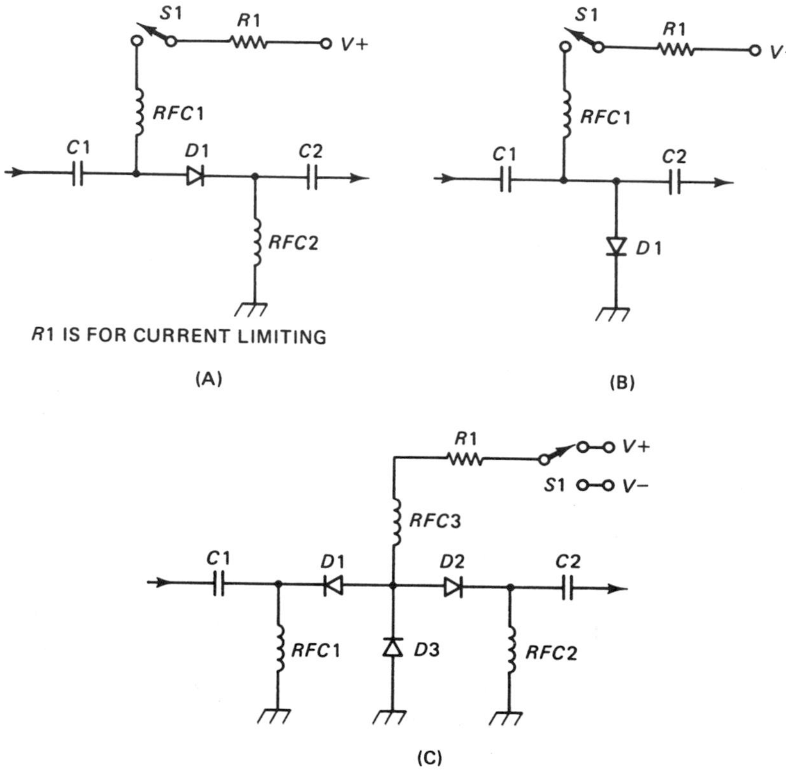

Figure 15-9 *PIN* diode switches.

But when *S*1 is in the off position, the opposite situation obtains. In this case, the applied potential is negative, so *D*1/*D*2 are reverse biased (and take on a high series resistance), while *D*3 is forward biased (and takes on a low series resistance). This circuit action creates a tremendous attenuation of the signal between input and output.

15-6.2 *PIN* Diode Applications

PIN diodes can be used either as variable resistors or as electronic switches for RF signals. In the latter case, the diode is basically a two-valued resistor, with one value being very high and the other being very low. These characteristics open several possible applications.

When used as switches, *PIN* diodes can be used to switch devices such as attenuators, filters, and amplifiers in and out of the circuit. It has become standard practice in modern radio equipment to switch dc voltages to bias *PIN* diodes, rather than directly switch RF/IF signals. In some cases, the *PIN* diode can be used to simply short out the transmission path to bypass the device.

Another application for *PIN* diodes is as voltage-variable *attenuators* in RF circuits. Because of its variable-resistance characteristic, the *PIN* diode can be used in a variety of attenuator circuits. Perhaps the most common is the bridge circuit, which is similar to a balanced mixer/modulator.

The *PIN* diode will also work as an *amplitude modulator*. In this application, a *PIN* diode is connected across a transmission line or inserted into one end of a piece of microwave waveguide. The audio-modulating voltage is applied through an RF choke to the *PIN* diode. When a CW signal is applied to the transmission line, the varying resistance of the *PIN* diode causes the signal to be amplitude modulated.

Our final application is shown in Fig. 15-10. Here we have a pair of *PIN* diodes used as a transmit-receive (TR) switch in a radio transmitter; models from low HF to microwave use this technique. Where you see a "relayless TR switch," it is almost certain that a *PIN* diode network such as Fig. 15-10 is in use. When switch S1 is open, diodes D1 and D2 are unbiased and so present a high impedance to the signal. Diode D1 is in series with the transmitter signal, so it blocks it from reaching the antenna; diode D2, on the other hand, is across the receiver input, so it does not attenuate the receiver input signal at all. But when switch S1 is closed, the opposite situation occurs: both D1 and D2 are now forward biased. Diode D1 is now a low resistance in series with the transmitter output signal, so the transmitter is effectively connected to the antenna. Diode D2 is also a low resistance and is across the receiver input, so it causes it to short out.

The isolation network can be a quarter-wavelength transmission line, microstrip line designed into the printed circuit board, or an *LC* pi-section filter. Transmitters up to several kilowatts have been designed using this form of switching. Almost all current VHF/UHF portable "handi-talkies" use *PIN* diode switching.

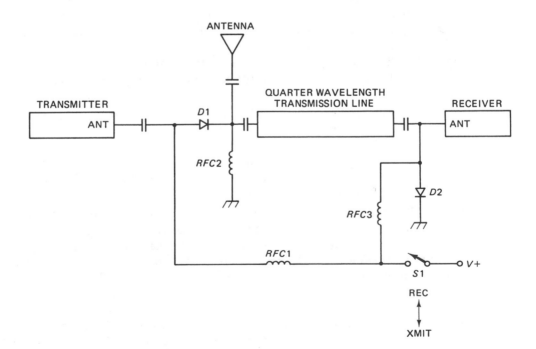

Figure 15-10 *PIN* diode TR switching in transceiver system.

15-7 TUNNEL DIODES

The *tunnel diode* was predicted by L. Esaki of Japan in 1958 and is often referred to as the *Esaki diode*. The tunnel diode is a special case of a *PN* junction diode in which carrier doping of the semiconductor material is on the order of 10^{19} or 10^{20} atoms per cubic centimeter, or about 10 to 1000 times higher than the doping used in ordinary *PN* diodes. The potential barrier formed by the depletion zone is very thin, on the order of 3 to 100 angstroms, or about 10^{-6} cm.

The tunnel diode operates on the quantum mechanical principle of *tunneling* (hence its name), which is a majority carrier event. Ordinarily, electrons in a *PN* junction device lack the energy required to climb over the potential barrier created by the depletion zone under reverse-bias conditions. The electrons are thus said to be trapped in a *potential well*. The current, therefore, is theoretically zero in that case. But if the carriers are numerous enough and the barrier zone thin enough, tunneling will occur. That is, electrons will disappear on one side of the barrier (inside the potential well) and then reappear on the other side of the barrier to form an electrical current. In other words, they appear to have tunneled under or through the potential barrier in a manner that is said to be analogous to a prisoner tunneling through the wall of the jailhouse.

The student should not waste a lot of energy attempting to explain the tunneling phenomenon, for indeed even theoretical physicists do not understand it very well. The tunneling explanation is only a metaphor for a quantum event that has no corresponding action in the nonatomic world. The electron does not actually go under, over, or around the potential barrier, but rather it disappears here and then reappears elsewhere. While the reasonable person is not comfortable with such seeming magic, until more is known about the quantum world we must be content with accepting tunneling as merely a construct that works for practical purposes. All that we can state for sure about tunneling is that, if the right conditions are satisfied, there is a certain *probability per unit of time* that tunneling will occur.

Three basic conditions that must be met to ensure a high probability of tunneling:

1. Heavy doping of the semiconductor material to ensure large numbers of majority carriers.
2. Very thin depletion zone (3 to 100 A).
3. For each tunneling carrier there must be a filled energy state from which it can be drawn on the source side of the barrier and an available empty state of the same energy level to which it can go on the other side of the barrier.

Figure 15-11 shows an *I* versus *V* curve for a tunnel diode. Note the unusual shape of this curve. Normally, in familiar materials that obey Ohm's law, the current (*I*) will increase with increasing potential (*V*), and decrease with decreasing potential.

Such behavior is said to represent *positive resistance* (+*R*) and is the behavior normally found in common conductors. But in the tunnel diode, there is a *negative-resistance zone* (NRZ in Fig. 15-11) in which *increasing potential forces a decreasing current*.

Points *A*, *B*, and *C* in Fig. 15-11 show us that tunnel diodes can be used in monostable (*A* and *C*), bistable (flip-flop from *A* to *C* or *C* to *A*), or astable (oscillate about *B*) modes of operation, depending on applied potential. These modes give us

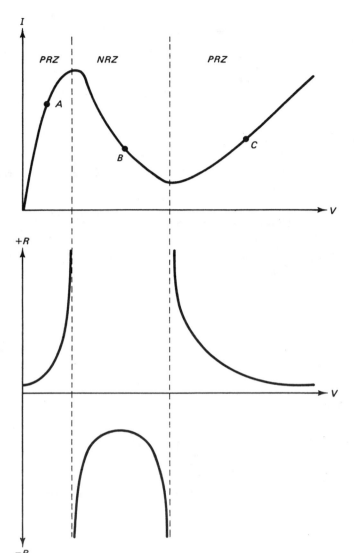

Figure 15-11
Tunnel diode current and resistance curves.

some insight into typical applications for the tunnel diode: *switching*, *oscillation*, and *amplification*. In Chapter 16, we will look at other negative-resistance oscillators, so keep in mind that a negative-resistance (-R) property indicates the ability to oscillate in *relaxation oscillator* circuits. Unfortunately, the power output levels of the tunnel diode are restricted to only a few milliwatts because the applied dc potentials must be less than the *bandgap potential* for the specific diode.

Figure 15-12A shows the usual circuit symbol for the tunnel diode, and Fig. 15-12B shows a simplified but adequate equivalent circuit. The tunnel diode can be modeled as a negative-resistance element shunted by the junction capacitance of the diode. The series resistance is the sum of contact resistance, lead resistance, and bulk material resistances in the semiconductor material, while the series inductance is primarily lead inductance.

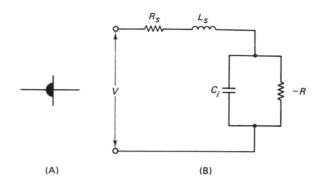

(A)   (B)

Figure 15-12
Tunnel diode equivalent circuit.

The tunnel diode will oscillate if the total reactance is zero ($X_L - X_C = 0$) and if the negative resistance balances out the series resistance. We can calculate two distinct frequencies with regard to normal tunnel diodes: *resistive frequency* (F_r) and the *self-resonant frequency* (F_s). These frequencies are given by

Resistive frequency

$$F_r = \frac{1}{2\pi RC_j} \sqrt{\frac{R}{R_s} - 1} \qquad (15\text{-}5)$$

Self-resonant frequency

$$F_s = \frac{1}{2\pi} \sqrt{\frac{1}{L_s C_j} - \frac{1}{(RC_j)^2}} \qquad (15\text{-}6)$$

where

F_r = resistive frequency in hertz (Hz)
F_s = self-resonant frequency in hertz (Hz)
R = absolute value of the negative resistance in ohms
R_s = series resistance in ohms
C_j = junction capacitance in farads
L_s = series inductance in henrys

EXAMPLE 15-4

Calculate the resistive frequency of a tunnel diode that has a −25-Ω negative resistance, 21 pF of junction capacitance, and 1.25 Ω of series resistance.

Solution

$$F_r = \frac{1}{2\pi RC_j} \sqrt{\frac{R}{R_s} - 1}$$

$$= \frac{1}{(2)(3.14)(25\,\Omega)(2.1 \times 10^{-11}\,F)} \times \sqrt{\frac{25\,\Omega}{1.25\,\Omega} - 1}$$

$$= \frac{1}{3.3 \times 10^{-19}} \sqrt{19}$$

$$= (3.03 \times 10^8)(4.4) = 1.33 \times 10\ Hz = 1330\ \text{MHz}$$

15-8 SUMMARY

1. The upper operating frequency of conventional *PN* diodes is limited by electron transit time, electron saturation velocity, and junction capacitance. However, point-contact construction allows operation at microwave frequencies.

2. The Schottky diode, also called the hot carrier diode, is a special case of the point-contact type of construction. By using majority carrier rectification phenomena, the Schottky diode greatly increases operating speed.

3. Varactors are voltage-variable capacitance diodes that can be used as tuning elements, frequency multipliers, and the like. The junction capacitance of the varactor varies under reverse-bias conditions.

4. The *PIN* diode is a multiregion device in which a lightly doped *intrinsic region* is sandwiched between *P* and *N* regions. The *PIN* diode can be used as a switch, attenuator, or amplitude modulator.

5. Tunnel diodes, also called Esaki diodes, are very heavily doped and have very thin depletion zones (3 to 100 angstroms) in order to encourage quantum mechanical tunneling currents to form. The tunnel diode can be used as an amplifier, one-shot multivibrator, or oscillator.

15-9 RECAPITULATION

Now return to the objectives and prequiz questions at the beginning of the chapter and see how well you can answer them. If you cannot answer certain questions, place a check mark by each and review the appropriate parts of the text. Next, try to answer the following questions and work the problems using the same procedure.

QUESTIONS AND PROBLEMS

1. List three factors that limit the operating frequency of common *PN* junction diodes such as the 1N4148 device.

2. What form of diode construction uses the cat's whisker structure to reduce junction capacitance?

3. Early microwave diodes were used as _____ stages in superheterodyne radar receivers.

4. _____ _____ velocity is a principal factor limiting the operating frequency of ordinary PN junction diodes.

5. The fact that minority carriers in PN junction diodes are in a storage state limits _____ time.

6. The _____ diode, also called the hot carrier diode, is generally more rugged and dissipates more power than point-contact diodes.

7. The hot carrier diode depends on _____ carrier rectification for fast switching speeds.

8. A certain silicon diode exhibits a capacitance of 50 pF when the applied voltage is zero ($V = 0$). If the barrier potential is 0.6 V, and for this material $m = 1/2$, what is the capacitance when the applied voltage (V) is –4 V dc? Assume $K = 1$.

9. A diode has a capacitance of 33 pF when $V = 0$. For this diode, the factor $m = 1/3$ and the barrier potential is 0.27 V. Find the capacitance at a potential of –10 V dc. Assume $K = 1$.

10. A varactor diode has a series resistance of 8 Ω and a junction capacitance of 36 pF at a potential of –10 V dc. Find the maximum operating frequency of this diode.

11. A certain variable-capacitance diode has a series resistance of 15 Ω and a junction capacitance of 8.2 pF. Find the maximum operating frequency.

12. An IF amplifier circuit must be varactor tuned to a frequency of 70 Mhz. If the inductance is 0.15 μH, what value of varactor capacitance must be used to resonate that circuit? Assume a stray capacitance of 4.4 pF.

13. Define Q as it applies to varactor diodes.

14. A varactor diode has a cutoff frequency of 2945 MHz. What is the Q of this diode at a frequency of 450 MHz?

15. A varactor diode with a cutoff frequency of 4.7 GHz is operated at a frequency of 901 MHz. What is the Q of this diode?

16. A varactor diode has a Q of 2215 at 1 MHz. Predict the maximum operating frequency.

17. A varactor diode has a measured Q of 2950 at a frequency of 1 MHz. What is the maximum operating frequency of this device?

18. A varactor diode has a capacitance range of 4.7 to 86 pF over the range of 0 to 30 V dc. Find the resultant capacitance range if this diode is placed in series with a dc-blocking capacitor of the following values:

    a. 100 pF, b. 0.001μF, c. 0.01 μF.

19. List several applications for varactor diodes.

20. A PIN diode has a lightly doped _____ region sandwiched between P and N regions.

21. PIN diodes have a _____ shift to RF signals because of a delay period for recombination of carriers in the center region of the diode.

22. There is always some small number of carriers in the _____ region of a PIN diode, so the resistance of that region is typically very low.

23. RF signals tend to see the PIN diode as a parallel _____ _____.

24. List two modes of switch action for PIN diodes.

25. List several applications for PIN diodes.

26. The Esaki diode depends on the quantum mechanical phenomenon called _____ for operation.

27. The doping level of the semiconductors in a tunnel diode is typically very much (higher/lower) than ordinary *PN* junction diodes.

28. List three basic conditions required for the operation of tunnel diodes.

29. Normal ohmic materials have a current characteristic that (increases/decreases) with increasing voltage.

30. Under some conditions, tunnel diodes exhibit (decreasing/increasing) current with increasing voltage; this phenomenon is called _____ resistance.

31. List three applications of the tunnel diode.

32. The maximum potential that can be applied to a tunnel diode in oscillator circuits must be less than the _____ potential of the diode.

33. List the conditions for oscillation required in a tunnel diode oscillator circuit.

34. A certain tunnel diode has a negative resistance of $-35\ \Omega$, a 33-pF junction capacitance, and $1.75\ \Omega$ of series resistance. Find the resistive frequency.

35. A tunnel diode shows 22 pF of junction capacitance and a negative resistance of $-20\ \Omega$. If the series resistance is $1.3\ \Omega$, what is the resistive frequency?

36. Find the self-resonant frequency of a tunnel diode if the series resistance is $1.4\ \Omega$, the junction capacitance is 18 pF, the negative resistance is $-25\ \Omega$, and the series inductance is $3.3\ \mu H$.

37. What is the self-resonant frequency of a tunnel diode that has a series inductance of $4.5\ \mu H$, a series resistance of $2.2\ \Omega$, a junction capacitance of 22 pF, and a negative resistance of $-30\ \Omega$?

KEY EQUATIONS

1. Capacitance of a varactor junction:

$$C_j = \frac{CK}{(V_b - V)^m}$$

2. Maximum varactor operating frequency:

$$F = \frac{1}{2\pi R_s C_d}$$

3. Varactor Q:

$$Q = \frac{F}{f}$$

4. Total capacitance of a dc-blocked varactor circuit:

$$C_t = \frac{C1 \times C_d}{C1 + C_d}$$

5. Tunnel diode frequencies
   a. Resistive frequency:

   $$F_r = \frac{1}{2\pi RC_j} \sqrt{\frac{R}{R_s} - 1}$$

   b. Self-resonant frequency:

   $$F_s = \frac{1}{2\pi} \sqrt{\frac{1}{L_s C_j} - \frac{1}{(RC_j)^2}}$$

CHAPTER 16

Microwave Diode Generators

OBJECTIVES

1. Learn the principal mechanism for two-terminal device oscillations.
2. Learn the properties and modes of operation for Gunn diodes.
3. Learn the theory of operation of IMPATT, TRAPATT, and BARITT diodes.
4. Learn the limitations and applications of two-terminal microwave generators.

16-1 PREQUIZ

These questions test your prior knowledge of the material in this chapter. Try answering them before you read the chapter. Look for the answers (especially those you answered incorrectly) as you read the text. After you have finished studying the chapter, try answering these questions again and those at the end of the chapter.

1. The efficiency of a Gunn diode operated in the transit time mode is typically less than _____ %.
2. List three modes of operation for a Gunn diode.
3. In Gunn transit time mode, the frequency of oscillation is a function of device _____.
4. In the _____ mode, the Gunn device operates without an internal space charge.

16-2 INTRODUCTION TO NEGATIVE RESISTANCE

In Chapter 15 you were introduced to the concept of *negative resistance* (-R), or *negative conductance* (-G = 1/-R). The negative resistance phenomenon, also called *negative differential resistance* (NDR), is a seemingly strange phenomenon in which

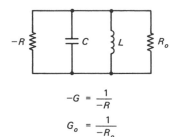

$$-G = \frac{1}{-R}$$

$$G_o = \frac{1}{-R_o}$$

CRITERION FOR OSCILLATION: $|-G| \geq G_o$

Figure 16-1
Negative-resistance oscillator equivalent circuit.

materials behave electrically contrary to Ohm's law. In ohmic materials (those that obey Ohm's law) we normally expect current to increase as electric potential increases. In these *positive-resistance* materials, $I = V/R$. In *negative-resistance* (NDR) devices, however, there may be certain ranges of applied potential in which *current decreases with increasing potential*.

In addition to the I versus V characteristic, certain other properties of negative-resistance materials distinguish them from ohmic materials. First, in ohmic materials we normally expect to find the voltage and current in phase with each other unless an inductive or capacitive reactance is also present. *Negative resistance*, however, causes *current and voltage to be 180° out of phase with each other*. This relationship is a key to recognizing negative-resistance situations. Another significant difference is that a positive resistance dissipates power, while a negative resistance generates power (that is, converts power from a dc source to another form). In other words, a $+R$ device *absorbs* power from external sources, while a $-R$ device *supplies* power (converted from dc) to the external circuit.

Two forms of oscillatory circuit are possible with NDR devices: *resonant* and *unresonant*. In the resonant type, depicted in Fig. 16-1, current pulses from the NDR device shock excite a high-Q LC tank circuit or resonant cavity into self-excitation. The oscillations are sustained by repetitive re-excitation. Unresonant oscillator circuits use no tuning. They depend on the device dimensions and the average charge carrier velocity through the bulk material to determine operating frequency. Some NDR devices, such as the Gunn diode of Section 16-3, will operate in either resonant or unresonant oscillatory modes.

16-3 TRANSFERRED ELECTRON (GUNN DIODE) DEVICES

Gunn diodes are named after John B. Gunn of IBM, who, in 1963, discovered a phenomenon since then called the *Gunn effect*. Experimenting with compound semiconductors such as gallium arsenide (GaAs), Gunn noted that the current pulse became unstable when the bias voltage increased above a certain crucial threshold potential. Gunn suspected that a negative-resistance effect was responsible for the unusual diode behavior. Gunn diodes are representative of a class of materials called *transferred electron devices* (TED).

Ordinary two-terminal, small-signal *PN* diodes are made from elemental semiconductor materials such as silicon (Si) and germanium (Ge). In the pure form before charge carrier doping is added, these materials contain no other elements. TED devices, on the other hand, are made from compound semiconductors; that is, materials that are chemical compounds of at least two chemical elements. Examples are gallium arsenide (GaAs), indium phosphide (InP), and cadmium telluride (CdTe). Of these, GaAs is the most commonly used.

N-type GaAs used in TED devices is doped to a level of 10^{14} to 10^{17} carriers per centimeter at room temperature (373 K). A typical sample used in TED/Gunn devices will be about 150×150 μm in cross-sectional area and 30 μm long.

16-3.1 Two-Valley TED Model

The operation of TED devices depends on a semiconductor phenomenon found in compound semiconductors called the *two-valley model* (for InP material there is also a three-valley model). The two-valley model is also called the Ridley-Watkins-Hilsum (RWH) theory. Figure 16-2 shows how this model works in TEDs. In ordinary elemental semiconductors, the energy diagram shows three possible bands: *valence band*, *forbidden band*, and *conduction band*. The forbidden band contains no allowable energy states and so contains no charge carriers. The difference in po-

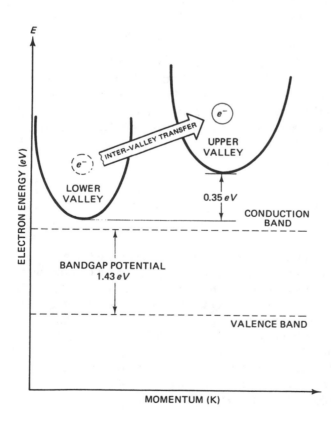

Figure 16-2
Transferred-electron two-valley model energy diagram.

tential between valence and conduction bands defines the forbidden band, and this potential is called the *bandgap voltage*.

In the two-valley model, however, there are two regions in the conduction band in which charge carriers (for example, electrons) can exist. These regions are called *valleys* and are designated the *upper valley* and the *lower valley*. According to the RWH theory, electrons in the lower valley have low effective mass (0.068) and consequently a high mobility (8000 cm²/V-s). In the upper valley, which is separated from the lower valley by a potential of 0.036 electron volts (eV), electrons have a much higher effective mass (1.2) and lower mobility (180 cm²/V-s) than in the lower valley.

At low electric field intensities (0 – 3.4 kV/cm), electrons remain in the lower valley and the material behaves ohmically. At these potentials, the material exhibits *positive differential resistance* (PDR). At a certain critical threshold potential (V_{th}), electrons are swept from the lower valley to the upper valley (hence the name transferred electron devices). For GaAs, the electric field must be about 3.4 kV/cm, so V_{th} is the potential that produces this strength of field. Because V_{th} is the product of the electric field potential and device length, a 10-μm sample of GaAs will have a threshold potential of about 3.4 V. Most GaAs TED devices operate at maximum dc potentials in the 7- to 10-V range.

The average velocity of carriers in a two-valley semiconductor (such as GaAs) is a function of charge mobility in each valley and the relative numbers of electrons in each valley. If all electrons are in the lower valley, then the material is in the highest average velocity state. Conversely, if all electrons were in the upper valley, then the average velocity is in its lowest state. Figure 16-3A shows drift velocity as a function of electric field, or dc bias.

In the PDR region, the GaAs material is ohmic, and drift velocity increases linearly with increasing potential. It will continue to increase until the saturation velocity is reached (about 10^7 cm/s). As the voltage increases above threshold potential, which creates fields greater than 3.4 kV/cm, more and more electrons are transferred to the upper valley, so the average drift velocity drops. This phenomenon gives rise to the negative resistance (NDR) effect. Figure 16-3B shows the NDR effect in the *I* versus *V* characteristic.

Figure 16-3C shows the *I* versus *T* characteristic of the Gunn diode operating in the NDR region. Ordinarily, we might expect a smooth current pulse to propagate through the material. But note the oscillations (Gunn's instabilities) superimposed on the pulse. It is this oscillating current that makes the Gunn diode useful as a microwave generator.

For the two-valley model to work, several criteria must be satisfied. First, the energy difference between lower and upper valleys must be greater than the thermal energy (*KT*) of the material; *KT* is about 0.026 eV, so GaAs with 0.036-eV differential energy satisfies the requirement. Second, to prevent hole-electron pair formation, the differential energy between valleys must be less than the forbidden band energy (that is, the bandgap potential). Third, electrons in the lower valley must have high mobility, low density of state, and low effective mass. Finally, electrons in the upper valley must be just the opposite: low mobility, high effective mass, and a high density of state. It is sometimes claimed that ordinary devices use so-called "warm" electrons (0.026 eV), while TED devices use "hot" electrons of greater than 0.026 eV.

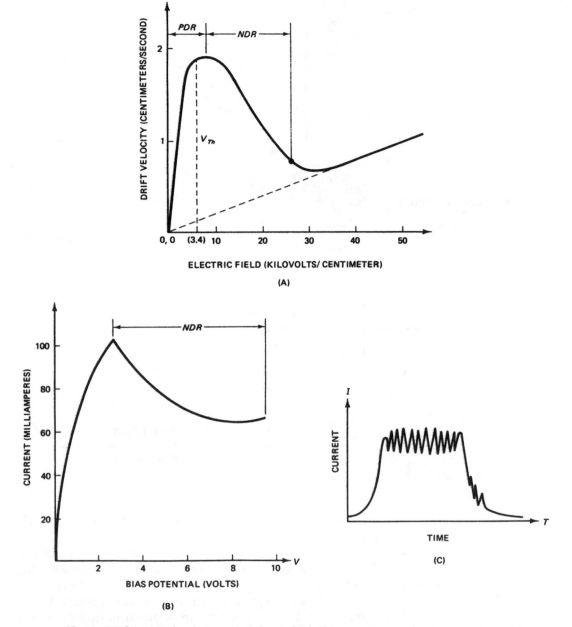

Figure 16-3 A) Drift velocity versus electric field; B) *I* versus *V* curve showing negative resistance region; C) Gunn oscillations.

16-3.2 Gunn Diodes

The Gunn diode is a transferred electron device that is capable of oscillating in several modes. In the unresonant *transit time* (TT) mode, frequencies between 1 and 18 GHz are achieved, with output powers up to 2 W (most are on the order of a few hundred milliwatts). In the resonant *limited space-charge* (LSA) mode, operating

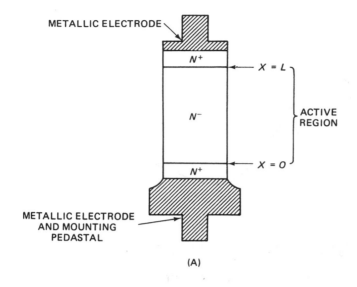

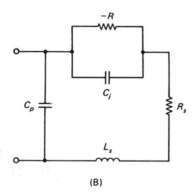

Figure 16-4
A) Gunn diode structure;
B) equivalent circuit.

frequencies to 100 GHz and pulsed power levels to several hundred watts (1% duty cycle) have been achieved.

Figure 16-4A shows a diagram for a Gunn diode, while an equivalent circuit is shown in Fig. 16-4B. The active region of the diode is usually 6 to 18 μm long. The N+ end regions are ohmic materials of very low resistivity (0.001 μ-cm) and are 1 to 2 μm thick. The function of the N+ regions is to form a transition zone between the metallic end electrodes and the active region. In addition to improving the contact, the N+ regions prevent migration of metallic ions from the electrode into the active region.

16-3.2.1 Domain Growth

The mechanism underlying the oscillations of a Gunn diode is the growth of *Ridley domains* in the active region of the device. An electron domain is created by a bunching effect (Fig. 16-5) that moves from the cathode end to the anode end of the active region. When an old domain is collected at the anode, a new domain forms at the cathode and begins propagating. The propagation velocity is close to the saturation velocity (10^7 cm/s). The time required for a domain to travel the length (L) of the material is called the *transit time* (T_t), which is

$$T_t = \frac{L}{V_s} \tag{16 - 1}$$

where

T_t = transit time in seconds (s)
L = length in centimeters (cm)
V_s = saturation velocity (10^7 cm/s)

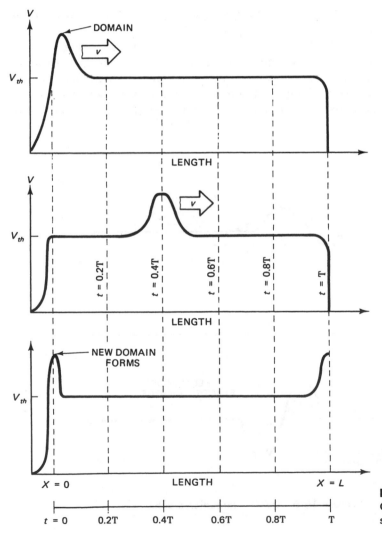

Figure 16-5
Growth of Ridley domains showing bunching effect.

EXAMPLE 16-1

Calculate the transit time for a 10-μm GaAs sample.

Solution

$$T_t = \frac{L}{V_s}$$

$$= \frac{(10 \, \mu m) \times \left(\dfrac{1 \, cm}{10^4 \, cm} \right)}{10^7 \, \dfrac{cm}{s}}$$

$$= \frac{10^{-3} \, cm}{10^7 \, cm} = 1 \times 10^{-10} \, s$$

Looking ahead to the next section, you might guess that a 10^{-10}-s transit time is comparable to a $1/10^{-10}$-Hz, or a 10-GHz, operating frequency.

Figure 16-6 graphically depicts domain formation. You may recognize that the domain shown here is actually a dipole, or double domain. The area ahead of the domain forms a minor depletion zone, while in the area of the domain there is a bunching of electrons. Thus, there is a difference in conductivity between the two regions, and this conductivity is different still from the conductivity of the rest of

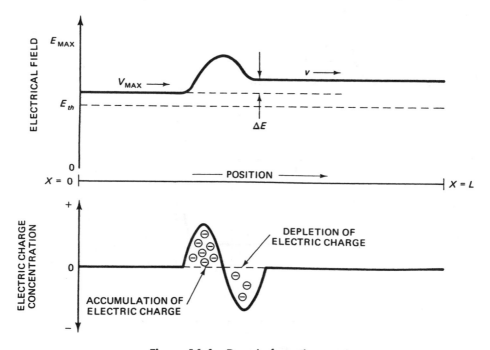

Figure 16-6 Domain-formation graph.

the active region. There is also a difference between the electric fields in the two domain poles and also in the rest of the material. The two fields equilibrate outside the domain. The current density is proportional to the velocity of the domain, so a current pulse forms.

16-3.3 Gunn Operating Modes

Transferred electron devices (Gunn diodes) operate in several modes and sub-modes. These modes depend in part on device characteristics and in part on external circuitry. For the Gunn diode, the operating modes are the *stable amplification* (SA) mode, *transit time* (TT) mode, *limited space-charge* (LSA) mode, and *bias circuit oscillation* (BCO) mode.

16-3.3.1 Stable Amplification (SA) Mode

In this mode, the Gunn diode will behave as an amplifier. We discussed this mode briefly in Chapter 13. The requirement for SA mode operation is that the product of doping concentration (N_o) and effective length of the active region (L) must be less than $10^{12}/cm^2$. Amplification is limited to frequencies in the vicinity of v/L, where v is the domain velocity and L is the effective length.

16-3.3.2 Transit Time (Gunn) Mode

The transit time (TT), or Gunn, mode is unresonant and depends on device length and applied dc bias voltage. The dc potential must be greater than the critical threshold potential (V_{th}). Because of the Gunn effect (Section 16-3), current oscillations in the microwave region are superimposed on the current pulse. Operation in this mode requires that the N_oL product be 10^{12} to $10^{14}/cm^2$. The operating frequency F_o is determined by device length, or rather the transit time of the pulse through the length of the material. Because domain velocity is nearly constant and is often close to the electron saturation velocity ($V_s = 10^7$ cm/s), length and transit time are proportional to each other. The operating frequency is inversely proportional to both length of the device and transit time.

The length of the active region in the Gunn diode determines the operating frequency. The frequency varies from about 6 GHz for an 18-μm sample (counting 1.5 μm for each $N+$ electrode region) to 18 GHz for a 6-μm sample. The operating frequency in the TT mode is approximately

$$F_o = \frac{V_{\text{dom}}}{L_{\text{eff}}} \qquad (16\text{-}2)$$

where

F_o = operating frequency in hertz (Hz)
V_{dom} = domain velocity in centimeters per second (cm/s)
L_{eff} = effective length in centimeters (cm)

EXAMPLE 16-2

A Gunn diode with an active region of 12 μM is operated in the TT mode. Assuming that the domain velocity is the saturation velocity (10^7 cm/s), what is the operating frequency?

Solution

$$F_o = \frac{V_{dom}}{L_{eff}}$$

$$= \frac{10^7 \text{ cm/s}}{12 \, \mu m \times \dfrac{10^{-4} \text{ cm}}{1 \, \mu m}}$$

$$= \frac{10^7 \text{ cm/s}}{1.2 \times 10^{-3} \text{ cm}} = 8.33 \times 10^9 \text{ Hz} = 8.33 \text{ GHz}$$

Operation in the TT mode provides efficiencies of 10% or less, with 4% to 6% being most common. Output powers are usually less than 1000 mW, although 2000 mW has been achieved.

16-3.3.3 Limited Space Charge (LSA) Mode

The LSA mode depends on shock exciting a high-Q resonant tank circuit or tuned cavity with current pulses from the Gunn diode. The LSA mode and its submodes are also called *accumulation mode, delayed domain mode,* and *quenched domain mode.* These various names reflect variations on the LSA theme. For the LSA mode, the N_oL product must be 10^{12}/cm² or higher and the N_o/F quotient between 2×10^5 and 2×10^4 s/cm³.

Figure 16-7A shows a simplified circuit for an LSA-mode Gunn oscillator, while Fig. 16-7B shows the waveforms. The circuit consists of the Gunn diode shunted by either an LC tank circuit (as shown) or a tuned cavity that behaves like a tank circuit. The criterion for resonant oscillation in a negative-resistance circuit is simple: the negative conductance ($-G = 1/-R$) must be greater than, or equal to, the conductance represented by circuit losses (see Fig. 16-1):

$$-G \geq G_o \tag{16-3}$$

At turn-on, transit-time current pulses (Fig. 16-7C) hit the resonant circuit and shock excite it into oscillations at the resonant frequency. These oscillations set up a sine wave voltage across the diode (Fig. 16-7B) that adds to the bias potential. The total voltage across the diode is the algebraic sum of the dc bias voltage and the RF sine wave voltage. The dc bias is set such that the negative swing of the sine wave forces the total voltage below the critical threshold potential V_{th}. Under this condition, the domain does not have time to build up, and the space charge dissipates. The domain quenches during the period when the algebraic sum of the dc bias and the sine wave voltage is below V_{th}.

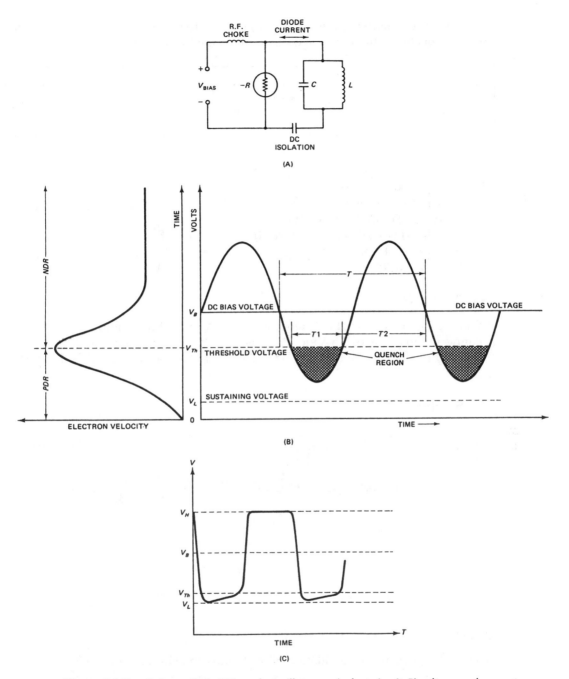

Figure 16-7 A) Gunn diode LSA mode oscillator equivalent circuit; B) voltage and current relationships; C) output pulse.

The LSA oscillation period ($T = 1/F$) is set by the external tank circuit and the diode adapts to it. The period of oscillation is found from

$$T = 2\pi \sqrt{LC} + \frac{L}{R(V_b/V_{th})} \qquad (16\text{-}4)$$

where

T = period in seconds (s)
L = inductance in henrys (H)
C = capacitance in farads (F)
R = low-field resistance in ohms (μ)
V_{th} = threshold potential
V_b = bias voltage

EXAMPLE 16-3

A Gunn diode with a 30-Ω resistance and a 3.4-V critical threshold potential (V_{th}) is connected to a bias of 2.5 V dc. What is the period of oscillation if the diode is connected to a tank circuit consisting of a 0.1-nH inductance and a 2.2-pF capacitor?

Solution

$$T = 2\pi \sqrt{LC} + \frac{L}{R(V_b/V_{th})} \text{ s}$$

$$= (2)(3.14)\sqrt{(10^{-10}\text{ H})(2.2 \times 10^{-12}\text{ F})} + \frac{10^{-10}\text{ H}}{(30)(2.53/3.4)} \text{ s}$$

$$= 6.28\sqrt{2.2 \times 10^{-22}} = \frac{10^{-10}}{22.06} \text{ s}$$

$$= 6.28(1.48 \times 10^{-11}) + (4.5 \times 10^{-12}) \text{ s}$$

$$= (9.29 \times 10^{-11}) + (4.5 \times 10^{-12})s = 9.7 \times 10^{-11}s$$

EXAMPLE 16-4

What is the frequency of oscillation of the Gunn diode in Example 16-3?

Solution

$$F = 1/T$$

$$= 1/9.7 \times 10^{-11}\text{ s}$$

$$= 1.03 \times 10^{10}\text{ Hz} = 10.3\text{ GHz}$$

The LSA mode is considerably more efficient than the transit time mode. The LSA mode is capable of 20% efficiency, so it can produce at least twice as much RF output power from a given level of dc power drawn from the power supply as transit time operation. At a duty factor of 0.01, the LSA mode is capable of delivering hundreds of watts of pulsed output power.

The output power of any oscillator or amplifier is the product of three factors: dc input voltage (V), dc input current (I), and the conversion efficiency (n, a decimal fraction between 0 and 1):

$$P_o = nVI \qquad\qquad (16\text{-}5A)$$

where

P_o = output power
n = conversion efficiency factor (0 to 1)
V = applied dc voltage
I = applied dc current

For the Gunn diode case, a slightly modified version of this expression is used:

$$P_o = n(MV_{th}L)(N_oeVA) \qquad\qquad (16\text{-}5B)$$

where

n = conversion efficiency factor (0 to 1)
v = average drift velocity
V_{th} = threshold potential (kV/cm)
M = multiple, V_{dc}/V_{th}
L = length in centimeters
n_o = donor ion concentration
e = electric charge (1.6×10^{-19} C)
A = area of the device (cm²)

EXAMPLE 16-5

A Gunn diode with a 3.4-V threshold potential is operated in an LSA circuit at 14 GHz. The diode has a cross-sectional area of 8×10^{-4} cm² and is 10 μm long. If the conversion efficiency is 8%, what is the output power if the dc bias of 7 V has a 6-V peak sine wave superimposed? Assume $v = 10^7$ cm/s and $N_o = 10^{15}$/cm³.

Solution

$$N = 0.08$$

$$M = \frac{7 + 6}{3.4} = 3.82$$

$$P_o = N \times (MV_{th}L) \times (N_oeVA)$$

$$= (0.08)\left[(3.82)(3.4\,\text{kv/cm})(10\,\mu m \times \frac{1\,\text{cm}}{10^4\,\mu m}\right]$$

$$\times [(10^{15}/\,\text{cm}^3)(1.6 \times 10^{-19}\,\text{C})(10^7\,\text{cm/s})]$$

$$\times (8 \times 10^{-4}\,\text{cm}^2)$$

$$= 0.08\,(1.3 \times 10^{-2}\,\text{kV})(1.28\,\text{C/s})$$

$$= 1.33 \times 10^{-3}\,\text{kw} = 1.33\,\text{W}$$

EXAMPLE 16-6

In Example 16-5, find the dc input power and current required to generate P_o.

Solution

$$\text{(a) } P_{dc} = P_o/N$$

$$= 1.33\,\text{W}/0.08$$

$$= 16.6\,\text{W}$$

$$\text{(b) } I = P/V$$

$$= 16.6\,\text{W}/7\,\text{V} = 2.37\,\text{A}$$

16-3.3.4 Bias Circuit Oscillation (BCO) Mode

This mode is quasi-parasitic in nature and occurs only during one of the normal Gunn oscillating modes (TT or LSA). If the product $F1$ is very, very small, then BCO oscillations can occur at frequencies from 0.01 to 100 MHz.

16-3.3.5 Gunn Diode Applications

Gunn diodes are often used to generate microwave RF power in diverse applications ranging from receiver local oscillators, to police speed radars, to microwave communications links. Figure 16-8 shows two possible methods for connecting a Gunn diode in a resonant cavity. In Fig. 16-8A, we see a cavity that uses a loop-coupled output circuit. The output impedance of this circuit is a function of loop size and position, with the latter factor dominating. The loop positioning is a trade-off between maximum output power and oscillator frequency stability.

In Fig. 16-8B, we see a Gunn diode mounted in a section of flanged waveguide. RF signal passes through an iris to be propagated through the waveguide to the load. In both Figs. 16-8A and 16-8B, the exact resonant frequency of the cavity is set by a tuning screw inserted into the cavity space.

16-4 IMPATT DIODES

The *avalanche phenomenon* is well known in *PN* junction diodes. If a reverse-bias potential exceeds a certain critical threshold, the diode breaks down and the reverse current increases abruptly from low leakage values to a very high value. The prin-

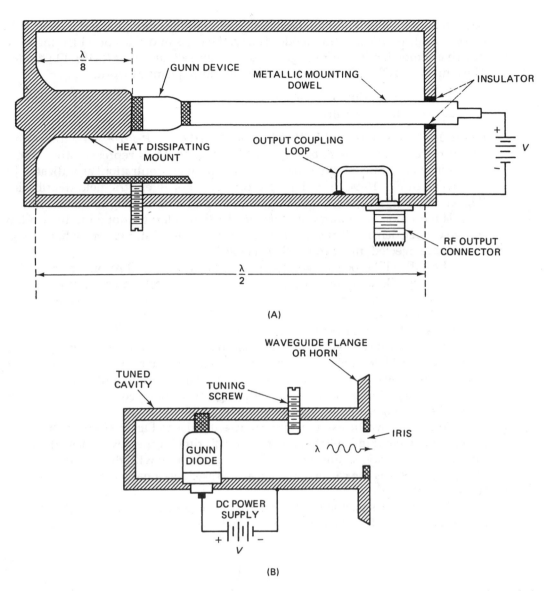

(A)

(B)

Figure 16-8   Gunn diode transmitters: A) closed resonant cavity using output link; B) iris-coupled oscillator.

cipal cause of this phenomenon is *secondary electron emission*; that is, charge carriers become so energetic as to be able to knock additional valence electrons out of the crystal lattice to form excess hole-electron pairs. The common zener diode works on this principle.

The onset of an avalanche current in a *PN* junction is not instantaneous, but rather there is a short phase delay period between the application of a sufficient breakdown potential and the creation of the avalanche current. In 1959, W. T. Read of Bell Telephone Laboratories postulated that this phase delay could create a negative resistance. It took until 1965 for others (C. A. Lee and R. L. Johnston) at Bell

Labs to create the first of these *Read diodes*. Johnston generated about 80 mW at 12 GHz in a silicon *PN* junction diode. Today, the class of diodes of which the Read device is a member is referred to as *impact avalanche transit time* (IMPATT) diodes. The name IMPATT reflects the two different mechanisms at work:

1. Avalanche (impact ionization)
2. Transit time (drift)

Figure 16-9 shows a typical IMPATT device based on a *N+-P-I-P+* structure. Other structures are also known, but the *NPIP* of Fig. 16-9A is representative. The + superscripts indicate a higher than normal doping concentration, as indicated by the profile in Fig. 16-9C. The doping profile ensures an electric field distribution (Fig. 16-9B) that is higher in the *P* region to confine avalanching to a small zone.

The *I* region is an *intrinsic semiconductor* that is lightly doped to have a low charge carrier density. Thus, the *I* region is a near insulator except when charge carriers are injected into it from other regions.

The IMPATT diode is typically connected in a high-Q resonant circuit (*LC* tank or cavity). Because avalanching is a very noisy process, noise at turn-on rings the tuned circuit and creates a sine wave oscillation at its natural resonant frequency (Fig. 16-9D). The total voltage across the *NPIP* structure is the algebraic sum of the dc bias and tuned circuit sine wave.

The avalanche current (I_o) is injected into the *I* region and begins propagating over its length. Note in Fig. 16-9E that the injected current builds up exponentially until the sine wave crosses zero and then drops exponentially until the sine wave reaches the negative peak. This current pulse is thus delayed 90° with respect to the applied voltage.

Compare now the external circuit current pulse in Fig. 16-9G with the tuned circuit sine wave (Fig. 16-9D). Note that transit time in the device has added additional phase delay, so the current is 180° out of phase with the applied voltage. This phase delay is the cause of the negative-resistance characteristic (see Section 16-2).

Oscillation is sustained in the external resonant circuit by successive current pulses reringing it. The resonant frequency of the external resonant circuit should be

$$F = \frac{V_d}{2L} \qquad (16\text{-}6)$$

where

F = frequency in hertz (Hz)
V_d = drift velocity in centimeters per second (cm/s)
L = length of the active region in centimeters (cm)

EXAMPLE 16-7

Find the resonant frequency of an external high-Q cavity if drift velocity is 10^7 cm/s and the length is 10 μm.

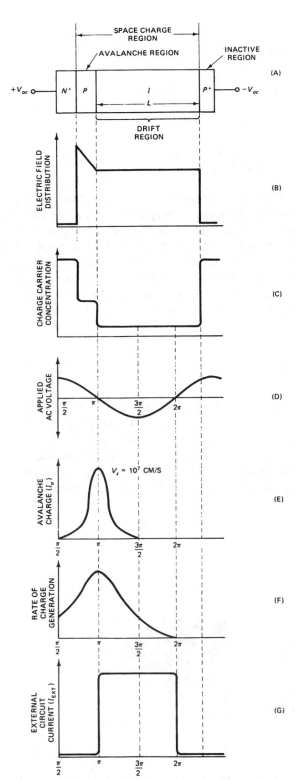

Figure 16-9
IMPATT diode and waveforms.

Solution

$$F = V_d / 2\,L$$

$$= \frac{10^7 \text{ cm/s}}{(2)\left[10\,\mu m \times \dfrac{1\,\text{cm}}{10^4\,\mu m} \right]}$$

$$= \frac{10^7 \text{ cm/s}}{2 \times 10^{-3}\,\text{cm}} = 5 \times 10^9 \text{ Hz} = 5 \text{ GHz}$$

IMPATT diodes typically operate in the 3- to 6-GHz region, although 100-GHz operation has been achieved. These devices typically operate at potentials in the 75- to 150-V dc range. Because an avalanche process is used, the RF output signal of the IMPATT diode is very noisy. Efficiencies of single-drift region devices (such as Fig. 16-9A) are about 6% to 15%. By using double-drift construction (Fig. 16-10), efficiencies can be improved to the 20% to 30% range. The double-drift device uses electron conduction in one region and hole conduction in the other. This operation is possible because the two forms of charge carrier drift approximately in phase with each other.

16-5 TRAPATT DIODES

Gunn and IMPATT diodes operate at frequencies of 3 GHz or above. Operation at lower frequencies (for example, 500 MHz to 3 GHz) was left to the transistor, which also limited available RF output power to a great extent. The reason why Gunn and IMPATT devices cannot operate at lower frequencies is that it proved difficult to increase transit time in those devices. One might assume that it is only necessary to lengthen the active region of the device to increase transit time. But

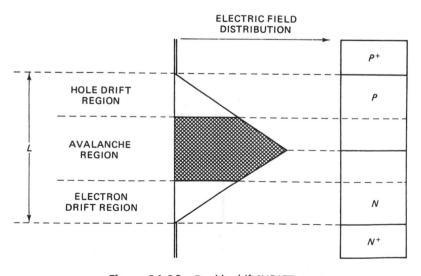

Figure 16-10 Double-drift IMPATT structure.

certain problems arose in long structures, domains were found to collapse, and sufficient fields were hard to maintain.

A solution to the transit time problem is found in a modified IMPATT structure that uses P^+-N-N^- regions (Fig. 16-11). The P^+ region is typically 3 to 8 μm, while the N region is typically 3 to 13 μm. The diameter of the device may be 50 to 750 μm, depending on the power level required. The first of these diodes was produced by RCA in 1967. The diode produced more than 400 W at 1000 MHz at an efficiency of 25%. Since then, frequencies as low as 500 MHz and powers to 600 W have been achieved. Today, efficiencies in the 60% to 75% range are common. One device was able to provide continuous tuning over a range of 500 to 1500 MHz.

The name of the P^+-N-N^- device reflects its operating principle: *trapped plasma avalanche-triggered transit* (TRAPATT). The method by which transit time is increased is by formation of a *plasma* in the active region. A plasma is a region of a large number of disassociated holes and electrons that do not easily recombine. If the electric

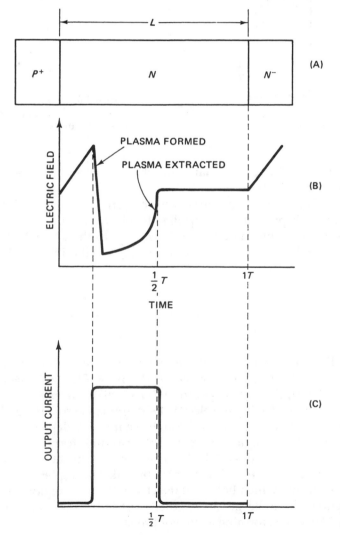

Figure 16-11
Modified IMPATT structure.

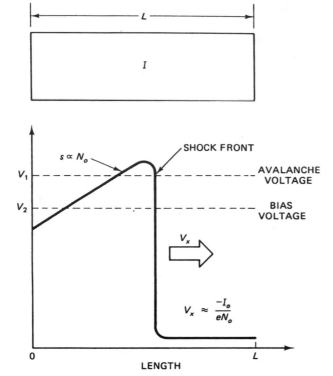

Figure 16-12
TRAPATT diode length and
associated shock front.

field is low, then the plasma is trapped. That is, the charge carriers are swept out of
the N region slowly (see Fig. 16-11B).

The output is a harmonic-rich, sharp-rise-time current pulse (Fig. 16-11C). To
become self-oscillatory, this pulse must be applied to a low-pass filter at the input
of the transmission line or waveguide that is connected to the TRAPATT. Harmonics are not passed by the filter and so are reflected back to the TRAPATT diode
to trigger the next current pulse.

16-6 BARITT DIODES

The BARITT diode (Fig. 16-13) consists of three regions of semiconductor material
forming a pair of abrupt PN junctions, one each P+-N and N-P+. The name of this
device comes from a description of its operation: *barrier injection transit time*. The
BARITT structure is designed so that the electric field applied across the end electrodes causes a condition at or near punch-through. That is, the depletion zone is
formed throughout the entire N region of the device. Current is formed by sweep-
ing holes into the N region of the device. Under ordinary circumstances, the P+-N
junction is forward biased and the N-P+ is reverse biased. The depleted N region
forms a potential barrier into which holes are injected into the N region from the
forward-biased junction. These charge carriers then drift across the N region at the
saturation velocity of 10^7 cm/s, forming a current pulse.

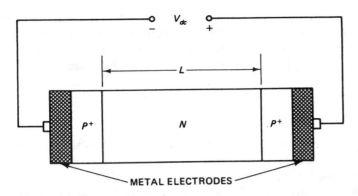

Figure 16-13
BARITT diode.

There are three conditions for proper BARITT operation:

1. The electrical field across the device must be great enough to force charge carriers in motion to drift at the saturation velocity (10^7 cm/s).
2. The electrical field must be great enough to create the punch-through condition.
3. The electrical field must not be great enough to cause avalanching to occur.

The normal circuit configuration for BARITT devices is in a resonant LC tank or cavity. At turn-on, noise pulses will initially shock excite the resonant circuit into

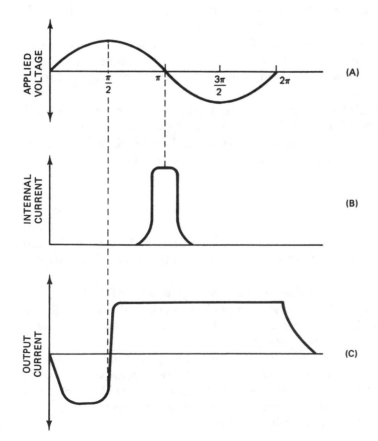

Figure 16-14
Oscillatory behavior of BARITT diode.

self-oscillation, and successive current pulses supply the energy to continue the oscillations. Figure 16-14 shows the relationship between the oscillatory sine wave voltage, the internal current, and the external current. As in the other diodes, the output energy comes from tapping the energy in the resonant circuit.

16-7 SUMMARY

1. Negative resistance, the basis for solid-state microwave diodes, is a condition under which the current decreases under increasing voltage. In a –R material, the current and voltage are 180° out of phase with each other.

2. Two forms of oscillator circuit are possible with –R devices: resonant and unresonant. In resonant circuits, noise pulses at turn-on shock excite the resonant tank or cavity into oscillation, and the RF sine wave voltage thus developed adds to or subtracts from the dc bias across the –R device. Unresonant circuits depend on the transit time of charge carriers through the bulk semiconductor media.

3. Transferred electron devices (TEDs) such as the Gunn diode depend on the fact that certain compound semiconductors (for example, GaAs, InP, and CdTe) have two regions, called valleys, in their conduction bands. The lower-valley electrons have high mobility and low effective mass, while in the upper valley the opposite is found: low mobility and high effective mass. As the electric field increases above a certain threshold potential, more and more electrons are swept into the upper valley, causing the average drift velocity to drop.

4. The mechanism underlying oscillations in the transit-time mode of the Gunn diode is the growth of Ridley domains, which is an electron bunching effect. The device transit time is the time required for the domain to propagate through the active region of the diode.

5. The Gunn diode has several operating modes or submodes: *stable amplification* (SA), *transit time* (TT), *limited space charge* (LSA), and *bias circuit oscillator* (BCO). Submodes to the LSA mode are called *accumulation mode, delayed domain mode,* and *quenched domain mode.* The transit time mode is unresonant, while the LSA mode requires an external resonant tank circuit or tuned cavity.

6. The IMPATT device operates on the principle that the phase delay between the onset of avalanching in a *PN* junction and the application of a voltage sufficient to cause avalanching creates a phase shift between current and voltage. Between the avalanche phase delay and the transit time of the device, the 180° phase difference between I and V required for –R operation is satisfied. IMPATT diodes operate in resonant oscillator circuits.

7. The TRAPATT diode is similar in many ways to the IMPATT, but it operates by providing a phase delay due to trapping of a plasma, that is, a high concentration of disassociated hole and electron charge carriers that do not easily recombine, in the active region of the device.

8. BARITT diodes are formed from a pair of abrupt *PN* junctions back to back, sharing the same *N* region. When the electric field approaches punchthrough, but is less than the avalanche potential, charge carriers are injected across one forward-biased junction and transit the *N* region at saturation velocity. The time delay thus created forms a negative resistance.

16-8 RECAPITULATION

Now return to the objectives and prequiz questions at the beginning of the chapter and see how well you can answer them. If you cannot answer certain questions, place a check mark by each and review the appropriate parts of the text. Next, try to answer the following questions and work the problems using the same procedure.

QUESTIONS AND PROBLEMS

1. List four types of microwave diode oscillator device.
2. The principal mechanism for microwave oscillation in two-terminal (diode) devices is

   _____ _____ .
3. In a _____ resistance, also called ohmic materials, the current and voltage are in phase with each other.
4. In a _____ resistance, the voltage and current are 180° out of phase with each other.
5. A _____ resistance absorbs power from the external circuit, while a _____ resistance delivers power to the external circuit.
6. List two forms of general oscillatory circuit for two-terminal microwave generators.
7. A transferred _____ device operates because some compound semiconductors have two energy valleys in the conduction band.
8. List three compound semiconductors that can be used for Gunn-type devices.
9. List the three energy bands in a semiconductor.
10. In a GaAs Gunn diode, the lower valley electrons have a (high/low) mobility and a (high/low) effective mass.
11. In a GaAs Gunn diode, the upper valley electrons have a (high/low) mobility and a (high/low) effective mass.
12. At electric field intensities below a critical threshold, Gunn device material acts as a _____ resistance, but above the threshold potential it is a _____ resistance.
13. The active region in a Gunn diode is 14 μm long, and the charge carriers drift at the saturation velocity (10^7 cm/s). What is the transit time?
14. The active region of a Gunn diode is 9×10^{-4} cm long. Find the transit time in seconds if the carrier drift is at saturation velocity.
15. In problem 14, calculate the operating frequency of the Gunn diode in the transit-time mode.
16. Find the operating frequency of a Gunn diode in the transit-time mode if the active region is 12.75 μm long and the charge carriers drift at saturation velocity.
17. List the various modes of operation for the Gunn diode.
18. Which of the modes asked for in question 17 are resonant? Unresonant?
19. A Gunn diode in the resonant mode sees losses of 14 Ω in the external circuit. What is the minimum negative-resistance value required for oscillation?
20. An external tank circuit is connected to a Gunn diode. For LSA-mode oscillation, what is the period of oscillation if the following parameters apply: $L = 0.12$ μH, $C = 1.9$ pF, $R = 27$ Ω, $V_{th} = 3.4$ V, bias voltage is 2.75 V?

21. In question 20, what is the operating frequency in gigahertz?

22. A certain device has an operating conversion efficiency of 0.14 and requires a dc current of 12 A when a potential of 7 V dc is applied. What is the RF output power?

23. A 10-μm Gunn diode with a cross-sectional area of 7.75×10^{-4} cm^2 has a conversion efficiency of 7.5 % and a threshold potential of 3.4 kV/cm. If the applied dc potential is 7 V and the resonant cavity produces a 6-V peak sine wave, what is the output power? Assume a charge carrier concentration of 10^{15}/cm^3.

24. The IMPATT diode operates because of the _____ phenomenon, which is caused by secondary electron emission.

25. List two factors that contribute to the negative resistance oscillation in an IMPATT diode.

26. In the TRAPATT diode, a trapped _____ increases the transit time to allow operation below 3 GHz.

27. A _____ is a condition in which a high concentration of charge carriers is disassociated and does not easily recombine.

28. A _____ diode is constructed of two abrupt PN junctions back to back, sharing the name N region.

KEY EQUATIONS

1. Electron transit time in a Gunn diode:

$$T_t = \frac{L}{V_s}$$

2. Operating frequency in the Gunn or transit-time mode:

$$F_o = \frac{V_{dom}}{L_{eff}}$$

3. Criterion for LSA oscillation:

$$-G \geq G_o$$

4. Period of oscillation in the LSA mode:

$$T = 2\pi(LC)^{1/2} + \frac{L}{R(V_b/V_{th})}$$

5. Output power from a Gunn diode:

$$P_o = n(MV_{th}L)(N_oeVA)$$

6. Resonant frequency of an IMPATT external tank circuit:

$$F = \frac{V_d}{2L}$$

CHAPTER 17

Transmitters

OBJECTIVES

1. Understand the function of the transmitter.
2. Learn the architecture of basic transmitters.
3. Learn the limitations and common problems associated with transmitters.

17-1 PREQUIZ

These questions test your prior knowledge of the material in this chapter. Try answering them before you read the chapter. Look for the answers (especially those you answered incorrectly) as you read the text. After you have finished studying the chapter, try answering these questions again and those at the end of the chapter.

1. A microwave transmitter operates on a frequency of 9410 MHz. The oscillator frequency of _____ MHz is passed through multiplication stages totalling 128 x.

2. A phase-locked-loop exciter operates on a frequency of 258.945 MHz. What is the microwave frequency at the output of the transmitter if the following multiplication stages are present: 2X, 3X, and 2X?

3. A heterodyne transmitter operates on 3355 GHz. Find the frequency of the modulated stage if the exciter frequency is 3280 MHz.

4. A transmitter in a relay system converts the receiver _____ signal to a microwave frequency for retransmission.

17-2 INTRODUCTION

A transmitter is a device that develops the RF signal, increases it to a useful power level, adds modulation where necessary for carrying information, and then delivers it to a load (usually an antenna). Beyond that general description, however, transmitters are so varied that common points are very few. As a result, this discussion must be highly generalized. We will discuss transmitters from a block diagram perspective.

17-3 TYPES OF TRANSMITTER

Classifying radio transmitters can be done along several lines: first, according to use (landmobile, broadcast, relay, telemetry link, and so on), and second, according to modulation type (CW, AM, SSBSC, DSBSC, FM, TDM, FDM, pulse, radar pulsed, and so on). We can also classify transmitters according to power level (milliwatt level, low power, medium power, high power, or superpower). Perhaps the most general approach to classification, however, is according to transmitter *architecture*, which will be our method in this chapter.

Transmitter architecture is dictated by any or all of several factors: *power level, modulation type, frequency, frequency tolerance required, frequency stability required, available technology, application,* and *physical requirements or limitations*. Some of these factors are more important than others when dealing specifically with microwave transmitters, but as technology progresses, some former distinctions are fading. For example, physical requirements for hand-held portable transmitters once did not apply to microwave transmitters. But today we already see 900-MHz and 1.3-GHz hand-held transceivers on the market, and there is a good prospect for even higher-frequency units in the near future.

17-4 SIMPLE TRANSMITTERS

Simple transmitters are those where the oscillator that generates the RF signal is coupled directly to the antenna (Fig. 17-1). Examples of this class of transmitter include reflex klystrons, magnetrons, and Gunn diode oscillators. The class includes 10-mW retail shoplifting detectors and megawatt-level pulsed radar transmitters.

A purely CW transmitter is unmodulated and consists only of the oscillator and antenna radiator element, plus whatever impedance-matching network is needed between the two. An example is the illuminator used in early (largely unsophisticated) antishoplifting system. Those systems transmitted a signal on 900 MHz

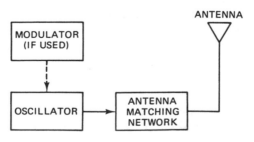

Figure 17-1
Simple transmitter.

to excite an 1800-MHz (second harmonic) dipole/diode combination on the hidden tag on the merchandise. The receiver section listens for a signal on 1800 MHz.

All non-CW transmitters require some form of *modulator* to alter the RF carrier signal in accordance with the information being transmitted. Modulator is a broad term with many meanings. The modulator could be a simple on-off switching device used in radio telegraphy, crude teletypewriter systems, and older noncoherent radar systems. The modulator could also be a complicated microcomputer-controlled device that combines a large number of modulation techniques.

The simple oscillator form of transmitter is appealing for its simplicity. There are, however, limitations to its use. First, certain complex modulation schemes are either not possible or are very difficult to achieve. For example, phase-coherent radar pulses are difficult or impossible in magnetron or reflex klystron transmitters. Second, many oscillators of this sort are not very frequency agile. Third, load variations can cause frequency shift in certain practical situations. As a result of these and other problems, the *master oscillator power amplifier* (MOPA) transmitter was designed.

17-5 MASTER OSCILLATOR POWER AMPLIFIER TRANSMITTERS

The MOPA transmitter, also called the *exciter-amplifier* transmitter, generates the basic RF signal at a low power level and then amplifies it to the required power level in following stages. The simplest form of MOPA transmitter is shown in Fig. 17-2. In this case, the oscillator generates the RF signal at the output frequency of the transmitter and passes it to the input of a power amplifier. See Chapters 11 through 16 for devices that amplify microwave signals. In some designs, one or more intermediate amplifier stages, usually called *driver* or *buffer* amplifiers, are used between the oscillator and the input of the final power amplifier.

The simple MOPA design of Fig. 17-2 provides three principal advantages. First, it provides isolation between the load (for example, the antenna) and the os-

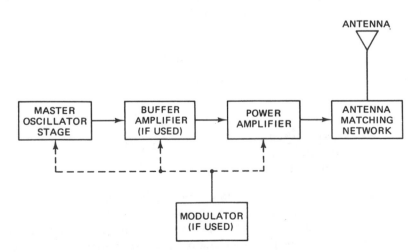

Figure 17-2 Master oscillator power amplifier (MOPA) transmitter.

cillator, resulting in an improvement over the stability available in the circuit of Fig. 17-1. Second, it allows flexibility in the types of modulation usable in the transmitter. The modulator may, depending on modulation type, be either high or low level. The former is modulation of the output power amplifier, while the latter is modulation of a stage prior to the power amplifier (a buffer or the oscillator itself). Finally, the MOPA transmitter design permits frequency agility because some of the devices used for RF signal generation at those power levels are inherently more flexible than high-level oscillators.

Some improvement in MOPA performance is available in modern transmitters from the fact that a microwave oscillator can be designed as part of a phase-locked loop (PLL). The PLL uses a voltage-controlled oscillator (VCO) on a microwave frequency. The output of the VCO is frequency divided and then compared with a lower-frequency crystal-controlled standard. The output of the comparison stage is a dc voltage that is used as an error signal to bring the VCO back on the correct frequency as it drifts. By adjusting either the comparison frequency or the division ratio of the VCO output frequency, we can create a large number of microwave channels.

17-5.1 Additional MOPA Designs

The simple MOPA of Fig. 17-2 is a basic form of transmitter, but it too suffers from problems. One such problem is that the oscillator must operate on the microwave output frequency. This requirement places some limits on stability, agility, and modulation types. A solution to these problems is to replace the simple MOPA with systems such as shown in Figs. 17-3 and 17-4.

In Fig. 17-3A, we see a transmitter in which the oscillator operates on a sub-harmonic of the microwave output frequency. Between the oscillator stage and the output stage are one or more *frequency multiplier* stages (doublers, triplers, quadruplers, and so on). An example of such a circuit is shown in Fig. 17-3B. The input circuit (tank 1) is tuned to the input frequency ($F1$). Diode $D1$ (or some other nonlinear element) distorts the $F1$ signal, thereby increasing its harmonic content. The output circuit (tank 2) will be tuned to a *harmonic* of $F1$ such that

$$F2 = N \times F1 \qquad\qquad (17\text{-}1)$$

where

> $F1$ = input frequency
> $F2$ = output frequency
> N = an integer $(2, 3, 4 \ldots, n)$

The harmonic content has a lower power level than the fundamental signal, so each multiplier may require an intermediate amplifier stage. In some frequency multiplier circuits, a single unbiased, common-emitter configuration transistor performs both the nonlinearity and amplification functions.

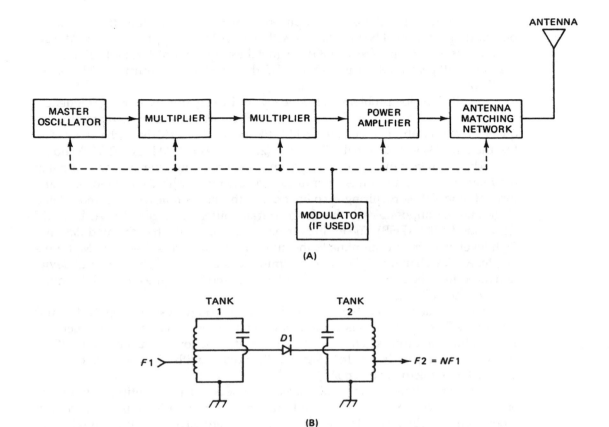

Figure 17-3 A) Frequency multiplier MOPA design; B) diode frequency multiplier.

EXAMPLE 17-1

A transmitter must operate on a frequency of 3345 MHz. The exciter section consists of an oscillator followed by three triplers and a doubler. Find the operating frequency of the master oscillator stage.

Solution

$$F_{MO} = \frac{F_{out}}{N1 \times N2 \times N3 \times N4}$$

$$= \frac{3345\,\text{MHz}}{(3) \times (3) \times (3) \times (2)} = \frac{3345\,\text{MHz}}{54} = 61.9444\,\text{MHz}$$

The MO frequency (61.9444 MHz) is well within the range of common crystal overtone oscillators and phase-locked-loop (PLL) circuits.

Some problems associated with this circuit are caused by the fact that frequency multiplication also multiplies errors. Consider *stability* and *accuracy* of the RF output frequency, for example. Each of these specifications is divided by the same multiplication factor as the output frequency. If, for example, the 3345-MHz output frequency must be maintained within ±0.001% (100 ppm), then the MO accuracy must be 100 ppm/54, or 1.85 ppm (a tight specification). Similarly, if the temperature drift permitted is, say, 2000 Hz/°C, then the MO drift must be 2000 Hz/54, or 37 Hz/°C.

Another major problem in frequency multiplier transmitters is frequency- and phase-modulated (FM/PM) noise. The FM/PM noise is exacerbated by frequency multiplication process. If the FM/PM noise of the MO is on the order of 20 Hz, then the 3345-MHz signal will exhibit (20 Hz × 54) = 1080 Hz of FM/PM noise.

The frequency multiplication system somewhat limits the available forms of modulation. Multiplication is a nonlinear process for amplitude variations, so any form of modulation requiring such linearity in the stages following the modulated stage may be impossible on this form of transmitter. Examples are AM, SSBSC (SSB), and DSBSC (DSB). The multiplier system works well when the modulation is high level or when it is immune to the multiplication process. FM and PM, for example, work well in multiplication transmitter designs, although the output deviation limits must be divided by the multiplication ratio to find the required deviation limits of the MO stage.

Some of these problems are relieved by the *heterodyne* design of Fig. 17-4. In this system the RF output signal is the sum or difference between two other frequencies. The modulation process generally takes place at some low frequency in the 30- to 100-MHz range, and the resultant signal is then mixed with a high-frequency signal in the UHF or microwave region.

The heterodyne method is also used in some frequency multiplier systems in order to gain channel flexibility while retaining certain advantages inherent in modulating a stable, fixed-frequency source. A microwave transmitter might consist of a 70-MHz modulated stage, a wide-range frequency synthesizer operating in the VHF/UHF region, a mixer/filter, and a chain of frequency multipliers that raises the output frequency to the desired microwave channel.

In Chapter 20 (microwave communications), we will discuss a variant of the heterodyne system that is used in microwave relay systems and frequency translators. The 70-MHz source in those systems is the IF amplifier of the receiver. The IF

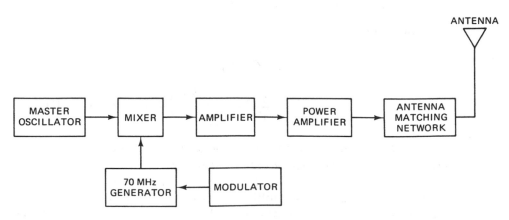

Figure 17-4   Heterodyne MOPA transmitter.

is mixed with an oscillator signal in order to re-transmit the signal on a frequency that is different from the received frequency.

17-6 SUMMARY

1. A transmitter is a device that develops an RF signal on a specified frequency, adds modulation to it as appropriate, amplifies it to the desired output power level, and then delivers to a load (usually an antenna).

2. Transmitter architecture is dictated by any or all of several factors: *power level, modulation type, frequency, frequency tolerance required, frequency stability required, available technology, application,* and *physical requirements or limitations*.

3. The simplest transmitters are oscillators coupled to the antenna circuit. While low cost and useful in some applications, these transmitters are generally inflexible and offer low stability.

4. The master oscillator power amplifier (MOPA) transmitter develops the signal in a low-power oscillator and then boosts it to the operational power level in one or more amplifier stages.

5. A multiplier-type transmitter develops a signal at a low frequency (usually in the 30- to 300-MHz range) and then passes it through a series of frequency multipliers to the desired UHF or microwave frequency.

6. A heterodyne transmitter develops two signals. The modulated stage is at a fixed frequency, often in the 30- to 100-MHz range, and then mixes it with another signal to produce a sum or difference frequency in the microwave range. The unmodulated oscillator may be a VCO in the microwave region, a secondary heterodyne system, or a frequency multiplier system.

17-7 RECAPITULATION

Now return to the objectives and prequiz questions at the beginning of the chapter and see how well you can answer them. If you cannot answer certain questions, place a check mark by each and review the appropriate parts of the text. Next, try to answer the following questions and work the problems using the same procedure.

QUESTIONS AND PROBLEMS

1. An RF exciter operates on a frequency of 58.85 MHz. Find the output frequency of the multiplier chain if the following stages are present: 2X, 3X, 3X, 2X, 4X.

2. A transmitter has a frequency multiplier chain with a multiplication ratio of 216. Find the oscillator frequency if the microwave channel is 8453 GHz.

3. List the factors that dictate transmitter architecture.

4. A _____ transmitter needs no modulator.

5. A MOPA transmitter generates the basic RF signal in a (low/high) level stage.

6. An FM transmitter operates on 910.55 MHz and has an X36 frequency multiplication ratio. Find

   a. the oscillator frequency, b. the oscillator stability needed to maintain 500 ppm, c. the oscillator deviation needed to produce an output deviation of 25 kHz.

7. A heterodyne transmitter operates on 3316.5 GHz; the modulated stage operates at 75 MHz. Find the oscillator frequency if the exciter chain has a 72X frequency multiplication factor, and its output is injected on the high side of the operating frequency.

8. A translator inputs on 4.234 GHz and outputs on 8.225 GHz. Calculate the local oscillator frequencies needed to down-convert the receiver to 70 MHz and to up-convert the IF to the transmit frequency. Assume low-side injection of the LO in each case.

KEY EQUATION

1. Output frequency of a multiplier

$$F2 = N \times F1$$

CHAPTER 18

UHF and Microwave Receivers

OBJECTIVES

1. Understand the principal receiver parameters, such as sensitivity, selectivity, intermodulation, 1-dB compression point, third-order intercept point, and dynamic range.
2. Be able to recognize and discuss the various receiver architectures.
3. Review the theory behind superheterodyne receivers.
4. Understand superheterodyne receiver problems such as noise and image response.
5. Understand the functions of the principal stages of a receiver.

18-1 PREQUIZ

These questions test your prior knowledge of the material in this chapter. Try answering them before you read the chapter. Look for the answers (especially those you answered incorrectly) as you read the text. After you have finished studying the chapter, try answering these questions again and those at the end of the chapter.

1. A microwave receiver operates on 4.445 GHz and has a 70-MHz IF amplifier section. Calculate the LO frequencies for high-side and low-side injection.
2. A receiver operates on 8.85 GHz and has an LO frequency of 9.0 GHz. Calculate the IF frequency.
3. In problem 2, calculate the image response frequency.
4. Does a doubly balanced mixer (DBM) suppress the even or odd harmonics of both LO and RF?

18-2 RECEIVERS AND RECEIVER ARCHITECTURES

Receivers are devices that intercept, detect, and demodulate (as needed) electromagnetic waves. The waves are initially intercepted by an antenna, and then fed to the input of a receiver. The receiver has two primary functions:

1. It must respond to signals that are desired.
2. It must reject signals that are not desired.

Most of the design issues regarding radio receivers are determined by these two requirements. Microwave and UHF receivers are used for radar, communications, electronic warfare, and a host of other applications. Many of these receivers have architectures that are common to lower-frequency receivers, while others are found only in the UHF or microwave regions.

Figure 18-1 shows the most basic processes in a receiver, regardless of the design. The antenna is used to capture the signal and convert it from an electromagnetic wave propagating in the atmosphere to an electrical signal oscillating in a conductor. The antenna is usually hooked to the receiver through either waveguide or transmission line. The receiver has three basic types of function in addition to the input and output: *front end*, *detector*, and *signal processing*. These categories are somewhat arbitrary, and often overlap.

The front end is the section of the receiver that receives the signal from the antenna. It will select the particular signal that is being received, or it may be broadbanded and limited to the band of signals is to be accepted. The front end in several types of receiver also converts the incoming signal RF frequency to another frequency; we will discuss this in more detail later in the chapter.

The detector may or may not be part of the front end, depending on the design. It might be something as simple as a diode envelope detector (rectifier and filter) for recovering amplitude modulation, or it may be frequency- or phase-sensitive for detecting frequency modulation. It may also be a product detector for single-sideband,

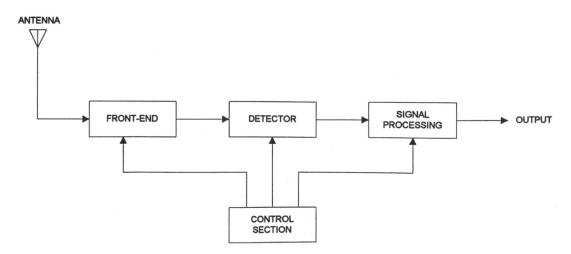

Figure 18-1   Broad view of a receiver block diagram.

CW, and radioteletype (RTTY) demodulation, or any of several detectors for various forms of pulse modulation detection.

The signal-processing section is found on most, but not all, receivers. This section may be as simple as an audio amplifier to build the signal up to a point where it can drive a loudspeaker or earphones; or it may contain filters to further restrict bandwidth or an analog-to-digital converter to convert the recovered signal to a form that a computer can handle. In some receivers, the signal-processing section may also contain functions such as *rectification, integration, differentiation,* or any of a host of others needed for specialized purposes.

The control section may be more than one function, but serves to regulate gain or control frequency, control the mode received, or distribute dc power to the various sections.

18-2.1 The Signal and Noise Problem

A receiver is little more than a device designed to separate electromagnetic signals from the background noise. Therefore, you are advised to be familiar with the noise material presented in Chapter 13. The goal of the receiver designer is to achieve a sufficient signal-to-noise ratio (SNR or S/N) to achieve the stated goal. For most applications, there is a specified SNR that must be achieved in order to provide reasonable service to the user. For voice communications, for example, an SNR of 10 dB is usually specified, although the modulation of the signal is certainly detectable at lower SNRs. This specification basically means that most users find 10-dB SNR a comfortable listening level, and that SNRs less than this provide progressively more difficult reception. Modes other than voice communications may have stricter or lesser SNR requirements.

Signal at the Receiver Input.   The signal level applied to a receiver input is usually specified as either a power level (watts), or as decibels (dBm, where 0 dBm = 1 mW dissipated in a 50-Ω resistive load). The power level at the receiver input (P_r) is a function of the *power density* (S) of the transmitted signal and the effective area of the receive antenna. The power density is a function of transmitter power output (P_t), and transmitter antenna gain (G_t), and is inversely proportional to the square of the distance between the transmitter and receiver (R):

$$S = \frac{G_t P_t}{4\pi R^2} \qquad (18\text{-}1)$$

The effective area of the antenna is a function of the receive antenna gain and the square of the wavelength of the received signal:

$$A_e = \frac{G_r \lambda^2}{4\pi} \qquad (18\text{-}2)$$

Combining these two equations gives the power level at the receiver input:

$$P_t = A_oS \qquad\qquad (18\text{-}3)$$

$$P_t = \frac{G_t\,P_t}{4\pi R^2}\,\frac{G_r\,\lambda^2}{4\pi} \qquad\qquad (18\text{-}4)$$

EXAMPLE 18-1

A transmitter and receiver are used in a one-way communications link. The transmitter produces an output power of 10 W at 1296 MHz, and drives an antenna with a gain of 10 dB. The receiver antenna, located 40 km away, has a gain of 30 dB. Calculate (a) the amount of power presented to the receiver's 50-Ω antenna input, assuming no loss in the transmission line between the antenna and the receiver; and (b) the rms signal voltage across the receiver's antenna terminals.

Solution

Given

$$P_t = 10 \text{ W}$$

$$G_t = 10 \text{ dB (X10)}$$

$$G_r = 30 \text{ dB (X1000)}$$

$$F = 1296 \text{ MHz}$$

$$R = 40,000 \text{ m}$$

$$R_{in} = 50 \text{ }\Omega$$

Calculate

$$l = 300/FMHz$$

$$l = 300/1,296 \text{ MHz} = 0.23 \text{ m}$$

a)

$$P_r = \frac{P_t G_t G_r \lambda^2}{16\pi^2\,R^2}$$

$$= \frac{(10 \text{ W})(10)(1000)(0.23 \text{ m})^2}{(16)(9.87)(40,000 \text{ m})^2}$$

$$= \frac{5290}{2.53 \times 10^{11}} = 2.09 \times 10^{-8} \text{ W}$$

b)

$$V = \sqrt{P_r P_\varepsilon}$$

$$= \sqrt{(2.09 \times 10^{-8} W)(50\Omega)} = 0.001V$$

Using Decibel Notation. It is often convenient to solve problems of this sort using decibel notation. Before re-solving the preceding example using decibels, let's review a few facts. First, all power levels are converted to dBm, that is, decibels relative to 1 mW dissipated in a 50-Ω resistive load. Second, the square terms can be handled by using 20 as the factor in the decibel equation rather than 10. For example, log R^2 is the same as 20 log R (either form can be used).

The goal when working in decibel notation is to add in all the power levels and gains, and subtract all losses. The gains are the transmit antenna gain (G_t) and the receive antenna gain (G_r). The loss term is the path loss, 10 log ($l^2/16p^2R^2$). The equation is

$$P_{rdBm} = P_{t\,(dBm)} + G_{t\,(dB)} + G_{rdB} - L_{dB} \qquad (18\text{-}5)$$

EXAMPLE 18-2

Work the previous example, using decibel notation.

1. Convert P_t (10 W) to dBm.

$$P_{t(dBm)} = 10 \log (10W/0.001W)$$

$$= 10 \log (10,000) = (10)(4) = 40 \text{ dBm}$$

2. Convert G_t (X10) to dB.

$$G_{t(dB)} = 10 \log (10) = (10)(1) = 10 \text{ dB}$$

3. Convert G_r (X1000) to dB.

$$G_{t(dB)} = 10 \log (1000) = (10)(3) = 30 \text{ dB}$$

4. Calculate path losses.

$$L_{dB} = 10 \log \left[\frac{\lambda^2}{16\pi^2 R^2} \right]$$

5. Calculate P_r.

$$P_{rdBm} = Pt\,(_{dBm}) + Gt\,(_{dB}) + G_{rdB} - L_{dB}$$

$$= 40 \text{ dBm} + 10 \text{ dB} + 30 \text{ dB} - 126.79 \text{ dB}$$

$$= -46.79 \text{ dBm}$$

If you convert −46.79 dBm to watts, you will find that it is the same number as the solution to the previous example. The procedure in this example looks longer than that of the previous example, but in reality the values are generally known in decibel form rather than as power levels and gain factors.

Noise Floor. The *noise floor* of the receiver is a statement of the amount of noise produced by the receiver's internal circuitry and directly affects the sensitivity. The noise floor is typically expressed in dBm. The noise floor specification is evaluated as follows: the more negative, the better. The best receivers have noise floor numbers of less than −130 dBm, while some very good receivers offer numbers of −115 dBm to −130 dBm.

The noise floor is directly dependent on the bandwidth used to make the measurement. Receiver advertisements usually specify the bandwidth, but be careful to note whether or not the bandwidth that produced the very good performance numbers is also the bandwidth that you'll need for the mode of transmission you want to receive. If, for example, you are interested only in weak 6-kHz-wide AM signals, and the noise floor is specified for a 250-Hz CW filter, then the noise floor might be too high for your use.

18-2.2 How Much Gain Is Required?

One way to view a receiver is as a frequency-selective gain block. The question therefore arises: "How much gain is required?" The answer depends on the output levels required and the input levels available. The required power gain is

$$G_p = \frac{P_{out}}{P_{in}} \tag{18 - 6A}$$

and the required voltage gain is

$$G_v = \frac{V_{out}}{V_{in}} \tag{18 - 6B}$$

Let's suppose a situation where 2×10^{-14} W is applied to the input of the receiver (this power level is the same as an input voltage of 1 mV across a 50-Ω load). The output is a power level of 250 mW (0.25 W) applied to an 8-Ω loudspeaker. The gain, therefore, is

$$G_p = P_{out}/P_{in} = (0.25 \text{ W})/(2 \times 10^{-14} \text{ W}) = 1.25 \times 10^{13}$$

Recalculated in terms of decibels,

$$G_{p \text{ (dB)}} = 10 \log (1.25 \times 10^{13}) = (10)(13.1) = 131 \text{ dB}$$

The gain of the receiver to produce a 250-mW output from an input of 2×10^{-14} W is 131 dB. In the sections to follow, we will examine some of the most common receiver architectures.

18-3 CRYSTAL VIDEO RECEIVERS

One of the earliest forms of radio receiver was the simple crystal set. And although the modern crystal video microwave receiver is at a much higher technological level, it is nonetheless little more than an outgrowth of that early technology. The early crystal set (Fig. 18-2) consisted of an antenna driving a crystal diode detector, and some output device, such as an earphone. A tuning network to select the operating frequency was optional. The crystal detector in those early sets was a natural mineral crystal called *galena* (a lead compound). The galena crystal usually sat in a metal cup (which formed an electrode), while a second electrode called a cat's whisker was used to probe the crystal surface to find a site that would provide radio detection. Research and development during World War II, however, led to a family of solid-state germanium and signal diodes (1N34, 1N60, 1N21, 1N23, and the like) that served as a radio detector. These *PN* junction diodes were the forerunners of all solid-state electronics parts used today, and the latter two were able to operate into the low microwave regions used during the war.

The single-diode crystal detector (as shown in Fig. 18-2) operates as a halfwave rectifier, while some multidiode circuits operate in the full-wave mode. The diode detector produces a pulsating dc level at its output that has a level proportional to the applied signal strength, which becomes a dc level when low-pass filtered, and because of this fact these simple detectors are called *envelope detectors*. The envelope detector is used for demodulating amplitude modulation (AM) signals.

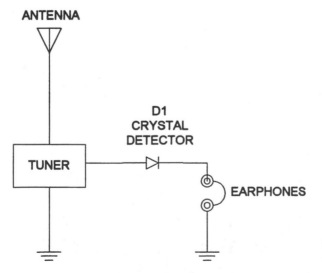

Figure 18-2
Crystal tuner receiver.

Several configurations of crystal video receiver as found in microwave applications are shown in Fig. 18-3. Nearly all of these are followed by a high-gain, wideband amplifier (similar or identical to a video amplifier); hence the name *crystal video* is used for these architectures even when no video signals are being processed.

The simplest form of crystal video receiver is shown in Fig. 18-3A. This type of receiver has the crystal detector between the antenna input and the video amplifier input. It is characterized by very wide bandwidth because the bandwidth of the detector is limited only by the antenna bandwidth, the transmission line bandwidth, and the natural limitations of the diode detector, whichever is less.

These receivers are also characterized as relatively low sensitivity because the signal levels that can be detected are limited by the junction potential of the crys-

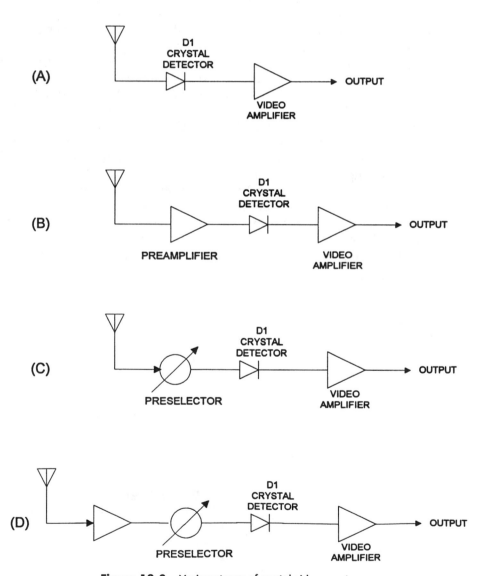

Figure 18-3   Various types of crystal video receiver.

tal diode (0.2 to 0.3 V for germanium and 0.6 to 0.7 V for silicon). Even if the video amplifier has a high gain, there is a limit to the level of signal that can be detected by the diode.

The sensitivity issue can be partially overcome by adding a preamplifier ahead of the crystal detector, as shown in Fig. 18-3B. The preamplifier will provide some gain, and may also provide some bandwidth-limiting effects. When combined with the gain of the video amplifier, this design is capable of considerably greater sensitivity.

Another problem with the two simple crystal video receivers of Figs. 18-3A and B is that they receive all frequencies at the same time. It is therefore relatively easy for a strong, but undesired, signal to dominate the receiver. Such signals will tend to diminish the receiver's dynamic range. A solution to this problem is to use a preselector ahead of the detector diode, as in Fig. 18-3C. The preselector passes some frequencies and rejects others. At lower frequencies, the preselector is usually an L-C filter of some sort, but at microwave frequencies tunable YIG filters are common.

The preselector may be a single-frequency tuner, or it may be a bandpass, low-pass, or high-pass filter. Combining the preselector and the preamplifier (Fig. 18-3D) provides both gain ahead of the crystal detector and frequency selection. Although only a single stage of preselection (shown following the preamplifier) is shown here, actual receiver designs may use the single stage of preselection ahead of the preamplifier. Also found are designs where there are two preselection stages, ahead of and following the preamplifier. In some cases, two preselection sections are different, one being a broad bandpass filter and the other being a single-frequency tuner.

18-4 TUNED RADIO FREQUENCY (TRF) RECEIVERS

The tuned radio frequency (TRF) receiver design is much like a crystal video, except that there are two or more stages of amplification ahead of the detector (Fig. 18-4), each of which can be modeled as shown in Fig. 18-3D. These receivers are capable of considerable sensitivity, although are rarely used. One of the problems facing the TRF designer is that they will work relatively well on a single frequency, but when made tunable they prove to be unstable, and work quite differently on different ends of the band. Although this design was common on broadcast and communications receivers prior to about 1928, it is used only rarely today. Some microwave examples are known, however.

18-5 BRAGG CELL RECEIVERS

The Bragg cell receiver (Fig. 18-5) uses the phenomenon of Bragg scattering in the structure of a crystal (not the same crystal used in crystal video receivers, but a natural crystalline solid). When excited by both a laser light source and a high-powered RF input signal, the Bragg cell crystal sets up internal compression lines with spacing that is proportional to the wavelength of the applied RF signal. These compression lines cause the laser light to diffract at an angle that depends on the

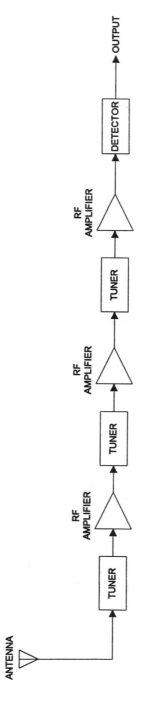

Figure 18-4 Tuned radio frequency (TRF) radio receivers.

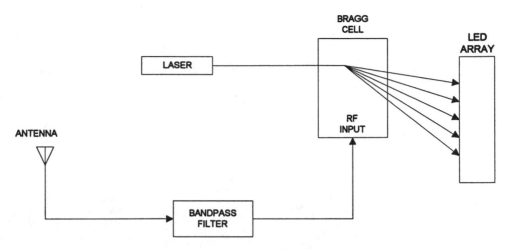

Figure 18-5   Bragg cell receiver.

wavelength (hence frequency) of the applied RF signal. These lines can be sensed and displayed, providing an instantaneous spectrum of the applied signal.

18-6 HOMODYNE DIRECT CONVERSION RECEIVERS

The homodyne receiver uses a *local oscillator* (LO) to produce a stable local signal ($F2$) that has the same frequency as the incoming RF signal ($F1$). When these signals are mixed together in a nonlinear circuit element, such as a crystal diode (Fig. 18-6A), a set of new frequencies is created (these will be discussed at length in the sections that follow). For our present purposes, however, consider that four frequencies will be produced: $F1$, $F2$, $F1+F2$, and $F1-F2$. If a low-pass filter is used at the output of the mixer, then only the difference signal ($F1-F2$) is passed.

Consider the example in Fig. 18-6B where an AM signal is presented at the input of the receiver. The AM signal consists of the carrier, which is $F1$ plus the two sidebands. The bandwidth of the sidebands is the same as the bandwidth of the

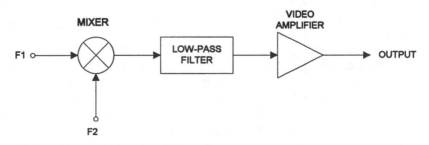

Figure 18-6A   Homodyne or heterodyne direct-conversion receiver.

Continued

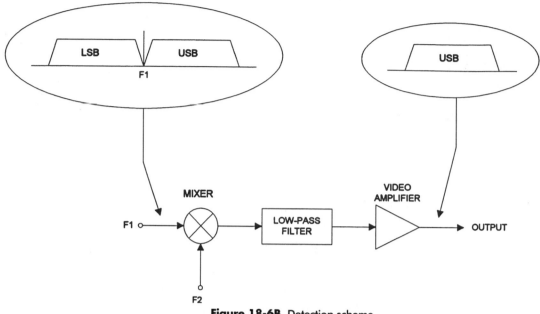

Figure 18-6B Detection scheme.

audio modulating signal, and the upper sideband (USB) and lower sideband (LSB) each contain 100% of the information carried on the radio signal. In the homodyne receiver, the local oscillator signal $F2$ is the same as $F1$, so by taking the difference frequency, only the sidebands lying close to $F1$ are recovered. Such receivers are also called *direct-conversion* receivers because they go direct from an incoming RF signal to baseband without using conventional envelope detection.

18-7 SUPERHETERODYNE RECEIVERS

Figure 18-7 shows the block diagram of a simple communications receiver. We will use this hypothetical receiver as the basic generic framework for evaluating receiver performance. The design in Fig. 18-7 is called a *superheterodyne* receiver and is representative of a large class of radio receivers; it covers the vast majority of receivers on the market.

Heterodyning.   The main attribute of the superheterodyne receiver is that it converts the radio signal's RF frequency to a standard frequency for further processing. Although today the new frequency, called the *intermediate frequency* or *IF*, may be either higher or lower than the RF frequencies, early superheterodyne receivers always down-converted to a lower IF frequency (IF < RF). The reason was purely practical, for in those days higher frequencies were more difficult to process than lower frequencies. Even today, because variable-tuned circuits still tend to offer different performance over the band being tuned, converting to a single IF

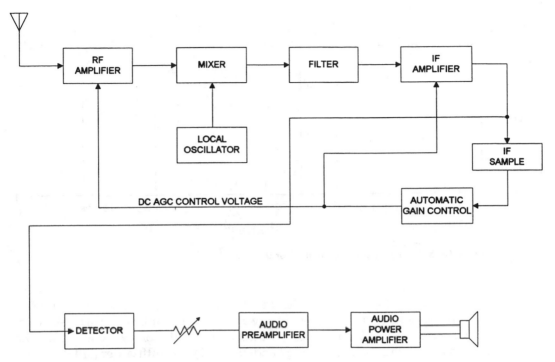

Figure 18-7   Superheterodyne receiver.

frequency, and obtaining most of the gain and selectivity functions at the IF, allows more uniform overall performance over the entire range being tuned.

A superheterodyne receiver works by frequency converting (*heterodyning* — the *super* part is 1920s vintage advertising hype) the RF signal. This occurs by non-linearly mixing the incoming RF signal with a *local oscillator* (LO) signal. When this process is done, disregarding noise, the output spectrum will contain a large variety of signals according to

$$F_o = mF_{RF} \pm nF_{LO} \tag{18-7}$$

where

F_{RF} = frequency of the RF signal
F_{LO} = frequency of the local oscillator
m and n = either zero or integers $(0, 1, 2, 3 \ldots n)$

Equation (18-7) means that there will be a large number of signals at the output of the mixer, although for the most part the only ones that are of immediate concern to understanding basic superheterodyne operation are those for which m and n are either 0 or 1. Thus, for our present purpose, the output of the mixer will be the fundamentals (F_{RF} and F_{LO}) and the second-order products ($F_{LO} - F_{RF}$ and $F_{LO} + F_{RF}$), as seen in Fig. 18-8. Some mixers, notably those described as *double-balanced mixers*

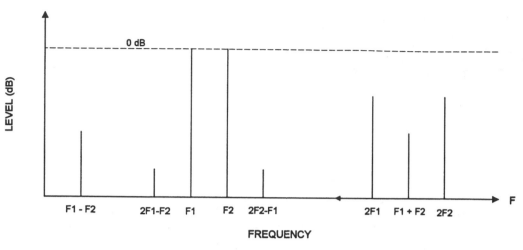

Figure 18-8 Frequency spectrum at mixer output.

(DBM), suppress F_{RF} and F_{LO} in the mixer output, so only the second-order sum and difference frequencies exist with any appreciable amplitude. This case is simplistic, and is used only for this present discussion. Later on, we will look at what happens when third-order ($2F1 \pm F2$ and $2F2 \pm F1$) and fifth-order ($3F1 \pm 2F2$ and $3F2 \pm 2F1$) become large.

Note that the local oscillator frequency can be either higher than the RF frequency (*high-side injection*) or lower than the RF frequency (*low-side injection*). There is ordinarily no practical reason to prefer one over the other except that it will make a difference whether the main tuning dial reads high-to-low or low-to-high.

The candidates for IF are the sum (LO + RF) and difference (LO – RF) of second-order products found at the output of the mixer. A high-Q tuned circuit following the mixer will select which of the two is used. Consider an example: suppose a receiver has an IF frequency of 70 MHz, and the tuning range is 540 to 1700 MHz. Because the IF is lower than any frequency within the tuning range, it will be the difference frequency that is selected for the IF. The local oscillator is set to be high-side injection, so will tune from (540 + 70) or 610 MHz, to (1700 + 70) or 1770 MHz.

Front-End Circuits. The principal task of the front-end section of the receiver in Fig. 18-7 is to perform the frequency conversion. In many radio receivers, however, there may be additional functions. In some cases (but not all), an RF amplifier will be used ahead of the mixer. Typically, these amplifiers have a gain of 3 to 10 dB, with 5 to 6 dB being very common. The tuning for the RF amplifier is sometimes a broad-bandpass fixed-frequency filter that admits an entire band. In other cases, it is a narrowband, but variable-frequency, tuned circuit.

Intermediate-Frequency Amplifier. The IF amplifier is responsible for providing most of the gain in the receiver, as well as the narrowest bandpass filtering. It is a high-gain, usually multistaged, single-frequency tuned radio frequency amplifier. For example, one receiver block diagram lists 120 dB of gain from antenna terminals to audio output, of which 85 dB are provided in the 70-MHz IF amplifier chain. In the example of Fig. 18-7, the receiver is a single conversion design, so there is only one IF amplifier section.

Detector. The detector demodulates the RF signal and recovers whatever audio (or other information) is to be heard by the listener. In a straight AM receiver, the detector will be an ordinary half-wave rectifier and ripple filter, and is called an *envelope detector*. In other detectors, notably double-sideband suppressed carrier (DSBSC), single-sideband suppressed carrier (SSBSC or SSB), or continuous-wave (CW or Morse telegraphy), a second local oscillator—usually called a *beat frequency oscillator* (BFO)—operating near the IF frequency is heterodyned with the IF signal. The resultant difference signal is the recovered audio. That type of detector is called a *product detector*. Many AM receivers today have a sophisticated *synchronous detector*, rather than the simple envelope detector.

Audio Amplifiers. The audio amplifiers are used to finish the signal processing. They also boost the output of the detector to a usable level to drive a loudspeaker or set of earphones. The audio amplifiers are sometimes used to provide additional filtering. It is quite common to find narrowband filters to restrict audio bandwidth, or notch filters to eliminate interfering signals that make it through the IF amplifiers intact.

There are three basic areas of receiver performance that must be considered. Although interrelated, they are sufficiently different to merit individual consideration: *noise, static,* and *dynamic*. We will look at all of these areas, but first let's look at the units of measure that we will use in this series.

18-8 INTERPRETING RADIO RECEIVER SPECIFICATIONS

There are a number of attributes of receivers that must be understood before they can be put into context, specified, and properly used. In the sections to follow you will find both static and dynamic specifications of receivers, and what they mean.

18-8.1 Units of Measure

Input Signal Voltage. Input signal levels, when specified as a voltage, are typically stated in either microvolts (mV) or nanovolts (nV); the volt is simply too large a unit for practical use on radio receivers. Signal input voltage (or sometimes power level) is often used as part of the sensitivity specification, or as a test condition for measuring certain other performance parameters.

There are two forms of signal voltage that are used for input voltage specification: *source voltage* (V_{EMF}) and *potential difference* (V_{PD}), as illustrated in Fig. 18-9 (after Dyer 1993). The source voltage (V_{EMF}) is the open terminal (no load) voltage of the signal generator or source, while the potential difference (V_{PD}) is the voltage that appears across the receiver antenna terminals with the load connected (the load is the receiver antenna input impedance, R_{in}). When $R_s = R_{in}$, the preferred "matched impedances" case in radio receiver systems, the value of V_{PD} is one-half V_{EMF}. This can be seen in Fig. 18-9 by noting that R_S and R_{in} form a voltage divider network driven by V_{EMF}, with V_{PD} as the output.

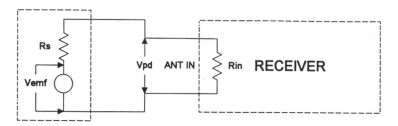

Figure 18-9 Receiver input equivalent circuit.

dBm. These units refer to decibels relative to one milliwatt (1 mW) dissipated in a 50-Ω resistive impedance (defined as the 0-dBm reference level), and are calculated from 10 log ($P_{\text{WATTS}}/0.001$) or 10 log (P_{MW}). In the noise voltage case calculated above, 0.028 mV in 50 Ω, the power is $V^2/50$, or 5.6×10^{-10} W, which is 5.6×10^{-7} mW. In dBm notation, this value is 10 log (5.6×10^{-7}), or –62.5 dBm.

dBmV. This unit is used in television receiver systems in which the system impedance is 75 Ω, rather than the 50 Ω normally used in other RF systems. It refers to the signal voltage, measured in decibels, with respect to a signal level of one millivolt (1 mV) across a 75-Ω resistance (0 dBmv). In many TV specs, 1 mV is the full quieting signal that produces no "snow" (noise) in the displayed picture.

dBμV. This unit refers to a signal voltage, measured in decibels, relative to one microvolt (1 mV) developed across a 50-Ω resistive impedance (0 dBmV). For the case of our noise signal voltage, the level is 0.028 mV, which is the same as –31.1 dBmV. The voltage used for this measurement is usually V_{EMF}, so to find V_{PD}, divide it by 2 after converting dBmV to mV. To convert dBmV to dBm, merely subtract 113; 100 dBmV = –13 dBm.

It requires only a little algebra to convert signal levels from one unit of measure to another. This job is sometimes necessary when a receiver manufacturer mixes methods in the same specifications sheet. In the case of dBm and dBmV, 0 dBmV is 1 mV V_{EMF}, or a V_{PD} of 0.5 mV, applied across 50 Ω, so the power dissipated is 5×10^{-15} W or –113 dBm.

18-9 STATIC MEASURES OF PERFORMANCE

The two principal static levels of performance for radio receivers are *sensitivity* and *selectivity*. Sensitivity refers to the level of input signal required to produce a usable output signal (variously defined). Selectivity refers to the ability of the receiver to reject adjacent channel signals (again, variously defined). Let's take a look at both of these factors. Keep in mind, however, that in modern high-performance radio receivers the static measures of performance may also be the least relevant, compared with the dynamic measures.

18-9.1 Sensitivity

Sensitivity is a measure of the receiver's ability to pick up (*detect*) signals, and is usually specified in microvolts (mV). A typical specification might be "0.5 mV sensitivity." The question to ask is: "Relative to what?" The sensitivity number in microvolts is meaningless unless the test conditions are specified. For most commercial receivers, the usual test condition is the sensitivity required to produce a 10-dB signal-plus-noise-to-noise (S+N/N) ratio in the mode of interest. For example, if only one sensitivity figure is given, one must find out what bandwidth is being used: 5 to 6 kHz for AM, 2.6 to 3 kHz for single sideband, 1.8 kHz for radioteletype, or 200 to 500 Hz for CW.

Bandwidth affects sensitivity measurements. Indeed, one place where creative spec writing takes place for commercial receivers is that advertisements will cite the sensitivity for a narrow-bandwidth mode (for example, CW), while the other specifications are cited for a more commonly used wider-bandwidth mode (for example, SSB). In one particularly egregious example, an advertisement claimed a sensitivity number that was applicable to the CW mode bandwidth only, yet the 270-Hz CW filter was an expensive option that had to be ordered separately!

The amount of sensitivity improvement is shown by some simple numbers. Recall that a claim of x mV sensitivity refers to some standard such as x mV to produce a 10-dB signal-to-noise ratio. Consider the case where the main mode for a receiver is AM (for international broadcasting), the sensitivity is 1.9 mV for 10-dB SNR, and the bandwidth is 5 kHz. If the bandwidth were reduced to 2.8 kHz for SSB, the sensitivity improves by the square root of the ratio, or Ö5/2.8. If the bandwidth is further reduced to 270 Hz (0.27 kHz) for CW, then the sensitivity for 10-dB SNR is Ö5/0.27. The 1.9-mV AM sensitivity therefore translates to 1.42 mV for SSB and 0.44 mV for CW. If only the CW version is given, then the receiver might be made to look a whole lot better, even though the typical user may never use the CW mode (note the differences in Fig. 18-10).

The sensitivity differences also explain why weak SSB signals can be heard under conditions when AM signals of similar strength have disappeared into the

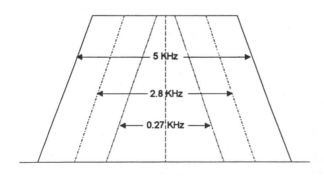

NOISE
PROPORTIONAL TO
AREA UNDER CURVE

Figure 18-10
Noise is proportional to receiver bandwidth.

noise, and why the CW mode has as much as 20 dB advantage over SSB, all other things being equal.

In some receivers, the difference in mode (AM, SSB, RTTY, CW, and the like) can conceivably result in sensitivity differences that are more than the differences in the bandwidths associated with the various modes. The reason is that there is sometimes a processing gain associated with the type of detector circuit used to demodulate the signal at the output of the IF amplifier. A simple AM envelope detector is lossy because it consists of a simple diode (1N60 and the like) and an R-C filter (a passive circuit). Other detectors (product detectors for SSB, synchronous AM detectors) have their own signal gain, so may produce better sensitivity numbers than the bandwidth suggests.

Another indication of sensitivity is *minimum detectable signal* (MDS), and is usually specified in dBm. This signal level is the signal power at the antenna input terminal of the receiver required to produce some standard S+N/N ratio, such as 3 dB or 10 dB (Fig. 18-11). In radar receivers, the MDS is usually described in terms of a single pulse return and a specified S+N/N ratio. Also, in radar receivers, the sensitivity can be improved by integrating multiple pulses. If N return pulses are integrated, then the sensitivity is improved by a factor of N if coherent detection is used, and ÖN if noncoherent detection is used.

Modulated signals represent a special case. For those sensitivities, it is common to specify the conditions under which the measurement is made. For example, in AM receivers the sensitivity to achieve 10-dB SNR is measured with the input signal modulated 30% by a 400- or 1000-Hz sinusoidal tone.

An alternate method is sometimes used for AM sensitivity measurements, especially in measuring consumer radio receivers (where SNR may be a little hard to measure with the equipment normally available to technicians who work on those radios). This is the standard output conditions method. Some manuals will specify the audio signal power or audio signal voltage at some critical point, when the 30% modulated RF carrier is present. In one automobile radio receiver, the sensitivity was specified as "X mV to produce 400 mW across 8-ohm resistive load substituted for the loudspeaker when the signal generator is modulated 30% with a 400-Hz audio tone." The cryptic note on the schematic showed an output sine wave across the loudspeaker with the label "400 mW in 8W (1.79 volts), @30% mod. 400 Hz, 1 mV RF." The sensitivity is sometimes measured essentially the same way, but the signal levels will specify the voltage level that will appear at the top of the volume control

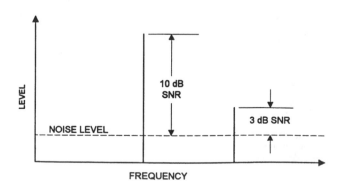

Figure 18-11
10-dB and 3-dB SNR situations.

or output of the detector/filter when the standard signal is applied. Thus, there are two ways of specifying AM sensitivity: *10-dB SNR* and *standard output conditions*.

There are also two ways to specify FM receiver sensitivity. The first is the 10-dB SNR method discussed above, that is, the number of microvolts of signal at the input terminals required to produce a 10-dB SNR when the carrier is modulated by a standard amount. The measure of FM modulation is *deviation* expressed in kilohertz. Sometimes the full deviation for that class of receiver is used; other times a value that is 25% to 35% of full deviation is specified.

The second way to measure FM sensitivity is the level of signal required to reduce the no-signal noise level by 20 dB. This is the 20-dB quieting sensitivity of the receiver. If you tune between signals on an FM receiver, you will hear a loud hiss signal, especially in the VHF/UHF bands. Some of that noise is externally generated, while some is internally generated. When an FM signal appears in the passband, that hiss is suppressed, even if the FM carrier is unmodulated. The *quieting sensitivity* of an FM receiver is a statement of the number of microvolts required to produce some standard quieting level, usually 20 dB.

Pulse receivers, such as radar and pulse communications units, often use the *tangential sensitivity* as the measure of performance, which is the amplitude of pulse signal required to raise the noise level by its own RMS amplitude (Fig. 18-12).

18-9.2 Selectivity

Although no receiver specification is unimportant, if one had to choose between sensitivity and selectivity, the proper choice most of the time would be selectivity.

Selectivity is the measure of a receiver's ability to reject adjacent channel interference. Put another way, it is the ability to reject interference from signals on frequencies close to the desired signal frequency.

In order to understand selectivity requirements, one must first understand a little bit of the nature of real radio signals. An unmodulated radio carrier theoretically has an infinitesimal (near-zero) bandwidth (although all real unmodulated carriers have a very narrow, but nonzero, bandwidth because they are modulated by noise and other artifacts). As soon as the radio signal is modulated to carry information, however, the bandwidth spreads.

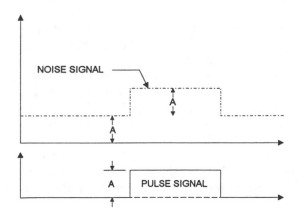

Figure 18-12
Tangential sensitivity.

An on/off telegraphy (CW) signal spreads out on either side of the carrier frequency an amount that depends on the sending speed and the shape of the keying waveform.

Figure 18-13 compares two radioteletype signals with 200-Hz (Fig. 18-13A) and 800-Hz (Fig. 18-13B) mark/space separation. Note how the sidebands spread out from the main mark and space signals.

An AM signal spreads out an amount equal to twice the highest audio-modulating frequencies. For example, a communications AM transmitter will have audio components from 300 to 3000 Hz, so the AM waveform will occupy a spectrum that is equal to the carrier frequency (*F*) plus/minus the audio bandwidth (*F* ± 3000 Hz in the case cited).

An FM carrier spreads out according to the *deviation*. For example, a narrowband FM landmobile transmitter with 5-kHz deviation spreads out ±5 kHz, while FM broadcast transmitters spread out ±75 kHz.

An implication of the fact that radio signals have bandwidth is that the receiver must have sufficient bandwidth to recover all of the signal. Otherwise, information may be lost and the output is distorted. On the other hand, allowing too much bandwidth increases the noise picked up by the receiver and thereby deteriorates the SNR. The goal of the selectivity system of the receiver is to match the bandwidth of the receiver to that of the signal. That is why receivers will use 270 or 500 Hz bandwidth for CW, 2 to 3 kHz for SSB, and 4 to 6 kHz for AM signals. They allow you to match the receiver bandwidth to the transmission type.

The selectivity of a receiver has a number of aspects that must be considered: *front-end bandwidth, IF bandwidth, IF shape factor,* and the *ultimate* (distant frequency) *rejection*.

Front-end Bandwidth.

The front end of a modern superheterodyne radio receiver is the circuitry between the antenna input terminal and the output of the first mixer stage. The reason that front-end selectivity is important is to keep out-of-band signals from afflicting the receiver. For example, AM broadcast band transmitters located nearby can easily overload a poorly designed shortwave receiver. Even if these signals are not heard by the operator (as they often are), they can desensitize a receiver, or create harmonics and intermodulation products that show up as birdies or other types of interference on the receiver. Strong local signals can take a lot of the receiver's dynamic range and thereby make it harder to hear weak signals.

In some crystal video microwave receivers, that front end might be wide open without any selectivity at all, but in nearly all other receivers there will be some form of frequency selection present.

Two forms of frequency selection are typically found. A designer may choose to use only one of them in a design. Alternatively, both might be used in the design, but separately (operator selection). Or finally, both might be used together. These forms can be called the *resonant frequency filter* (Fig. 18-14A) and the *bandpass filter* (Fig. 18-14B) approaches.

The resonant frequency approach uses L-C elements tuned to the desired frequency to select which RF signals reach the mixer. In some receivers, these L-C elements are designed to track with the local oscillator that sets the operating frequency.

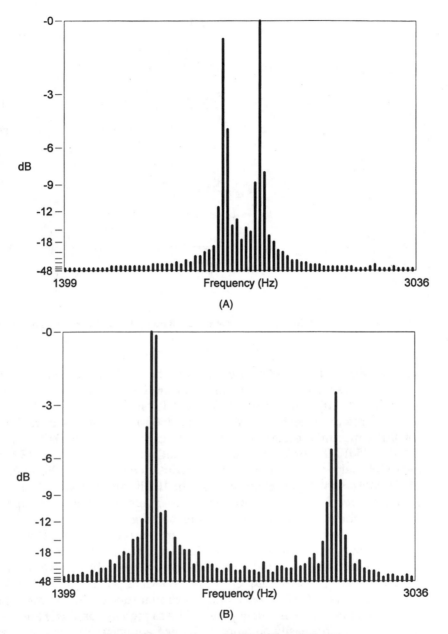

Figure 18-13   A) 200-Hz RTTY; B) 800-Hz RTTY.

The other approach uses a suboctave bandpass filter to admit only a portion of the RF spectrum into the front end. For example, a shortwave receiver that is designed to take the HF spectrum in 1-MHz pieces may have an array of RF input bandpass filters that are each 1 MHz wide (for example, 9 to 10 MHz).

In addition to the reasons cited above, front-end selectivity also helps improve a receiver's *image rejection* and *1st IF rejection* capabilities.

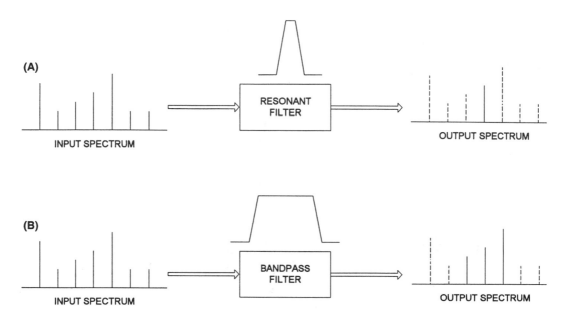

Figure 18-14 A) –3-dB bandwidth; B) ratio of –3 dB and –60 dB form the skirt factor.

Image Rejection. An *image* in a superheterodyne receiver is a signal that appears at twice the IF distance from the desired RF signal, located on the opposite side of the LO frequency from the desired RF signal. In Fig. 18-15, a superheterodyne microwave receiver operates with a 450-MHz IF, and is turned to 2400 MHz (F_{RF}). Because this receiver uses low-side LO injection, the LO frequency F_{LO} is (2400 to 450) MHz or 1950 MHz. If a signal appears at twice the IF below the RF (900 MHz below F_{RF}), and reaches the mixer, then it too has a difference frequency of 450 MHz, so will pass right through the IF filtering as a valid signal. The image rejection specification tells how well this image frequency is suppressed. Normally, anything over about 70 dB is considered good.

Tactics to reduce image response vary with the design of the receiver. The best approach, at design time, is to select an IF frequency that is high enough that the image frequency will fall outside the passband of the receiver front end. Some modern receivers use IF frequencies of 8.83 MHz, 9 MHz, 10.7 MHz, 45 MHz, 70 MHz, 270 MHz, 450 MHz, or something similar (depending on the use). For good image rejection, the IF frequency should be as high as possible. However, selectivity is usually more easily obtained at lower frequencies. A common trend is to do *double conversion* in order to take advantage of both attributes. In most such designs, the first IF frequency is considerably higher than the RF, being in the range 250 to 1200 MHz. This high IF makes it possible to suppress microwave images with a simple low-pass filter.

Consider an example. If the 2400-MHz signal (above) were first converted to 70 MHz (2470-MHz LO) in a single-conversion receiver, the image would be found at 2540 MHz. This frequency may be well within the passband of the front end of the receiver, and thus be heard in the output. But if the first conversion were to 450 MHz, then the LO frequency is 2850 MHz, and the image would be 3300 MHz. The

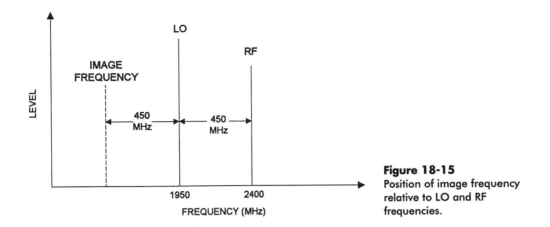

Figure 18-15
Position of image frequency relative to LO and RF frequencies.

frequency 3300 MHZ is much more likely to be outside the front-end passband than 2540 MHz. The second conversion brings the 450-MHz first-IF down to one of the frequencies mentioned above, most likely 70 MHz.

In triple-conversion receivers, the second-IF signal will be converted to still another IF. The lower frequencies are preferable in narrowband cases to 70 MHz for bandwidth selectivity reasons, because good-quality crystal, ceramic, or mechanical filters in those ranges are easily available.

1st IF Rejection.   The 1st IF rejection specification refers to how well a receiver rejects radio signals operating on the receiver's first IF frequency. For example, if your receiver has a first IF of 70 MHz, it must be able to reject radio signals operating on that frequency when the receiver is tuned to a different frequency. Although the shielding of the receiver is also an issue with respect to this performance, the front-end selectivity affects how well the receiver performs against 1st IF signals. If there is no front-end selectivity to discriminate against signals at the IF frequency, then they arrive at the input of the mixer unimpeded. Depending on the design of the mixer, they then may pass directly through to the high-gain IF amplifiers and be heard in the receiver output.

IF Bandwidth.   Most of the selectivity of the receiver is provided by the filtering in the IF amplifier section. The filtering might be L-C filters (especially if the principal IF is a low frequency like 50 kHz), a ceramic resonator, a crystal filter, or a mechanical filter. Of these, the mechanical filter is usually regarded as best, with the crystal filter and ceramic filters coming in next.

The IF bandwidth is expressed in kilohertz and is measured from the points on the IF frequency response curve where gain drops off –3 dB from the mid-band value (Fig. 18-16). This is why you will sometimes see selectivity referred to in terms such as "6 kHz between –3-dB points."

The IF bandwidth must be matched to the bandwidth of the received signal for best performance. If a too-wide bandwidth is selected, then the received signal will be noisy, and SNR deteriorates. If too narrow, then you might experience difficulties recovering all the information that was transmitted. For example, an AM

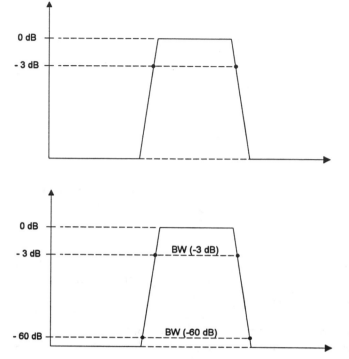

Figure 18-16
Bandwidth is measured between the –3 dB points, while the shape factor is the ratio of the –3 dB and –60 dB bandwidths.

broadcast band radio signal has audio components to 5 kHz, so the signal occupies up to 10 kHz of spectrum space ($F \pm 5$ kHz). If a 2.8-kHz SSB IF filter is selected, then it will tend to sound mushy and distorted.

IF Passband Shape Factor.   The shape factor is a measure of the steepness of the receiver's IF passband and is taken by measuring the ratio of the bandwidth at –6 dB to the bandwidth at –60 dB. The general rule is that the closer these numbers are to each other, the better the receiver. Anything in the 1:1.5 to 1:1.9 region can be considered high quality, while anything worse than 1:3 is not worth looking at for serious receiver uses. If the numbers are between 1:1.9 and 1:3, then the receiver could be regarded as being middling, but useful.

The importance of shape factor is that it modifies the notion of bandwidth. The cited bandwidth (for example, 2.8 kHz for SSB) does not take into account the effects of strong signals that are just beyond those limits. Such signals can easily punch through the IF selectivity if the IF passband skirts are not steep. After all, the steeper they are, the closer a strong signal can be without adversely affecting the receiver's operation. Thus, selecting a receiver with a shape factor as close to the 1:1 ideal as possible will result in a more usable radio.

Distant-Frequency ("Ultimate") Rejection.   This specification tells something about the receiver's ability to reject very strong signals that are located well outside the receiver's IF passband. This number is stated in negative decibels, and the higher the number the better. An excellent receiver will have values in the

–60- to –90-dB range, a middling receiver will see numbers in the –45- to –60-dB range, and a terrible receiver will be –44 dB or worse.

18-9.3 Stability

The *stability* specification measures how much the receiver frequency drifts as time elapses or temperature changes. The LO drift sets the overall stability of the receiver. This specification is usually given in terms of *short-term drift* and *long-term drift* (for example, from crystal aging). The short-term drift is important in daily operation, while the long-term drift ultimately affects general dial calibration.

If the receiver is VFO controlled, or uses partial-frequency synthesis (which combines VFO with crystal oscillators), then the stability is dominated by the VFO stability. In fully synthesized receivers, the stability is governed by the master reference crystal oscillator. If either an *oven-controlled crystal oscillator* (OCXO) or a *temperature-compensated crystal oscillator* (TCXO) is used for the master reference, then stabilities on the order of 1 part in $10^8/°C$ are achievable.

For most users, the short-term stability is what is most important, especially when tuning SSB, ECSS, or RTTY signals. A common spec for a good receiver will be 50 Hz/hr after a 3-hr warmup, or 100 Hz/hr after a 15-min warmup. The smaller the drift the better the receiver.

The foundation of good stability is found at design time. The local oscillator, or VFO or crystal oscillator portions of a frequency synthesizer, must be operated in a cool, temperature-stable location within the equipment, and must have the correct types of components. Capacitor temperature coefficients are often selected in order to cancel out temperature-related drift in inductance values.

Post-design time changes can also help, but these are less likely to be possible today than in the past. The chief cause of drift problems is heat. In the days of tube oscillators, the heat of the tubes produced lots of heat that created drift. One popular receiver of the 1950s was an excellent performer except for the drift of the VFO/LO. A popular modification was to insert metal panels backed with insulating material between the heat source (an IF amplifier chain) and the VFO case. Rather dramatic improvement resulted. In most modern solid-state receivers, however, space is not so easily found for such modifications.

18-9.4 AGC Range and Threshold

Modern communications receivers must be able to handle signals over a range of about 1,000,000:1. Tuning across a band occupied by signals of wildly varying strengths is hard on the ears and hard on the receiver's performance. As a result, most modern receivers have an *automatic gain control* (AGC) circuit that smooths out these changes. The AGC will reduce gain for strong signals and increase it for weak signals (AGC can be turned off on most HF communications receivers). The AGC range is the change of input signal (in $dB\mu V$) from some reference level (for example, $1 \mu V_{EMF}$) to the input level that produces a 2-dB change in output level. Ranges of 90 to 110 dB are commonly seen.

The AGC threshold is the signal level at which the AGC begins to operate. If it is set too low, the receiver gain will respond to noise and irritate the user. If it is set too high, the user will experience irritating shifts of output level as the band is tuned. AGC thresholds of 0.7 to 2.5 μV are common on decent receivers, with the better receivers being in the 0.7- to 1-μV range.

Another AGC specification sometimes seen deals with the speed of the AGC. Although sometimes specified in milliseconds, it is also frequently specified in subjective terms like fast and slow. This specification refers to how fast the AGC responds to changes in signal strength. If it is set too fast, then rapidly keyed signals (for example, CW) or noise transients will cause unnerving large shifts in receiver gain. If it is set too slow, then the receiver might as well not have an AGC. Many receivers provide two or more selections in order to accommodate different types of signals.

Now let's examine what some authorities believe are the most important specifications for receivers used on crowded bands, or in high-electromagnetic interference (EMI) environments: the dynamic measures of performance. These include *intermodulation distortion, 1-dB compression point, third-order intercept point,* and *blocking*. We will also look at a couple of important additional measures.

18-10 DYNAMIC PERFORMANCE

The dynamic performance specifications of a radio receiver are those that deal with how the receiver performs in the presence of very strong signals, either cochannel or adjacent channel. Until about the 1960s, dynamic performance was somewhat less important than static performance for most users. However, today the role of dynamic performance is probably more critical than static performance because of crowded band conditions.

There are at least two reasons for this change in outlook (Dyer 1993). First, in the 1960s receiver designs evolved from tubes to solid-state. The new solid-state amplifiers were somewhat easier to drive into nonlinearity than tube designs. Second, there has been a tremendous increase in radio frequency signals on the air. The UHF and microwave bands were nearly empty in those days. There are far more transmitting stations than ever before, and there are far more sources of *electromagnetic interference* (EMI – pollution of the air waves) than in prior decades. With the advent of new and expanded wireless services available to an ever-widening market, the situation can only worsen. For this reason, it is now necessary to pay more attention to the dynamic performance of receivers than in the past.

18-10.1 Intermodulation Products

Understanding the dynamic performance of the receiver requires knowledge of *intermodulation products* (IP) and how they affect receiver operation. Whenever two signals are mixed together in a nonlinear circuit, a number of products are created, according to the rule $mF1 \pm nF2$, where m and n are either integers or zero. Mixing can occur in either the mixer stage of a receiver front end or in the RF amplifier (or any outboard preamplifiers used ahead of the receiver) if the RF amplifier is overdriven by a strong signal.

The spurious IP signals are shown graphically in Fig. 18-17. Given input signal frequencies of $F1$ and $F2$, the main IPs are

Second-order:	$F1 \pm F2$
Third-order:	$2F1 \pm F2$
	$2F2 \pm F1$
Fifth-order:	$3F1 \pm 2F2$
	$3F2 \pm 2F1$

When an amplifier or receiver is overdriven, the second-order content of the output signal increases as the square of the input signal level, while the third-order responses increase as the cube of the input signal level.

Consider the case where two VHF signals, $F1 = 100$ MHz and $F2 = 150$ MHz, are mixed together. The second-order IPs are 50 and 250 MHz; the third-order IPs are 50, 200, 350, and 400 MHz; and the fifth-order IPs are 0, 250, 600, and 650 MHz. If any of these are inside the passband of the receiver, they can cause problems. One such problem is the emersion of *phantom* signals at the IP frequencies. This effect often occurs when two strong signals ($F1$ and $F2$) exist and can affect the front end of the receiver, and one of the IPs falls close to a desired signal frequency, F_d.

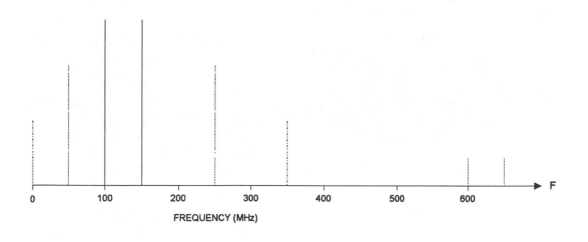

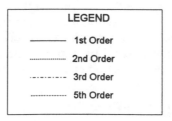

Figure 18-17   Intermodulation products.

If a receiver were tuned to 50 MHz, for example, a spurious signal would be found from the $F1$-$F2$ pair.

Another example is related to strong in-band, adjacent channel signals. Consider a case where the receiver is tuned to a station at 9610 MHz and there are also very strong signals at 9600 MHz and 9605 MHz. The near (in-band) IP products are

Third-order:	9595 MHz ($\Delta F = 15$ MHz)
	9610 MHz ($\Delta F = 0$ MHz) [ON CHANNEL!]
Fifth-order:	9590 MHz ($\Delta F = 20$ MHz)
	9615 MHz ($\Delta F = 5$ MHz)

Note that one third-order product is on the same frequency as the desired signal and could easily cause interference if the amplitude is sufficiently high. Other third- and fifth-order products may be within the range where interference could occur, especially on receivers with wide bandwidths.

The IP orders are theoretically infinite because there are no bounds on either m or n. However, in practical terms, because each successively higher-order IP is reduced in amplitude compared with its next lower order mate, only the second-order, third-order, and fifth-order products usually assume any importance. Indeed, only the third-order product is normally used in receiver specifications sheets.

18-10.2 The –1-dB Compression Point

An amplifier produces an output signal that has a higher amplitude than the input signal. The transfer function of the amplifier (indeed, any circuit with output and input) is the ratio OUT/IN, so for the power amplification of a receiver RF amplifier, it is P_o/P_{in} (or, in terms of voltage, V_o/V_{in}). Any real amplifier will saturate given a strong enough input signal (see Fig. 18-18). The dotted line represents the theoretical output level for all values of input signal (the slope of the line represents the gain of the amplifier). As the amplifier saturates (solid line), however, the actual gain begins to depart from the theoretical at some level of input signal (P_{in1}). The –1-dB compression point is that output level at which the actual gain departs from the theoretical gain by –1 dB.

The –1-dB compression point is important when considering either the RF amplifier ahead of the mixer (if any), or any outboard preamplifiers that are used. The –1-dB compression point is the point at which intermodulation products begin to emerge as a serious problem. Also, harmonics are generated when an amplifier goes into compression. A sine wave is a pure signal because it has no harmonics (all other waveshapes have a fundamental plus harmonic frequencies). When a sine wave is distorted, however, harmonics arise. The effect of the compression phenomenon is to distort the signal by clipping the peaks, thus raising the harmonics and intermodulation distortion products.

18-10.3 Third-Order Intercept Point

It can be claimed that the third-order intercept point (TIP) is the single most important specification of a receiver's dynamic performance because it predicts per-

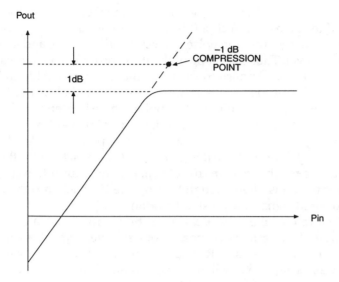

Figure 18-18
-1-dB compression point.

formance with regard to intermodulation cross-modulation, and blocking desensitization.

Third-order (and higher) intermodulation products (IP) are normally very weak and don't exceed the receiver noise floor when the receiver is operating in the linear region. But as input signal levels increase, forcing the front end of the receiver toward the saturated nonlinear region, the IPs emerge from the noise (Fig. 18-19) and begin to cause problems. When this happens, new, spurious signals appear on the band and self-generated interference begins to arise.

Figure 18-20 shows a plot of the output signal versus fundamental input signal. Note the output compression effect that was shown in Fig. 18-18. The dotted

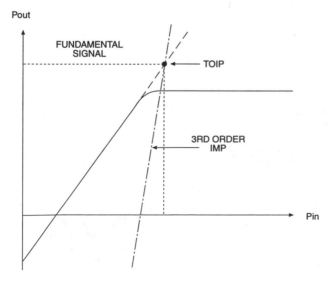

Figure 18-19
Third-order intercept point.

gain line continuing above the saturation region shows the theoretical output that would be produced if the gain did not clip. It is the nature of third-order products in the output signal to emerge from the noise at a certain input level and increase as the cube of the input level. Thus, the slope of the third-order line increases 3 dB for every 1-dB increase in the response to the fundamental signal. Although the output response of the third-order line saturates similarly to that of the fundamental signal, the gain line can be continued to a point where it intersects the gain line of the fundamental signal. This point is the third-order intercept point.

Interestingly enough, one receiver feature that can help reduce IP levels back down under the noise is a *front-end attenuator* (or *input attenuator*). Because the fundamental output changes much less than the intermodulation output, even a few decibels of input attenuation is often enough to drop the IPs back into the noise, while afflicting the desired signals only a small amount.

Other effects that reduce the overload caused by a strong signal also help. Sometimes the apparent third-order performance of a receiver improves dramatically when a lower-gain antenna is used. Rotating a directional antenna away from the direction of the interfering signal will also often accomplish this effect.

Add-on preamplifiers are popular receiver accessories, but can often reduce rather than enhance performance. Two problems commonly occur (assuming that the preamplifier is a low-noise device; if not, then there are three). The best-known problem is that the preamplifier amplifies noise as much as signals, and while it makes the signal louder, it also makes the noise louder by the same amount. Since it is the signal-to-noise ratio that is important, that does not improve the situation. Indeed, if the preamplifier itself is noisy, it will deteriorate the SNR. The other problem is less well known, but potentially more devastating. If the increased signal levels applied to the receiver drive the receiver nonlinear, then IPs begin to emerge.

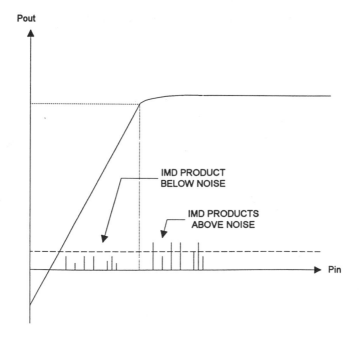

Figure 18-20
Effect of compression point on Intermodulation Distortion (IMD) products.

When evaluating receivers, remember: a TIP of +5 to +20 dBm is excellent performance, while up to +27 dBm is relatively easily achievable, and +35 dBm has been achieved with good design; around +50 dBm is attainable. Receivers are still regarded as reasonably good performers in the 0 to +5 dBm range, and middling performers in the –10 to 0 dBm range. Anything below –10 dBm is not a wonderful machine to own. A general rule is to buy the best third-order intercept performance that you can afford, especially if there are strong signal sources in your vicinity.

18-11 DYNAMIC RANGE

The *dynamic range* of a radio receiver is the range from the minimum discernible signal to the maximum allowable signal (measured in decibels). While this simplistic definition is conceptually easy to understand, in the concrete it is a little more complex. Several definitions of dynamic range are used (Dyer 1993).

One definition of dynamic range is that it is the input signal difference between the sensitivity figure (for example, 0.5 μV for 10-dB S+N/N) and the level that drives the receiver far enough into saturation to create a certain amount of distortion in the output. This definition was common on consumer broadcast band receivers at one time (especially automobile radios, where dynamic range was somewhat more important due to mobility). A related definition takes the range as the distance in decibels from the sensitivity level and the –1-dB compression point. There is still another definition, the *blocking dynamic range*, which is the range of signals from the sensitivity level to the blocking level.

A problem with the preceding definitions is that they represent single-signal cases, so do not address the receiver's dynamic characteristics. Dyer (1993) provides both a loose definition and a more formal one that is somewhat more useful, and is at least standardized. The loose version is that dynamic range is the range of signals over which dynamic effects (for example, intermodulation) do not exceed the noise floor of the receiver. Dyer's recommendation for HF receivers is that the dynamic range is two-thirds the difference between the noise floor and the third-order intercept point in a 3-kHz bandwidth. Dyer's more formal definition of dynamic range is that it is the difference between the fundamental response input signal level and the third-order intercept point along the noise floor, measured with a 3-kHz bandwidth. For practical reasons, this measurement is sometimes made not at the actual noise floor (which is sometimes hard to ascertain), but rather at 3 dB above the noise floor.

Magne and Sherwood (1987) provide a measurement procedure that produces similar results (the same method is used for many amateur radio magazine product reviews). Two equal-strength signals are input to the receiver at the same time. The frequency difference has traditionally been 20 kHz for HF and 30 to 50 kHz for VHF receivers (modern band crowding may indicate a need for a specification at 5-kHz separation on HF). The amplitudes of these signals are raised until the third-order distortion products are raised to the noise floor level. For 20-kHz spacing, using the two-signal approach, anything over 90 dB is an excellent receiver, while anything over 80 dB is decent.

Defining the difference between the single-signal and two-signal (dynamic) performance is not merely an academic exercise. Besides the fact that the same receiver can show as much as a 40-dB difference between the two measures (favoring the single-signal measurement), the most severe effects of poor dynamic range show up most in the dynamic performance.

18-11.1 Blocking

The blocking specification refers to the ability of the receiver to withstand very strong off-tune signals that are at least 20 kHz (Dyer) away from the desired signal, although some use 100-kHz separation (Magne and Sherwood). When very strong signals appear at the input terminals of a receiver, they may desensitize the receiver, that is, reduce the apparent strength of desired signals over what they would be if the interfering signal were not present.

Figure 18-21 shows this blocking behavior. When a strong signal is present, it takes up more of the receiver's resources than normal, so there is not enough of the output power budget to accommodate the weaker desired signals. But if the strong undesired signal is turned off, then the weaker signals receive a full measure of the unit's power budget.

The usual way to measure blocking behavior is to input two signals, a desired signal at 60 dBμV and another signal 20 (or 100) kHz away at a much stronger level. The strong signal is increased to the point where blocking desensitization causes a 3-dB drop in the output level of the desired signal. A good receiver will show ≥ 90 dBμV, with many being considerably better. An interesting note about modern receivers is that the blocking performance is so good that it is often necessary to specify the input level difference (dB) that causes a 1-dB drop, rather than a 3-dB drop, of the desired signal's amplitude (Dyer 1993).

The phenomenon of blocking leads us to an effect that at first appears paradoxical. Many receivers are equipped with front-end attenuators that permit fixed attenuations of 1 dB, 3 dB, 6 dB, 12 dB, or 20 dB (or some subset of same) to be inserted into the signal path ahead of the active stages. When a strong signal that is capable of causing desensitization is present, *adding attenuation often increases the level of the desired signals in the output,* even though overall gain is reduced. This occurs because the overall signal that the receiver front end is asked to handle is below the threshold where desensitization occurs.

18-11.2 Cross-Modulation

Cross-modulation is an effect in which amplitude modulation (AM) from a strong, undesired signal is transferred to a weaker, desired signal. Testing is usually done (in HF receivers) with a 20-kHz spacing between the desired and undesired signals, a 3-kHz IF bandwidth on the receiver, and the desired signal set to 1000 μV_{EMF} (-53 dBm). The undesired signal (20 kHz away) is amplitude modulated to the 30% level. This undesired AM signal is increased in strength until an unwanted AM output 20 dB below the desired signal is produced.

A cross-modulation specification ≥ 100 dB would be considered decent performance. This figure is often not given for modern HF receivers, but if the receiver

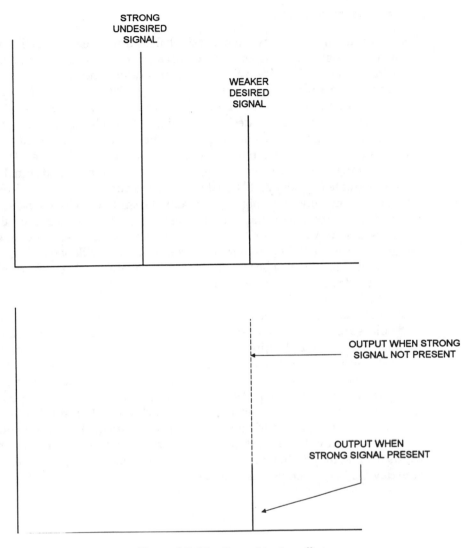

Figure 18-21   Desensitization effect.

has a good third-order intercept point, then it is likely to also have good cross-modulation performance.

Cross-modulation is also said to occur naturally, especially in transpolar and North Atlantic radio paths where the effects of the aurora are strong. According to one legend there was something called the Radio Luxembourg Effect discovered in the 1930s. Modulation from very strong broadcasters appeared on the Radio Luxembourg signal received in North America. This effect was said to be an ionospheric cross-modulation phenomenon. If anyone has any direct experience with this effect, or a literature citation, then the author would be interested in hearing from him or her.

18-11.3 Reciprocal Mixing

Reciprocal mixing occurs when noise sidebands from the local oscillator (LO) signal in a superheterodyne receiver mix with a strong, undesired signal that is close to the desired signal. Every oscillator signal produces noise, and that noise tends to amplitude modulate the oscillator's output signal. It will thus form sidebands on either side of the LO signal. The production of phase noise in all LOs is well known, but in more recent designs the digitally produced synthesized LOs are prone to additional noise elements as well. The noise is usually measured in –dBc (decibels below carrier, or, in this case, dB below the LO output level).

In a superheterodyne receiver, the LO beats with the desired signal to produce an intermediate frequency (IF) equal to either the sum (LO + RF) or difference (LO – RF). If a strong unwanted signal is present, it might mix with the noise sidebands of the LO to reproduce the noise spectrum at the IF frequency (see Fig. 18-22). In the usual test scenario, the reciprocal mixing is defined as the level of the unwanted signal (in decibels) at 20 kHz required to produce a noise sideband 20db down from the desired IF signal in a specified bandwidth (usually 3 kHz on HF receivers). Figures of 90 dBc or better are considered good.

The importance of the reciprocal mixing specification is that it can seriously deteriorate the observed selectivity of the receiver, yet is not detected in the normal static measurements made of selectivity (it is a *dynamic selectivity* problem). When the LO noise sidebands appear in the IF, the distant-frequency attenuation (>20 kHz off-center of a 3-kHz bandwidth filter) can deteriorate 20 to 40 dB.

The reciprocal mixing performance of receivers can be improved by eliminating the noise from the oscillator signal. Although this sounds simple, in practice it is often quite difficult. A tactic that will work well, at least for those designing their own receivers, is to add high-Q filtering between the LO output and the mixer input. The narrow bandwidth of the high-Q filter prevents excessive noise sidebands from getting to the mixer. Although this sounds like quite an easy solution, "the devil is in the details," as they say.

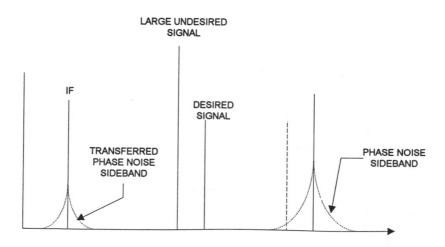

Figure 18-22   Transfer of LO noise sidebands to IF signal.

18-12 OTHER IMPORTANT SPECIFICATIONS

18-12.1 IF Notch Rejection

If two signals fall within the passband of a receiver, they will both compete to be heard. They will also heterodyne together in the detector stage, producing an audio tone equal to their carrier frequency difference. For example, suppose we have a 1295-MHz receiver with a 50-kHz bandwidth and a 70-MHz IF. If two signals appear on the band, one at an IF of 70.01 MHz and the other at 70.02 MHz, then both are within the receiver passband and both will be heard in the output. However, the 0.01-MHz (10-kHz) difference in their carrier frequency will produce a 10-kHz heterodyne difference signal in the output of the envelope detector.

In some receivers, a tunable high-Q (narrow and deep) notch filter is in the first- or second-IF amplifier circuit. This tunable filter can be turned on and then adjusted to attenuate the unwanted interfering signal, reducing the irritating heterodyne. Attenuation figures for good receivers vary from –35 to –65 dB or so (the more negative the better).

There are some trade-offs in notch filter design. First, the notch filter Q is more easily achieved at low IF frequencies (50 kHz to 500 kHz) than at high IF frequencies (9 MHz and up). Also, the higher the Q the better the attenuation of the undesired squeal, but the touchier it is to tune. Some happy middle ground between the irritating squeal and the touchy tune is mandated here.

Some receivers use audio filters rather than IF filters to help reduce the heterodyne squeal. In the AM broadcast band, channel spacing is typically 10 kHz and the transmitted audio bandwidth (hence the sidebands) are 5 kHz. Designers of AM BCB consumer grade receivers usually insert an R-C low-pass filter with a –3-dB point just above 4 or 5 kHz right after the detector in order to suppress the 10-kHz heterodyne. This R-C filter is called a *tweet filter* in the slang of the electronic service and repair trade.

Another audio approach is to sharply limit the bandpass of the audio amplifiers. A voice-grade communications channel typically needs a bandpass of 3000 or 3300 Hz. The audio amplifier of such a receiver may be designed for a highly limited bandpass of 300 to 3300 Hz.

18-12.2 Internal Spurii

All receivers produce a number of internal, spurious signals that sometimes interfere with the operation. Both old and modern receivers have spurious signals from assorted high-order mixer products, from power supply harmonics, parasitic oscillations, and a host of other sources. Newer receivers with either synthesized local oscillators and digital frequency readouts (or both) produce noise and spurious signals in abundance (low power digital chips with slower rise times—CMOS, NMOS, and the like—are generally much cleaner than higher-power, fast-rise-time chips like TTL devices).

With appropriate filtering and shielding, it is possible to hold the "spurs" down to –100 dB relative to the main maximum signal output, or within about 3 dB of the noise floor, whichever is lower.

When Harry Helms (author of *All About Ham Radio* and the *Shortwave Listener's Guide*) did comparisons of basic spur/noise levels on several high-quality receivers, including tube models and modern synthesized models, his results were surprising. A high-quality tube receiver from the 1960s appeared to have a lower noise floor than the modern receivers. Helms attributed the difference to the internal spurii from the digital circuitry used in the modern receivers.

18-13 RECEIVER IMPROVEMENT STRATEGIES

There are several strategies that can improve the performance of a receiver. Some of them are available for after-market use with existing receivers, while others are only practical when designing a new receiver.

One method for preventing dynamic effects is to reduce the signal level applied to the input of the receiver at all frequencies. A front-end attenuator will help here. Indeed, many modern receivers have a switchable attenuator built into the front-end circuitry. This attenuator will reduce the level of all signals, backing the largest away from the compression point and, in the process, eliminating many dynamic problems.

Another method is to prevent the signal from ever reaching the receiver front end. This can be achieved by using any of several tunable filters, bandpass filters, high-pass filters, or low-pass filters (as appropriate) ahead of the receiver front end. These filters may not help if the intermodulation products fall close to the desired frequency, but otherwise are quite useful.

When you are designing a receiver project, there are some things that you can do. If you use an RF amplifier, then use either field-effect transistors (MOSFETs are popular) or a relatively high-powered bipolar (NPN or PNP) transistor intended for cable-TV applications, such as the 2N5109 or MRF-586. Gain in the RF amplifier should be kept minimal (4 to 8 dB). It is also useful to design the RF amplifier with as high a dc power supply potential as possible. This tactic allows a lot of head room as input signals become larger.

You might also wish to consider deleting the RF amplifier altogether. One design philosophy holds that one should use a high dynamic range mixer with only input tuning or suboctave filters between the antenna connector and the mixer's RF input. It is possible to achieve noise figures of 10 dB or so using this approach, and this is sufficient for HF work (although it may be marginal for VHF and up). The tube-design 1960s-vintage Squires-Sanders SS-1 receiver used this approach. The front-end mixer tube was the 7360 double-balanced mixer.

The mixer in a newly designed receiver project should be a high dynamic range type, regardless of whether an RF amplifier is used. Popular with some designers is the double-balanced switching-type mixer (Makhinson 1993). Although examples can be fabricated from MOSFET transistors and MOS digital switches, there are also some IC versions on the market that are intended as MOSFET switching mixers. If you opt for diode mixers, then consult a source such as *Mini-Circuits* (1992) for DBMs that operate up to RF levels of +20 dBm.

Anyone contemplating the design of a high dynamic range receiver will want to consult Makhinson (1993) and DeMaw (1976a/b) for design ideas before starting.

18-14 SUMMARY

1. Most microwave receivers are superheterodynes in which the RF signal is beat against a local oscillator (LO) to produce an intermediate frequency (IF) that is either the sum or the difference of LO and RF. In most common receivers, the IF is the difference product of LO and RF.

2. Receiver sensitivity is a measure of the weakest signal that the receiver can detect, and must be measured against the noise produced by the receiver circuits. Selectivity is the measure of a receiver's ability to separate signals close to each other in frequency. The IF amplifier is the principal factor in setting both sensitivity and selectivity.

3. Dynamic range and AGC control range are related parameters, but are different from each other. The dynamic range refers to the ratio of the largest signal that does not cause significant distortion to the minimum discernible signal. AGC control range refers to the ability of the receiver to control its gain over a wide range of input signal strengths.

4. Single-conversion receivers use one mixer and have one IF frequency; double-conversion receivers use two mixers and have two IF frequencies; triple-conversion receivers have three mixers and employ three IF frequencies.

5. The dynamic receiver performance specifications are often more important than the static performance measures, especially in crowded bands.

18-15 RECAPITULATION

Now return to the objectives and prequiz questions at the beginning of the chapter and see how well you can answer them. If you cannot answer certain questions, place a check mark by each and review the appropriate parts of the text. Next, try to answer the following questions and work the problems using the same procedure.

QUESTIONS AND PROBLEMS

1. A receiver delivers 10^{-6} W to the IF amplifier input when the RF signal applied to the mixer is 10^{-5} W. Calculate the conversion loss in decibels.
2. A receiver uses an LO frequency of 2477 MHz when the receiver is tuned to 2400 MHz. What is the IF frequency if the mixer output filter selects the difference product?
3. A receiver is tuned to 5.555 GHz. What are the possible LO frequencies if the IF frequency is 110 MHz?
4. A receiver is tuned to 4500 MHz and has an IF of 70 MHz. How many possible LO frequencies are there? What are they?
5. A receiver produces acceptable levels of distortion only when the RF input signal is 300 μV or less. If the dynamic range is 100 dB, what is the minimum discernible signal?
6. List several potential problems on a receiver with poor dynamic range.
7. A receiver with a symmetrical passband uses 30-MHz IF frequencies. What is the bandwidth of the receiver if the IF response curve is −3 dB down from the center band level at 29.9 and 30.1 MHz?

8. Define third-order intercept point.

9. Define –1-dB compression point.

10. Two frequencies, $F1$ = 2400 MHz and $F2$ = 4000 MHz, are presented to the input of a receiver. Calculate the first-, second-, third-, fifth-, and seventh-order intermodulation products.

11. Why might placing an attenuator in the front end of a receiver improve performance, rather than deteriorate it?

12. Draw the block diagram of a crystal video receiver.

13. Draw the block diagram of a TRF receiver.

14. Draw the block diagram for a Bragg cell receiver.

15. The output of a Bragg cell receiver is a _____.

18-16 REFERENCES

Carr, Joseph J. "Be a RadioScience Observer – Part 1." *Shortwave* (November 1994): 32–36.

Carr, Joseph J. "Be a RadioScience Observer – Part 2." *Shortwave* (December 1994): 36–44.

Carr, Joseph J. "Be a RadioScience Observer – Part 3." *Shortwave* (January 1995) 40–44.

DeMaw, Doug. "His Eminence — The Receiver Part 1." *QST* (June 1976, American Radio Relay League, Newington, CT): 27ff.

DeMaw, Doug. "His Eminence — The Receiver Part 2." *QST* (July 1976, American Radio Relay League, Newington, CT): 14ff.

Dyer, Jon A. "Receiver Performance: Descriptions and Definitions of Each Performance Parameter." *Communications Quarterly* (Summer 1993): 73–88.

Gruber, Mike. "Synchronous Detection of AM Signals: What Is It and How Does It Work?" *QEX* (September 1992): 9–16.

Helms, Harry. *All About Ham Radio* and *Shortwave Listeners Guide*. Solana Beach, CA: High-Text Publications, Inc. Personal telephone conversation.

Kinley, R. Harold. *Standard Radio Communications Manual: With Instrumentation and Testing Techniques*. Englewood Cliffs, NJ: Prentice-Hall, 1985.

Kleine, G. "Parameters of MMIC Wideband RF Amplifiers." *Elektor Electronics* (UK). (February 1995): 58–60.

Magne, Lawrence and J. Robert Sherwood. *How to Interpret Receiver Specifications and Lab Tests*. 2nd ed. A Radio Database International white paper. Penn's Park, PA: International Broadcasting Services, Ltd., February 18 1987.

Makhinson, Jacob. "A High Dynamic Range MF/HF Receiver Front-End." *QST* (February 1993): 23–28.

Mini-Circuits RF/IF Designers Handbook (1992). Mini-Circuits (P.O. Box 350166, Brooklyn, NY 11235-0003, USA. In UK: Mini-Circuits/Dale, Ltd., Dale House, Wharf Road, Frimley Green, Camberley, Surrey, GU16 6LF, England).

Poole, Ian G3YWX. "Specification — The Mysteries Explained." *Practical Wireless* (April 1995): 41.

Poole, Ian G3YWX. "Specification — The Mysteries Explained." *Practical Wireless* (March 1995): 57.

Rohde, Ulrich L. and T.T.N. Bucher. *Communications Receivers: Principles & Design*. New York: McGraw-Hill, 1988.

Tsui, James B. *Microwave Receivers and Related Components*. Los Altos, CA: Peninsula Publishing, 1985.

CHAPTER 19

Radar Systems

OBJECTIVES

1. Learn the applications of radar systems.
2. Understand the radar range equation.
3. Learn the types of equipment needed for radar systems.
4. Recognize the most common radar display modes.

19-1 PREQUIZ

These questions test your prior knowledge of the material in this chapter. Try answering them before you read the chapter. Look for the answers (especially those you answered incorrectly) as you read the text. After you have finished studying the chapter, try answering these questions again and those at the end of the chapter.

1. List the principal functional building blocks for a simple radar system.
2. A police speed radar works by virtue of the _____ principle.
3. An ___-scan display shows return echo amplitude versus range.
4. A PPI display shows _____ and _____.

19-2 INTRODUCTION

Although by common usage now regarded as a noun, the word *radar* was originally an acronym (RADAR) meaning *radio detection and ranging*. Thus, the radar set is an electronic device that can detect distant objects and yield information about the distance to these objects. With appropriate design, the radar can also give us either relative or compass bearing to an object, and in the case of aircraft, the height of the object. In radar jargon, the object detected is called a *target*.

Modern radar sets can be landbased, shipbased, aircraftbased, or spacecraft-based. They are used for weather tracking, air traffic control, marine and aeronautical navigation, ground mapping, scientific studies, velocity measurement, military applications, and remote sensing from outer space. Readers with an unfortunate love of driving fast cars will also recognize at least one velocity measurement application: police speed radar guns.

Scientists and engineers were working on radar in the 1920s and 1930s, but World War II caused the sudden surge of development effort that led to radar as we know it today. One of the earliest experiments in radar was accidental. In 1922, scientists at the Naval Research Laboratory (NRL) in Washington, D.C., were experimenting with 60-MHz communications equipment. A transmitter was set up at Anacostia Naval Air Station, and a receiver was installed at Haines Point on the Potomac River (a distance of about one-half mile). Signal fluctuations were noted when a wooden vessel, the *U.S.S. Dorchester*, passed up the river.

Further experiments at 300 MHz showed that *reflections* of the radio waves by passing ships were the cause of those signal fluctuations. Distances of 3 miles were achieved in those trials. The first practical radars were fielded by a British team under Sir Robert Watson-Watt in the 1930s.

The basic principle behind radar is simple: a short burst of RF energy is transmitted, and then a receiver is turned on to listen for the returning echo of that pulse as reflected from the target (Fig. 19-1A). The transmitted pulse (Fig. 19-1B) is of short duration, or width (T), with a relatively long receiver time between pulses. It is during the receiver time that the radar listens for echoes from the target. The number of pulses per second is called the *pulse repetition rate* (PRR) or *pulse repetition frequency* (PRF), while the time between the onset of successive pulses is called the *pulse repetition time* (PRT). Note that

$$\text{PRF} = \frac{1}{\text{PRT}} \qquad (19\text{-}1)$$

The pulse propagates at the speed of light (c), which is a constant, so we can measure *range* indirectly by actually measuring the *time* required for the echo to return:

$$R = \frac{ct}{2} \qquad (19\text{-}2)$$

where

R = range to the target in meters
c = speed of light (3×10^8 m/s)

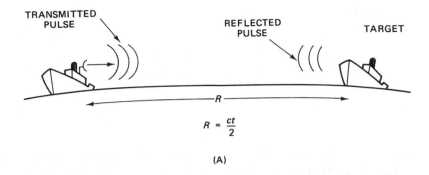

$$R = \frac{ct}{2}$$

(A)

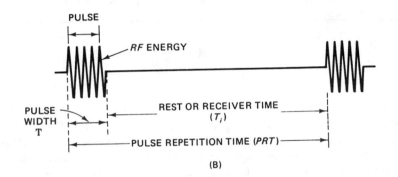

(B)

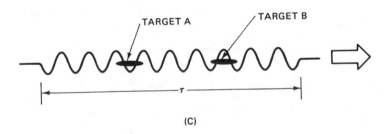

(C)

Figure 19-1 A) Radar geometry; B) pulse waveforms; C) why resolution depends on pulse width.

t = time in seconds between the original pulse transmitted and arrival of the echo from the target

The factor 1/2 in Eq. (19-2) reflects the fact that the echo must make a round trip to and from the target, so the apparent value of t is twice the correct value.

EXAMPLE 19-1

EX = A target return echo is noted 98.6 microseconds (μs) after the transmit pulse. Calculate the range (R) to the target.

Solution

$$R = \frac{ct}{2}$$

$$= \frac{(3 \times 10^8 \text{ m/s})(9.86 \times 10^{-6} \text{ s})}{2}$$

$$= 14,790 \text{ m}$$

A popular radar rule of thumb derives from the fact that a radio wave travels one nautical mile (nmi), or 6000 ft, in 6.18 μs. Because of the round-trip problem, radar range can be measured at 12.36 μs/nmi. In equation form,

$$R = \frac{T}{12.36} \text{ nmi} \qquad (19\text{ - }3)$$

Unfortunately, real radar waves do not behave so well as we have implied for at least two reasons. First, the pulse might re-reflect and thus present a "second time around" error echo for the same target. Second, the maximum unambiguous range is limited by the interval between pulses. Echoes arriving late will fall into the receive interval of the *next* transmitted pulse, so the second (erroneous) target appears much closer than it really is. If T_i is the interpulse interval (or receiver period, as we called it earlier), then the maximum unambiguous range (R_{max}) is

$$R_{max} = \frac{cT_i}{2} \qquad (19\text{ -}4)$$

where

R_{max} = maximum unambiguous range in meters
c = speed of light (3×10^8 m/s)
T_i = interpulse interval in seconds

EXAMPLE 19-2

A radar transmits a 2-μs pulse at a PRF of 10 kHz. What is the unambiguous range?

Solution

(a) Find the pulse repetition time.

$$\text{PRT} = \frac{1}{\text{PRF}}$$

$$= \frac{1}{10^4 \text{ Hz}} = 10^{-4} \text{ s}$$

(b) Calculate T_i.

$$T_i = PRT - T$$

$$= (10^{-4}\,s) - (2 \times 10^{-6}\,s) = 9,8 \times 10^{-5}\,s$$

(c) Calculate R_{max}.

$$R_{max} = \frac{cT_i}{2}$$

$$= \frac{(3 \times 10^8\ \text{m/s})(9.8 \times 10^{-5}\ \text{s})}{2} = 14,700\ \text{m}$$

The pulse width (T) represents the time the transmitter is on and emitting. During this period, the receiver is shut off and so cannot hear any echoes. As a result, there is a *dead zone* in front of the antenna in which no targets are detectable. The dead zone is a function of transmitter pulse width and the transmit/receive (T-R) switching time. If T-R time is negligible, then

$$R_{dead} = \frac{cT}{2} \qquad (19\text{-}5)$$

Obviously, Eq. (19-5) requires very short pulse widths for nearby targets to be detectable.

Another radar property that depends on pulse width is *range resolution* (ΔR). Simply defined, ΔR is the smallest distance along the same azimuth line that two targets must be separated before the radar can differentiate between them. At distances less than the minimum, two different targets appear as one.

Consider an example (Fig. 19-1C). Each pulse has a width T, and because it propagates through space with velocity c, it also has *length*. Let's assume a 1-μs pulse on a frequency of 1.2 GHz. The wavelength of this signal is 0.25 m, and in 1 μs the transmitter sends out a pulse containing $(1.2 \times 10^9$ cps$) \times (1 \times 10^{-6}$ s$)$, or 1200 cycles, each of which is 0.25 m long. Thus, the *pulse length* is $(1,200 \times 0.025$ m$)$, or 300 m. Two targets must be 300 m apart in order to not fall inside the same pulse length at the same time. Because of the round-trip phenomenon, which means that the result is divided by 2, we can express range resolution as

$$\Delta R = \frac{cT}{2} \qquad (19\text{-}6)$$

where

ΔR = range resolution in meters
c = speed of light (3×10^8 m/s)
T = pulse width in seconds

Range resolution can be improved by a process called *pulse compression*. One of the most common forms of compression is *FM chirp*, in which the radar carrier frequency is swept over a fixed range by a ramp function. If B is the change of frequency and T is the pulse width, then the compressed range resolution is the simple range resolution, Eq. (19-6), divided by BT:

$$\Delta R_c = \frac{\Delta R}{BT} \qquad (19\text{-}7A)$$

or

$$\Delta R_c = \frac{c}{2B} \qquad (19\text{-}7B)$$

19-2.1 Pulse Properties

The radar transmits relatively short-duration (T) pulses and then pauses to listen for the echo. The actual number of cycles in the pulse is the product of pulse width (T) and operating frequency. For example, a 0.5-μs pulse at 1 GHz sends out 500 cycles of RF energy per pulse. For radar pulses, we need to know the *peak power* (P_p), *average power* (P_a), and the *duty cycle* (D).

The output power of the transmitter is usually measured in terms of peak power (P_p), and it is that peak power that we will use shortly to develop the radar range equation. The average power, which affects transmitter design (see Chapter 17), is found from either

$$P_a = \frac{P_p T}{PRT} \qquad (19\text{-}8)$$

or

$$P_a = P_p T(PRF) \qquad (19\text{-}9)$$

where

P_a = average power in watts
P_p = peak pulse power in watts
T = pulse width in seconds
PRF = pulse repetition frequency in hertz
PRT = pulse repetition time in seconds

EXAMPLE 19-3

A radar transmitter produces a 15-kW peak power pulse that is 1 μs long and has a PRT of 500 μs (0.0005 s). Calculate the average output power of the transmitter.

Solution

$$P_a = \frac{P_p T}{\text{PRT}}$$

$$= \frac{(1.5 \times 10^4 \text{ W})(1 \times 10^{-6} \text{ s})}{5 \times 10^{-4} \text{ s}} = 30 \text{ W}$$

Note in Example 19-3 that the average power is very much smaller than the peak power. This difference accounts for the fact that radar transmitters seem to pack a large peak power into a small space. In other words, the *duty cycle* is low on radar transmitters. The duty cycle is defined as

$$D = \frac{P_a}{P_p} \tag{19 - 10}$$

or, in percent,

$$D = \frac{P_a (100\%)}{P_p} \tag{19 - 11}$$

If you compare the duty cycle and average power equation, you might deduce that the duty cycle is also

$$D = \frac{T}{\text{PRT}} \tag{19 - 12}$$

or

$$D = T \times PRF \tag{19 - 13}$$

19-3 RADAR RANGE EQUATION

Basic to understanding radar is understanding the *radar range equation*. We can best see this equation by building it up in parts from first principles: the *transmitted energy, reflected energy, system noise,* and *system losses.* The maximum RF energy on target is essentially the effective radiated power (ERP) of the radar reduced by the inverse square law $(1/R^2)$. The ERP of a pulsed transmitter is the product of pulse width (T), peak power (P_p), and antenna gain (G): P_pTG. We need to calculate the power density produced by this energy over an area (A in Fig. 19-2). This is found from

$$p = \frac{P_p TG}{4\pi R^2} \qquad (19\text{-}14)$$

The reflected energy is only a fraction of the incident energy (p), so can be calculated from Eq. (19-14) and a factor called the *radar cross section* (RCS) of the target. In essence, RCS (σ) is a measure of how well radio waves reflect from the target. RCS determination is very difficult because it depends on the properties of the target. A mid-sized ocean liner will have an RCS of about 5000 m², while a commercial airliner might have an RCS of 20 to 25 m². The reflected energy reaching the receiver antenna is also reduced by the inverse square law. With this information, we can rewrite Eq. (19-12) in the form

$$S = \frac{P_p TGA_e \sigma}{16\pi^2 R^4} \qquad (19\text{-}15)$$

where
    S = signal energy in watt-seconds (W-s)
    P_p = peak power in watts (W)
    T = pulse width in seconds (s)
    G = antenna gain
    A_e = receive antenna effective aperture in square meters (m²)
    σ = radar cross-section (RCS) of the target in square meters (m²)
    R = range to the target in meters (m)

For an ideal radar system Eq. (19-13) suffices, although in the practical world losses and noise become important.

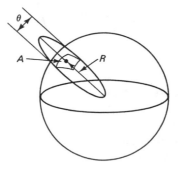

Figure 19-2
Development of directivity from an isotropic source.

EXAMPLE 19-4

A radar transmits a 3-μs, 10-kW pulse through an antenna with a power gain of 30 dB ($G = 1000$) and an effective aperture of 0.19 m². Calculate the signal energy returned from a 1000-m² ship 10 km (10,000 m) from the radar site.

Solution

$$S = \frac{P_p TGA_e \sigma}{16\pi^2 R^4}$$

$$= \frac{(10^4 \text{ W})(3 \times 10^{-6} \text{ s})(1000)(0.19 \text{ m}^2)(1000 \text{ m}^2)}{(16)(\pi^2)(10,000 \text{ m}^2)}$$

$$= \frac{5.7 \times 10^3 \text{ W} - s}{5.03 \times 10^9} = 1.1 \times 10^{-6} \text{ W} - s$$

In the real world, radars are not ideal, so we must also account for system noise and various losses (L). Examples of miscellaneous losses include weather losses as signal is attenuated in moist air and transmission lines. When these factors are added into Eq. (19-15), we have the classical *radar range equation*:

$$\frac{S}{N} = \frac{P_p TGA_e \sigma}{16\pi^2 R^4 KTL} \qquad (19\text{-}16)$$

where

S/N = signal-to-noise ratio
K = Boltzmann's constant (1.38×10^{-23} J/K)
T = noise temperature in kelvins (K)
L = sum of all losses in the system

All other terms are as defined for Eq. (19-15).

19-4 RADAR HARDWARE

A radar set must transmit a burst of RF energy, turn on a receiver to listen for the echo of that energy if any reflects from a target, process the echo signal to extract information, and display the extracted information in a form that a human operator can use. In summary then, the radar must *transmit, receive, process*, and *display*.

Two forms of system architecture are used: *monostatic* and *bistatic*. The monostatic system is, by far, the most common form of radar. The monostatic radar has both receive and transmit hardware in the same location. Indeed, not only are the receiver and transmitter colocated, but they most often also use the same antenna for both functions. The bistatic radar has the two functions significantly separated in distance. One site transmits the pulse while another site receives the echo reflected from the target. A timing and control signal must be transmitted via data

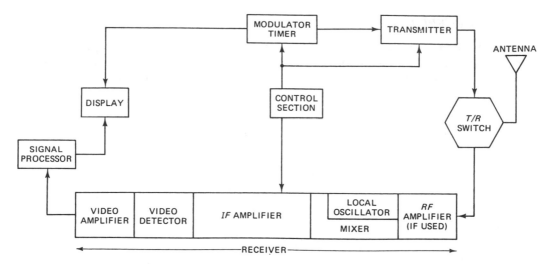

Figure 19-3   Simple radar system.

link between the two sites. Bistatic radars are used in security surveillance work, but most other applications use the monostatic design.

Figures 19-3 and 19-4 show two forms of radar hardware design. The system in Fig. 19-3 is representative of early designs and the simpler modern designs. The antenna is shared between functions, so a *transmit-receive* (T/R) switch is placed in the line to the antenna. The T/R switch must operate very rapidly, so devices like gas discharge tubes are used. In other cases, *duplexers* are used. These devices allow transmission of RF energy between ports, but provide a very high degree of isolation between transmitter and receiver.

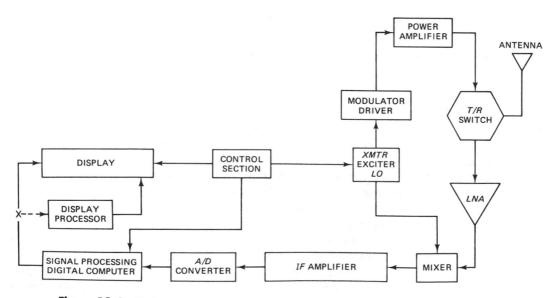

Figure 19-4   Radar system.

The transmitter and modulator form the RF pulse that is output to the antenna. The transmitter on older (and simpler) systems is a magnetron or reflex klystron. Some systems, however, might use an exciter-amplifier design. The purpose of the modulator is to initiate and terminate the pulse by turning the transmitter on and off.

The receiver section is a *superheterodyne* design, which means that the RF signal received from the antenna is down-converted to a lower *intermediate frequency* (IF) by heterodyning with the signal from a stable local oscillator (STALO). It is at the IF that most of the signal amplification and bandpass filtering take place. Following the IF amplifier are the video detector and video amplifier sections. These sections extract the information from the target return echo.

The signal processor may be either analog or digital, depending on the radar design age or mission. Its purpose is to further refine the information extracted from the signal. Signal processing may be as little as integration or filtering to improve S/N ratio or very complex. The need for signal processing is shown in one form in Fig. 19-5. The input signal is very noisy, and the return echo may exceed the critical trigger threshold only by a small amount. Some noise signals may also exceed the threshold and appear on the display as a false target. If there is a wide band between average noise level and target returns, then the operator can adjust the threshold level to eliminate false alarms. In other cases, low-pass filtering (integration or time averaging) will be used to improve S/N ratio. By integrating multiple pulses, the target return is enhanced and the noise is diminished, so the S/N ratio rises.

The display unit on most radars is a cathode ray tube (CRT) that shows targets as a function of range and/or bearing and/or elevation (see Section 19-5). In simple radar devices, the display might be numerical, such as on police speed radars and airborne radar altimeters.

The block diagram shown in Fig. 19-4 is more sophisticated and represents modern digital radars. There are differences in both transmitter and receiver sections. The transmitter uses a power amplifier driven by an exciter, which is in turn driven by an oscillator or frequency synthesizer. This design allows generation of *coherent* output signals, that is, those for which constant phase relationships are maintained over time. Coherent radars can offer *moving target indication* (MTI) and *clutter cancellation* more easily than noncoherent radars.

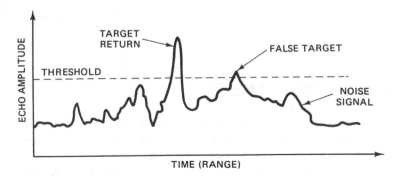

Figure 19-5   Detection of target return.

The radar of Fig. 19-4 also uses a superheterodyne receiver, but instead of a local oscillator integral to the receiver it derives the STALO signal from the exciter or synthesizer. The STALO in any superheterodyne receiver operates on a frequency equal to either the sum or difference of the IF and received signal (that is, transmitter) frequencies.

Some modern radars use a conventional video detector and video amplifier; others do not. In either case, however, the receiver output signal must be converted into binary numbers by an *analog-to-digital* (A/D) converter stage. The output of the A/D converter is input to a programmable digital computer, called in this illustration the *radar signal processor*. There may be a second computer called a *display processor* in some designs, especially in systems that use digital display oscilloscopes.

In both Figs. 19-3 and 19-4, a *control section* is used to allow the operator to work the radar set. This section may be switch and/or relay logic in older designs, digital IC logic in more recent designs, and a microcomputer in the latest designs.

19-4.2 Radar Antennas

The radar antenna is critical to the success of the design. Chapter 9 discusses microwave antennas in detail, so the reader is referred to that material. Radars may use parabolic dish antennas, flat-plate antennas, or a variety of array antennas. The selection depends on application, gain required, receive aperture required, and so forth. Two critical antenna requirements are, however, *beamwidth* and *sidelobe level*.

Beamwidth determines the *azimuth resolution* of the radar (in other words, the ability of the radar to discriminate between two targets at the same range but separated in azimuth). The narrower the beamwidth, the finer the azimuth resolution.

The antenna sidelobe performance is also critical. High sidelobes allow interference between cochannel systems in close proximity. Radar systems are further affected by sidelobes because a sidelobe will impinge on a target and cause a reflection that will be seen as a target. The same target will also reflect signal in the main beam, so the target will be seen twice (at two different azimuths); only the main beam reflection is valid.

The radar antenna must also be *steerable*. In most systems, the beam must be swept over an area to allow wide coverage while retaining narrow azimuth resolution. Some antennas steer a 360° circular track, while others only steer a limited track (for example, ±45°) producing a pie-slice-shaped track. Both electronic and mechanical steering are used, although mechanical is at this point the most common.

19-5   RADAR DISPLAY MODES

Figure 19-6 shows several common display schemes used on radar sets. The oldest form is the *A-scan* of Fig. 19-6A in which the vertical axis shows the target echo *amplitude*, and the horizontal axis shows the *range* to the target. The first blip on the left is the transmitter pulse, while the smaller blips are target echo returns. Range is measured by the time-base distance between transmitter pulse and target pulse.

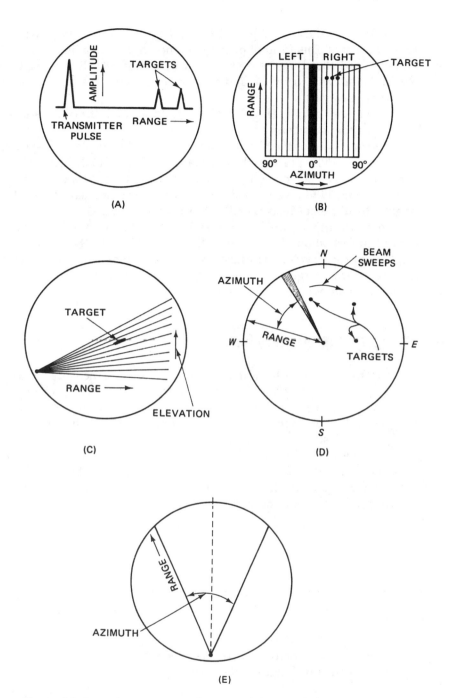

Figure 19-6 Radar scan types: A) A-scan; B) B-scan; C) elevation scan; D) planned position indicator (PPI); E) airborne radar PPI.

The *B-scan* shown in Fig. 19-6B shows *range* versus *azimuth*, with range along the vertical axis. This form of display (as shown) covers a limited azimuth extent, with 0° (dead ahead) in the center of the screen. Azimuth is along the horizontal axis, and displays 0 to 90° left and right of dead ahead. B-scans can be used in navigation applications.

The *E-scan* display (Fig. 19-6C) is used for height-finding radars and shows *height* (altitude) versus *range*.

Two versions of the *plan position indicator* (PPI) display are shown in Fig. 19-6D and E. The PPI is perhaps the most familiar display to the general public. In Fig. 19-6D, we see a 360° coverage PPI such as used on shipboard and land-based radars. In Fig. 19-6E, we see a limited azimuth extent such as found on airborne radars. The 0° point (dead ahead) is the aircraft flight path.

In a PPI display, the beam sweeps across the extent of coverage, providing azimuth information. The distance from the origin to the target blip gives the range. Very rough velocity indications (measurement is not a good term here) can be deduced from target position changes from scan to scan. Accurate velocity measurement requires *doppler shift* measurement.

19-6 VELOCITY MEASUREMENT

As many speeders ruefully admit, radar can be used to measure velocity. The basic principle is the *doppler effect*. The apparent frequency of waves changes if there is a nonzero relative velocity between the wave source and the observer. The classic example is the train whistle. As the train approaches the track-side observer, the pitch of the whistle rises an amount proportional to the train's speed. As it passes, however, the pitch abruptly shifts to a point lower than its zero-velocity value. Thus, the direction of the relative velocity determines whether frequency increases or decreases, while velocity determines how great a change takes place. The *doppler frequency* (F_d) is the amount of frequency shift:

$$F_d = \frac{2v}{\lambda} \qquad (19\text{-}17)$$

where

F_d = doppler shift in hertz
v = relative velocity in meters per second
λ = radar wavelength in meters

19-7 GROUND-MAPPING RADAR

Ground-mapping radar is carried aboard both aircraft and spacecraft for both civil and military purposes. Space-based *synthetic aperture radars* (SARs) launched by NASA have mapped considerable portions of the Earth. In some of those maps, over North Africa and the Middle East, dark river-like lines appeared in the displays that were not on any maps. It was noted, however, that the locations were approximately where ancient sources (the Bible, plus Arab and Greek historians) indicated rivers once were located, but are now gone. Further research showed that the lines indeed represented buried ancient riverbeds! This *backscatter* seen by the radar was caused by the harder texture of the riverbed compared with the sand overburden. Just as happens in optics when light passes from one material to another, the radar signal was partially absorbed and partially reflected by the harder material that had once been a riverbed.

One principal application for ground-mapping radar is aircraft navigation. Critical landmarks and waypoints can be located using radar a great distance further than by eye. Radar ground maps can be made by a real-beam radar, which displays ground-return video with little or no processing. In general, depending on what's being mapped, the radar should have a resolution that is one-fifth to one-tenth the dimension of the smallest object being mapped.

Significant improvements in range and cross-range resolution are realized by using SAR techniques. Figure 19-7 shows an example of an SAR ground map. Real-beam radars use the ground-return data (range, azimuth, and amplitude of the return) to form a map on the video display unit. SARs also use a doppler shift component caused by the motion of the aircraft as it flies past the ground swath being mapped.

On real-beam radars, the resolution of the map displayed is a function of the range and the cross-range resolution of the radar (Fig. 19-8). The cross-range resolution is a function of the radar antenna azimuth resolution or aperture. The term synthetic aperture refers to the fact that the radar provides a resolution that is far greater than the natural aperture of the antenna. It gains this extra resolution because the radar takes a number of sequential "snapshots" of the ground area being

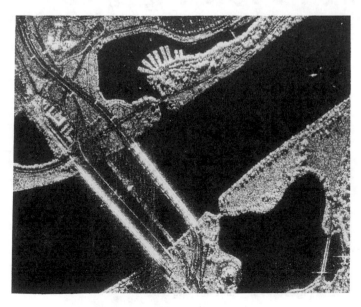

Figure 19-7
Radar ground map image.

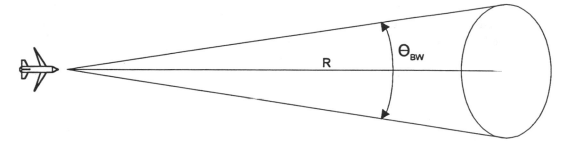

Figure 19-8   Real-beam radar mapping.

mapped as it flies along a path (Fig. 19-9), and these snapshots are processed to make it appear as if the radar antenna has an apparent aperture approximating the length of the flight path.

Real-beam radars have a cross-range resolution that is set by the range to the object being mapped and the beamwidth of the antenna:

$$\delta_{CR} = R\theta_{BW} \qquad\qquad (19\text{-}18)$$

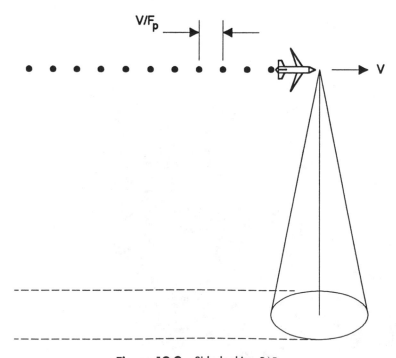

Figure 19-9   Side-looking SAR.

where

δ_{CR} = cross-range resolution
R = range
q_{BW} = antenna's –3-dB beamwidth

The natural beamwidth of the antenna is given by

$$\theta_{BW} = \frac{k\lambda}{D} \tag{19-19}$$

where

θ_{BW} = antenna's –3-dB beamwidth
λ = wavelength
D = antenna diameter
k = a factor dependent on the antenna's current illumination

The factor k is often 1, so for our purposes we will keep it simple and make $k = 1$ for the following discussion.

The cross-range resolution can therefore be written by combining Eqs. (19-18) and (19-19) into the following form:

$$\delta_{CR} = \frac{R\lambda k}{D} \tag{19-20}$$

For the SAR, with an effective synthetic array length Le, the synthetic beam width becomes

$$\theta_s = \frac{k\lambda}{2L_e} \tag{19-21}$$

where the factor 2 is due to the necessity for accounting for the round trip of the radar signal. A constraint on Le is that it can be no larger than the width of the illuminated region, or

$$L_e \leq R\theta_{BW} \tag{19-22}$$

19-7.1 Unfocused SAR

In the unfocused SAR mode, the aperture size is limited because it must be such that the phase front of the advancing wave looks like a plane wave front, that is, all segments very nearly in phase with each other. For effective length Le, the effective cross-range resolution is

$$\delta_{CR} = \frac{\sqrt{R\lambda}}{2} \qquad\qquad (19\text{-}23)$$

A phase correction at each range must be provided:

$$\Delta\phi = \frac{2\pi x^2}{\lambda R} \qquad\qquad (19\text{-}24)$$

where

$\Delta\phi$ = phase change
x = distance from the center of the synthetic aperture

The other terms are as previously defined.

Focused SAR results when the phase correction is applied for each range, in which case,

$$\delta_{CR} = R\theta_s = \frac{D}{2} \qquad\qquad (19\text{-}25)$$

The preceding equations and examples assume that the radar is side-looking at a 90° angle from the flight path. While this approach will make ground maps, it is often more useful to provide a *squint angle look* (Fig. 19-11); in other words, look ahead and to the side of the aircraft flight vector by angle ϕ. In that case, the cross-range resolution becomes

$$\delta_a = \frac{\lambda R}{2vT \sin\theta} \qquad\qquad (19\text{-}26)$$

where

δ_a = cross-range resolution
v = aircraft velocity
θ = squint angle
T = signal integration time

Note that the denominator of Eq. (19-26) contains the sine of the squint angle (see Fig. 19-11 for the squint angle). As this angle gets larger, the sine gets smaller, so the integration time T gets larger to keep the resolution constant. The consequence of

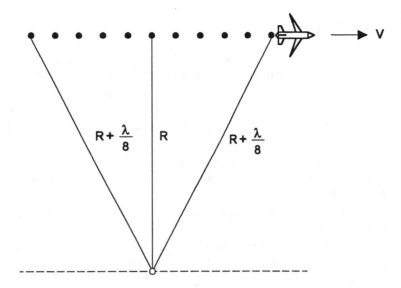

Figure 19-10   Longer apparent antenna formed by taking successive looks at the target.

this situation is that map resolution drops markedly around the aircraft nose. Some airborne radars leave the notch at ±5° or 10° around straight-ahead blank, while others fill the space with real-beam radar data. A practical consequence of this problem for aircraft using the SAR for navigation is that the aircraft must fly at an angle with respect to the direction where it wants to go for at least long enough to build the map. If the SAR is being used only for map data collection, then the antenna can be pointed at 90° from the flight path and the resolution is maximum.

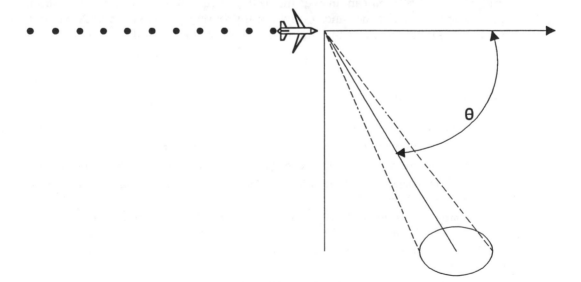

Figure 19-11   Squint-angle SAR.

19-8 SUMMARY

1. Radar (radio detection and ranging) uses reflections of RF pulses from a target to determine its location, range, azimuth, height, velocity, or other parameters.

2. The radar signal is a train of microwave RF pulses of *T*-pulse width (duration). The *pulse repetition rate* (PRR) or *pulse repetition frequency* (PRF) is the number of pulses per second, while *pulse repetition time* is the period between the onset of successive pulses.

3. The factors in the radar range equation are transmitted energy, reflected energy, system noise, and system losses. These factors determine the operation of a radar system.

4. The four functions of the basic radar system are transmit, receive, process, and display.

5. Radar antennas may be any type of microwave antenna, but most require high power gain, high effective aperture, narrow beamwidth, and low side-lobe levels.

6. The basic display modes are *A-scan* (amplitude versus range), *B-scan* (range versus azimuth left and right), *E-scan* (height versus range), and *plan position indicator* (PPI) (range versus azimuth).

7. Velocity measurement is made by measuring the frequency shift caused by the doppler effect on the moving target.

19-9 RECAPITULATION

Now return to the objectives and prequiz questions at the beginning of the chapter and see how well you can answer them. If you cannot answer certain questions, place a check mark by each and review the appropriate parts of the text. Next, try to answer the following questions and work the problems using the same procedure.

QUESTIONS AND PROBLEMS

1. List four applications for radar systems.
2. Radar operates by evaluating the energy _____ by the target.
3. A radar transmits pulses at a PRF of 12,000 Hz. What is the pulse repetition time?
4. A target is detected at a time of 78.4 μs after the pulse is transmitted. What is the range of the target in meters?
5. A target is located exactly 34 km from the radar. Calculate the round-trip transit time required by the pulse.
6. A radar pulse is found to return in 45.8 μs. What is the range to the target in nautical miles?
7. Calculate the round-trip transit time of a pulse if the target is 27 nmi from the radar.

8. Calculate the unambiguous range in meters of a target if the interpulse interval is 100 μs.

9. What is the unambiguous range in meters if the radar pulse is 3 μs long and the pulse repetition frequency is 10 kHz?

10. How wide is the dead zone in front of the radar if the pulse width (T) is 2.5 μs?

11. Calculate the pulse duration of a radar if the dead zone is 250 m.

12. Calculate the range resolution in meters of a radar if the pulse width is 1.5 μs.

13. Calculate the range resolution of a radar if the pulse width is 5 μs.

14. A radar transmitter outputs a peak power of 24 kW at a PRF of 1000 Hz and a pulse width of 2.5 μs. Calculate the average power of this pulse.

15. Calculate the peak power of a transmitter of 150 W average power when the PRF is 20 kHz and the pulse width is 5 μs.

16. Calculate the duty cycle (D) of the transmitters in problems 14 and 15.

17. A transmitter has a pulse width of 0.5 μs and a PRF of 50 kHz. Calculate the duty cycle.

18. Calculate the power density of a radar transmitter at a distance of 30 km if the peak power is 25 kW, the pulse width is 3 μs, and the antenna gain is 20 dB.

19. The radar set in question 18 is retrofitted with a 45-dB gain antenna. Calculate the power density at 30 km.

20. Calculate the signal energy in watt-seconds of the echo if the target is 25 km from a radar with the following specifications: peak power = 10 kW, pulse width = 10 μs, antenna gain = 30 dB, receiver aperture = 0.2 m^2, and the radar cross-section of the target is 500 m^2.

21. A radar is known to have 30-kW peak power into a parabolic dish antenna with a power gain of 45 dB. Calculate the signal level (S) of the echo from a 100-m^2 target that is 22,000 m from the radar. (*Hint:* You will need information from a previous chapter.)

22. In question 26, it is found that the radar system has a noise temperature of 90 K and system losses total 17 dB. Calculate the S/N ratio.

23. A _____ radar has both receiver and transmitter at the same site, often sharing the same antenna.

24. A _____ radar requires a data link between receiver and transmitter.

25. Two advantages of a radar with _____ pulses are moving target indication and clutter cancellation.

26. List desirable features of a radar antenna.

27. List four forms of display scan used on radar sets.

28. The ____-scan system displays azimuth versus range.

29. A _____ display shows azimuth and range.

30. An _____ antenna beam pattern is often used for making ground maps using airborne radar.

31. _____ _____ radar is a technique for obtaining very high-resolution ground maps by making the processor act as if the antenna were very large.

32. A _____ _____ radar uses the motion of the aircraft to increase cross-range resolution by taking sequential shots, thus making the antenna appear as if it were larger.

RECOMMENDATIONS FOR FURTHER READING

Skolnik, Merrill I. *Introduction to Radar Systems*. 2nd Edition. New York: McGraw-Hill Book Co., 1962, 1980.

Toomay, J.C. *Radar Principles for the Non-Specialist*. Belmont, CA: Lifetime Learning Publications, 1982.

KEY EQUATIONS

1. Relationship of PRF and PRT:

$$PRF = \frac{1}{PRT}$$

2. Range to target as a function of elapsed time:

$$R = \frac{ct}{2}$$

3. Radar range rule of thumb:

$$R = \frac{T}{12.36} \, \text{nmi}$$

4. Maximum unambiguous range:

$$R_{max} = \frac{cT_i}{2}$$

5. Extent of dead zone in front of radar antenna:

$$R_{dead} = \frac{cT}{2}$$

6. Simple range resolution:

$$\Delta R = \frac{cT}{2}$$

7. Compressed range resolution:

$$\Delta R_c = \frac{\Delta R}{BT}$$

or

$$\Delta R_c = \frac{c}{2B}$$

8. Average power as a function of PRT and pulse width:

$$P_a = \frac{P_p T}{\text{PRT}}$$

9. Average power as a function of PRF and pulse width:

$$P_a = P_p T(PRF)$$

10. Duty cycle as a decimal fraction:

$$D = \frac{P_a}{P_p}$$

11. Duty cycle in percent:

$$D = \frac{P_a(100\%)}{P_p}$$

12. Duty cycle as a function of PRT:

$$D = \frac{T}{\text{PRT}}$$

13. Duty cycle as a function of PRF:

$$D = T \times PRF$$

14. Incident transmitter energy on the target at range R:

$$p = \frac{P_p TG}{4\pi R^2}$$

15. Signal energy at the receiver due to reflected echo:

$$S = \frac{P_p TGA_e \sigma}{16\pi^2 R^4}$$

16. Classical radar range equation:

$$\frac{S}{N} = \frac{P_p TGA_e \sigma}{16\pi^2 R^4 KTL}$$

17. Doppler shift frequency:

$$F_d = \frac{2v}{\lambda}$$

CHAPTER 20

Microwave Communications

OBJECTIVES

1. Learn the modulation forms used in microwave communications.
2. Understand how time- and frequency-domain multiplexing work.
3. Understand the basic configuration of terrestial microwave communications systems.
4. Learn the basics of space satellite microwave communications systems.

20-1 PREQUIZ

These questions test your prior knowledge of the material in this chapter. Try answering them before you read the chapter. Look for the answers (especially those you answered incorrectly) as you read the text. After you have finished studying the chapter, try answering these questions again and those at the end of the chapter.

1. A _____-domain multiplex system *interleaves* the various voice channels, rather than being truly simultaneous.
2. A microwave system allows two-way communications, but only in one direction at a time. This system is a ____-_____ system.
3. A geosynchronous, or geostationary, satellite orbits at a height of approximately _____ miles.
4. An audio channel with a 0- to 4-kHz baseband modulates a 60-kHz subcarrier. The (upper/lower) sideband occupies 56 to 60 kHz.

20-2 INTRODUCTION TO MICROWAVE COMMUNICATIONS

A microwave RF signal is just like any other radio carrier, except for the frequency range. Because those frequencies are very high, a great deal of information can be transmitted over a single microwave carrier. A very large number of voice-grade telephone channels can be superimposed on the microwave carrier. Some microwave links carry a mix of voice telephone, radio broadcast, television broadcast, and digital data channels.

Microwave communications systems can take the form of point-to-point, one-way radio relay links, one-way studio-to-transmitter links, remote TV pickups, Earth satellite links, ship-to-shore links, two-way voice links, and a variety of other services. The system may be a single-channel design or it may have dozens of channels multiplexed onto the same microwave carrier. In this chapter we will examine multiplex methods, modulation methods, and system architecture.

20-3 MULTIPLEX METHODS

Multiplex, according to Webster, means "manifold" and refers to the transmission of several messages over the same channel at the same time. There are two general forms of multiplexing: *time-domain multiplex* (TDM) and *frequency-domain multiplex* (FDM). The FDM method is truly simultaneous, while TDM is actually interleaved rather than simultaneous.

Figure 20-1 shows a simple four-channel TDM system. Each of four voice telephony channels (*VC*1 through *VC*4) are filtered and then applied to a switch (*TS*1-*TS*4). A timing or synchronization section rapidly actuates all four transmit switches in sequence such that a brief sample of each channel is transmitted. Thus, a continuous burst of channel samples is sent over the single-channel output to the microwave transmitter.

At the receiver end, the process is reversed. In this circuit, a bank of sequentially actuated switches alternately connects the common input line to the receive filters. As long as the receiver switches are in step with the transmit switches, the message will be reconstituted properly.

Timing can be either *synchronous* or *asynchronous*. In a synchronous system, either a clock pulse train is transmitted along with the information signal or handshaking signals are transmitted on a parallel circuit. In the asynchronous variation, the clocks at either end are maintained to a very tight tolerance of a certain standard frequency.

The audio voice signals input to the multiplexer are chopped by the sampling process. The signal can be reconstituted by the integration provided by the receiver low-pass filtering, provided that the sampling rate is high enough. According to the *Nyquist criterion*, the sampling rate per channel must be twice the highest Fourier component frequency present in the input signal. For a 4-kHz voice channel, therefore, the sampling rate must be 8 kHz or more. Because four channels are "muxed" together, however, the overall sampling speed must be 4×8 kHz, or 32 kHz.

In actual practice, Nyquist's rate of two times the highest frequency is not always satisfactory. In many systems it is found that good fidelity requires a sam-

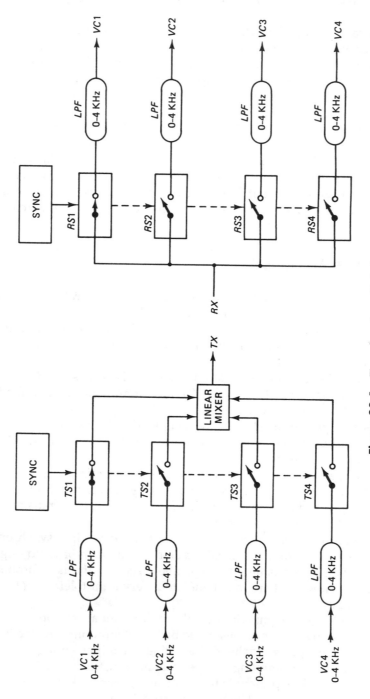

Figure 20-1 Time-domain multiplexing.

pling rate of four or five times the highest frequency, or two to two-and-a-half times greater than the Nyquist rate.

A four-channel example of FDM is shown in Fig. 20-2. The simplified circuit is shown in Fig. 20-2A, while the spectrum is shown in Fig. 20-2B. Each voice telephony channel has a 0- to 4-kHz *baseband*. When modulated onto a carrier or subcarrier, the spectrum will be baseband plus and minus the carrier frequency. But when single-sideband suppressed carrier (SSBSC) is used, the signal occupies the spectrum of the carrier plus the baseband (for upper sideband, USB) or carrier minus baseband (for lower sideband, LSB). In Fig. 20-2B, we see a series of SSBSC signals displaced in carrier frequency from each other by 4 kHz. By using bandpass filters at the receiver end, each voice channel can be recovered and reconstituted. An SSBSC signal is demodulated in a *product detector* by injecting a replacement locally generated carrier at the same frequency and phase as the suppressed carrier.

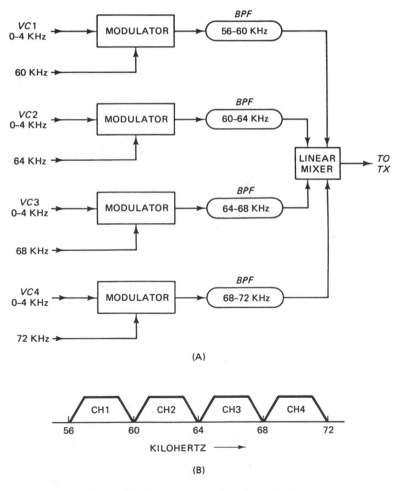

Figure 20-2 Frequency-domain multiplexing.

20-4 MODULATION METHODS

From the multiplexer section the composite signal (whether TDM or FDM) is sent to the modulator of the transmitter. At this point, any of several possible modulation types is employed or a combination of methods may be used. Examples include *amplitude, frequency, phase,* or *pulse modulation.*

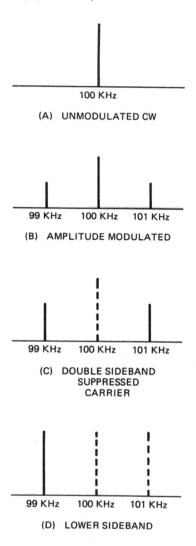

(A) UNMODULATED CW

(B) AMPLITUDE MODULATED

(C) DOUBLE SIDEBAND
SUPPRESSED
CARRIER

(D) LOWER SIDEBAND

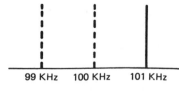

(E) UPPER SIDEBAND

Figure 20-3
A) Unmodulated signal; B) amplitude-modulated signal; C) DSSB-SC signal; D) LSB-SC signal; E) Upper sideband.

Amplitude modulation (AM) was discussed briefly in the previous section and is the oldest form of telephony. Variants on the AM theme include double-sideband suppressed carrier (DSBSC), single-sideband suppressed carrier (SSBSC), which includes LSB and USB, and vestigial sideband (VSB). Figure 20-3 shows the frequency spectrum for these modulation schemes when a 1-kHz audio sine wave (Fig. 20-3A) modulates a 100-kHz sinusoidal subcarrier (Fig. 20-3B).

The standard AM signal (Fig. 20-3B) has a spectrum consisting of the 100-kHz carrier and the sum and difference signals (99 and 101 kHz) produced by the modulation process (that is, the *sidebands*). On a 100% modulated AM signal, 33.3% of the RF power is distributed to the sidebands and the remainder is in the RF carrier.

A standard AM signal is wasteful of power because all the information can be conveyed in either sideband. As a result, removing the carrier and one sideband provides a much more efficient system. The carrier is removed in a balanced modulator, the output of which is a DSBSC signal, as shown in Fig. 20-3C. Bandpass filters pass either lower or upper sideband, as in Figs. 20-3D and 20-3E, respectively. In each case the filter rejects the undesired sideband.

Vestigial sideband is an AM signal in which one sideband is reduced in amplitude, but not eliminated. The VSB method is used for the video portion of a television signal.

Figure 20-4 shows a variant of Fig. 20-2A in which a single carrier frequency can support two FDM channels. Each 4-kHz baseband telephony channel has its own balanced modulator to produce a DSBSC output. Following each modulator is a bandpass filter that can pass different sidebands. For *VC*1 the USB is used, and for *VC*2 the LSB is used.

Frequency modulation (FM) and *phase modulation* (PM) are similar forms of angular modulation. Figure 20-5 shows the FM system and its relationship to the modulating audio sine wave signal. With the exception that phase is involved instead of frequency, the same concepts apply equally to PM. In the example of Fig. 20-5, a 70-MHz IF carrier is modulated by a sine wave signal. When the modulating signal *V*1 is zero, the RF carrier frequency is exactly 70 MHz. As *V*1 goes positive, however, the frequency of the carrier increases to an upper limit. The reverse happens on negative excursions of *V*1: the carrier frequency decreases to the lower limit. The overall difference between limits is the *swing*, while the difference between the limits and the unmodulated carrier is called the *deviation* of the signal. In

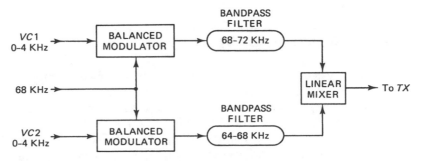

Figure 20-4   Using upper and lower sidebands to transmit two different audio signals.

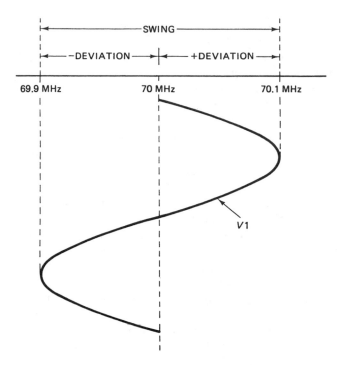

Figure 20-5
Frequency modulation.

the case shown in Fig. 20-5, positive and negative deviation are both 100 kHz, while the swing is 200 kHz.

Microwave communications systems also use various forms of *pulse modulation*, as shown in Figs. 20-6 and 20-7. *Pulse-amplitude modulation* (PAM) is shown in Fig. 20-6. In PAM, the pulse heights are proportional to the instantaneous value of the modulating signal (Fig. 20-6A). There are several forms of PAM signal: *normal monophasic* (Fig. 20-6B), *flat-top monophasic* (Fig. 20-6C), *normal biphasic* (Fig. 20-6D), and *flat-top biphasic*.

Two other forms of pulse modulation are shown in Fig. 20-7. *Pulse-position modulation* (PPM), Fig. 20-7A, varies the position of the pulse in accordance with the modulating signal. *Pulse-width modulation* (PWM), Fig. 20-7B, varies the pulse duration (width) in accordance with the modulating signal.

20-5 TERRESTIAL MICROWAVE SYSTEMS

A terrestial system is one that operates on Earth's surface (as opposed to satellite systems, which are in space). Terrestial systems may carry communications (telephony, data), television, or control system/data acquisition signals. Three general forms are used: *simplex*, *half-duplex*, and *full-duplex*. In the simplex system, communications occur in only one direction; half-duplex systems allow two-way communications but not at the same time; full-duplex systems allow two-way communication simultaneously.

(A)   AUDIO SIGNAL

(B)   NATURAL SAMPLING

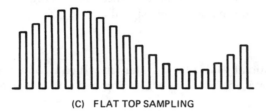

(C)   FLAT TOP SAMPLING

(D)   BIPHASIC

Figure 20-6
Sampled signals.

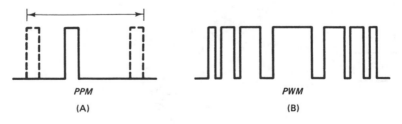

PPM

(A)

PWM

(B)

Figure 20-7   Pulse-modulated signals.

Figure 20-8 shows a microwave simplex system involving a single transmitter and single receiver. This system is perhaps the simplest possible architecture. The performance of the system can be evaluated through *Friis's transmission equation* for antennas aligned with respect to each other for maximum power transfer:

$$\frac{P_r}{P_t} = \frac{G_t G_r \lambda^2}{16\pi^2 d^2} \qquad (20\text{-}1)$$

where

P_r = power density at the receiver site
P_t = power density at the transmitter site
G_t = gain of the transmitter antenna
G_r = gain of the receiver antenna
d = distance between transmitter and receiver antennas
λ = wavelength of the microwave signal

EXAMPLE 20-1

Calculate the P_r/P_t ratio at 9 GHz for a system in which two identical 30-dB antennas ($G = 1000$) are spaced 25 km apart, facing each other for maximum power transfer.

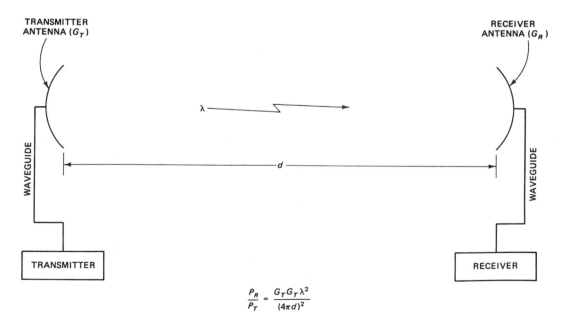

Figure 20-8 Simplex communication system.

Solution

$$\frac{P_r}{P_t} = \frac{G_t\,G_r\,\lambda^2}{16\pi^2\,d^2}$$

$$= \frac{(1000)(1000)(300/9000\text{ MHz})^2}{[(16)(\pi^2)(25,000\text{ m})^2]}$$

$$= \frac{1111}{9.9\times10^{10}} = 1.12\times10^{-8}$$

By algebraic manipulation of the Friis equation, Eq. (20-1), we can select parameters for the system. For example, the required power density at the receiver is a function of receiver sensitivity and noise figure, as well as the antenna gain. From that information, it is possible to calculate the required transmitter power levels.

EXAMPLE 20-2

In Example 20-1, it was discovered that the P_r/P_t ratio is 1.12×10^{-8}, so the P_t/P_r ratio is $1/1.12\times10^{-8} = 8.9\times10^9$. Thus, the transmitter power density is

$$P_t = \frac{16P_r\,\pi^2\,d^2}{G_t\,G_r\,\lambda^2} \tag{20-2}$$

From Eq. (20-2) we could calculate the transmitter power density required. For example, suppose the required receive power density is on the order of 10^{-7} W/m². The transmit power density is therefore on the order of

$$P_t = (8.9\times10^9)\times P_r$$

$$= (8.9\times10^9)(10^{-7}\text{ W/m}^2)$$

$$= 890\text{ W/m}^2$$

If we know the antenna specifics, including gain, then we could calculate the required transmitter output power. Given a 30-dB power gain, we conclude that less than 1 W of RF power is needed into the antenna.

The Friis equation can also be expressed in terms of decibels:

$$10\log\frac{P_r}{P_t} = G_{t\,(\text{dB})} + G_{r\,(\text{dB})} + 10\log\left(\frac{\lambda}{4\pi d}\right)^2 \tag{20-3}$$

These equations are approximations because they do not include error budgets or system losses. A major loss is atmospheric attenuation (see Fig. 20-9). The attenuation in decibels is a function of frequency, with special problems showing up at 22 and 64 GHz. These spikes are caused by *water vapor* and *atmospheric oxygen* absorp-

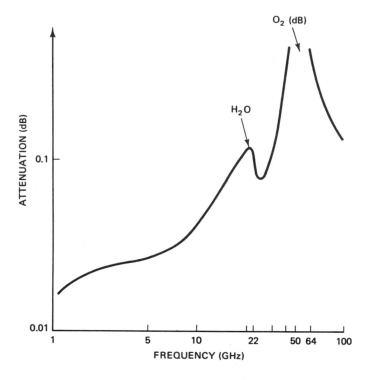

Figure 20-9
Atmospheric absorption of microwave energy.

tion of microwave energy, respectively. Operation of any microwave frequency requires consideration of atmospheric losses, but operation at the two spike frequencies poses special problems. At 22 GHz, for example, about 1 dB of loss must be calculated for the system.

Microwave systems are limited in range principally because of the line-of-sight propagation phenomena. As a result, distances beyond 50 km (approximately 30 miles) must be covered by a series of *relay stations*. Such stations are typically located on top of tall buildings, tall towers, or mountain peaks. Figure 20-10 shows a typical relay station.

It is common practice to use two frequencies at each station in order to avoid intrasystem interference (that is, between transmitter and receiver at the same site). Typically, receive and transmit frequencies are separated by about 2 GHz. For a relay system consisting of more than one station, the frequencies will alternate in a pattern of RX/TX, $F1/F2$, $F2/F1$, $F1/F2$, . . ., TX/RX.

In the simple system of Fig. 20-10, the received signal $F1$ is picked up on the receive antenna and then routed through waveguide to the receiver. In some systems, a low-noise amplifier (LNA) is part of the antenna system at the feed end of the waveguide (often as part of the antenna feed assembly). The demodulated information is routed to the modulator input of the transmitter in very simple systems (see Section 20-5.1 for an exception). The output of the transmitter is fed to the transmit antenna via waveguide.

It is also possible to use reflectors instead of antennas and waveguides (see Fig. 20-11). The incoming microwave signal $F1$ is reflected to the receiver horn at the base of the tower. The recovered information signal is then transmitted through another horn and reflected in the direction of the next relay station.

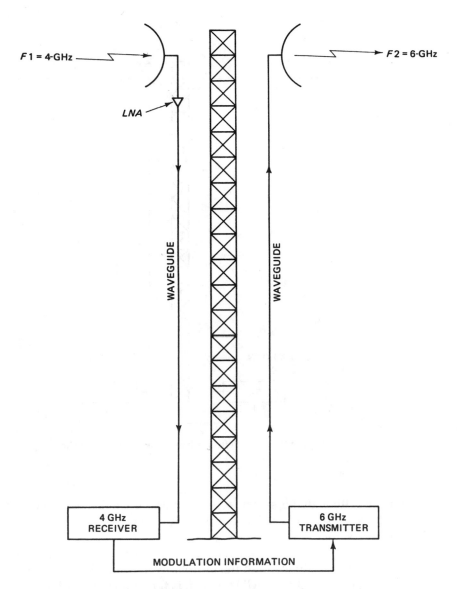

Figure 20-10 Simple microwave relay station.

In a few sites a *passive repeater* can be used. These installations have neither receiver nor transmitter and consist of a reflector used to redirect a signal around an
obstacle. An example is a mountain-top passive repeater to direct signal down in
a valley or further down the slope of the hill.

20-5.1 Relay Repeater Designs

In the previous examples, the microwave signal was demodulated, the recovered
modulating information signal fed to the modulator input of the transmitter, and
then retransmitted on another frequency that is displaced from the original receive

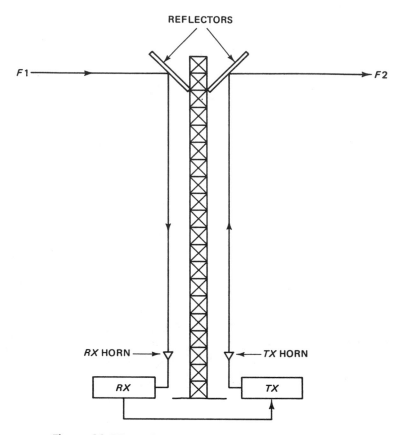

Figure 20-11   Reflector-based microwave relay station.

frequency. Although this system works for simple single-channel voice or data signals, a better system is available, especially where a TDM/FDM subcarrier is the modulating signal.

Figure 20-12 shows a relay repeater station architecture that couples receiver to transmitter directly through the mutual IF amplifier chain. The received signal $F1$ is down-converted to 70 MHz, which is then up-converted to the transmit frequency $F2$. This architecture eliminates considerable demodulation and remodulation circuitry, as well as the nonlinearities normally associated with demod/mod processes. Thus, IF frequency translators are inherently more efficient and less complex.

20-5.2 Multistation Relays

The simple systems shown thus far are capable of communication at either line-of-sight distances or about twice that far through a single intervening relay station. In Fig. 20-13, however, we see a multistation relay system that can be generalized to almost any number of relay points. Such chains can relay messages over thousands of miles.

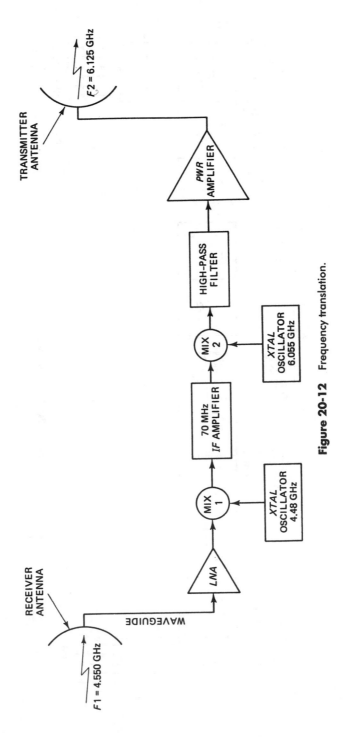

Figure 20-12 Frequency translation.

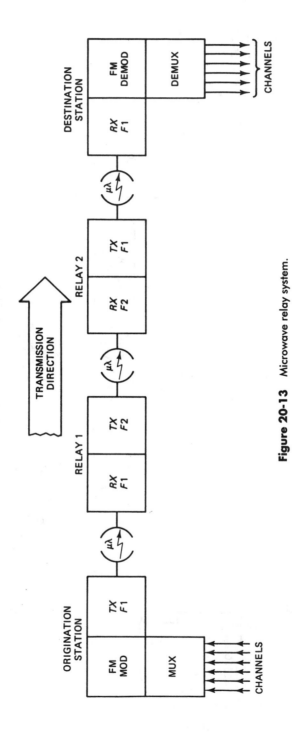

Figure 20-13 Microwave relay system.

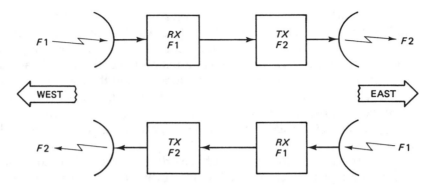

Figure 20-14 Bidirectional relay system.

The system in Fig. 20-13 is reduced to the most primitive simplex case, which allows transmission in only one direction. The origination station transmits on $F1$, and the first relay receives $F1$ and translates it to $F2$ to avoid intrasystem interference. Each successive relay reverses the RX/TX order, so the pattern will be $F1/F2$, $F2/F1$, $F1/F2$, and so on. The destination station receives the final signal (in this case on frequency $F1$) and demodulates it to recover the information.

Half-duplex and the more common full-duplex can be accommodated by providing two channels, shown in Fig. 20-14 as east and west. The same frequency translation takes place, except that the order is reversed in the two directions. In some systems, where the frequencies are too close together, it might be necessary to use four frequencies. For example, the east circuit would use $F1/F2$, while the west circuit uses $F3/F4$, with the frequency pairs in each direction alternating as before.

20-6 SATELLITE COMMUNICATIONS

In the 1950s, Russian and American engineers managed to launch the first artificial Earth satellites and thereby launched a new era. Within a few years, the first communications satellite was aloft, and it revolutionized both the communications and the broadcasting industries. Telephone users today have little concept of what it was like to place international calls in earlier decades. All we have to do is dial the correct country code and the individual phone number, and the phone rings on the other end of the line.

Although the first transatlantic telegraph line was laid under the ocean in 1858, it was not until 1955 that the first transatlantic telephone line was laid, and it could only handle a few calls at a time. Users had to wait hours, or even days, to make a call. Until 1955, all calls across the Atlantic Ocean were carried by point-to-point shortwave radio circuits, and even after 1955 a large amount of the traffic went by radio. But those links were subject to the vagaries of ionospheric propagation and would go silent for hours and even days as solar radiation storms interrupted the shortwaves. Earth satellites not only put up many more circuits, which allow many more calls to be placed, but they are also less subject to solar storms and other problems.

The Earth satellite also offers the advantage that it can handle other services as well as voice telephony. The satellite acts as a stationary relay station that can handle radio broadcasts, television broadcasts, telex or teletypewriter transmissions, and digital data. Several newspapers transmit the entire text of each day's edition through a satellite to regional Earth stations around the country, where local printing is performed. Thus a *USA Today* or *Wall Street Journal* is truly a national newspaper because of satellite communications. The cable TV industry simply would not exist in any size to speak of without the programming available from a host of satellite sources.

Earth satellites in close orbit have orbital times that are too fast for communications purposes. Nearby satellites may orbit Earth every two hours or so, meaning that the time a satellite is in view of any given Earth station is measured in hours or even minutes, hardly satisfactory for communications purposes. The communications satellite must have an orbital period equal to that of Earth itself in order to appear stationary above a given point on the ground. The period of Earth is 23 hours, 56 minutes (the *sidereal day* as opposed to the 24-hour civil day), so an Earth satellite in *geosynchronous* (also called *geostationary*) orbit must have that orbital period. Because orbital period is a function of altitude, the satellite must be approximately 36,000 km (22,300 miles) above the equator.

Geostationary satellites also have another advantage over closer satellites. On low-orbit satellites, the transmission frequency varies according to the doppler shift of the moving satellite. On geostationary satellites, however, the relative velocity between the satellite relay station and the ground station is zero, so the doppler shift disappears.

The satellite tends to drift off the correct parking spot in space because of various forces acting on it. As a result, ground controllers must activate thrusters that periodically correct the location of the satellite to the required ± 0.1° of arc azimuth and elevation tolerance. This correction is called *station keeping* in satellite terminology.

Figure 20-15 shows a typical application of Earth communications satellites: maritime satellite communications. MARISAT is the system of three maritime satellites, parked one each above the Atlantic, Pacific, and Indian oceans. These three satellites cover virtually all points on Earth's oceans. Earth stations connect the microwave shore receivers to the land-line telephone and telex systems. The MARISAT system has all but eliminated the need for traditional Morse-code-trained radio operators on ships.

Because of the large number of potential users, and the limited number of satellite uplink and downlink channels, the shore station operates as a control and gatekeeper. When the ship wants to make a telephone call or send a telex, an operator enters a three-digit access code into the shipboard terminal and transmits it as an *access request* burst transmission to the satellite. The request identifies the user and states which form of communications channel (voice or telex) is needed. The shore station constantly listens for these access requests, and on receiving one assigns a communications channel (uplink and downlink frequencies, as well as multiplex slots) to the requester. Following the receipt of the *assignment message*, the shipboard terminal is activated and the circuit closed.

When a shore user wants to call the ship, a message is sent through the satellite that alerts the shipboard terminal and establishes the linkup. In the early com-

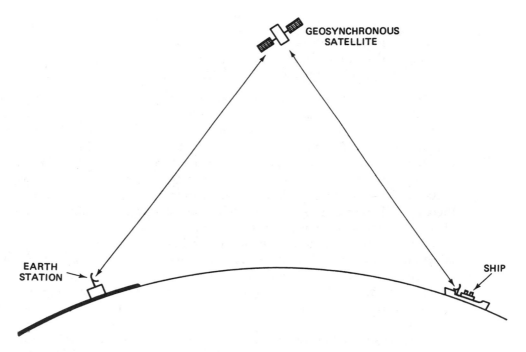

Figure 20-15   Satellite communications system.

munications satellites, the FDM multiplex system was used, but now time-domain multiplex is used for both the access channels (TDMA) and main communications channels (TDM).

20-6.1 Technical Problems in Satellite Communications

Recall from our previous discussion that terrestial communication was limited to about 50 km. In satellite communications, we need to communicate over a 36,000-km path. To make matters worse, the satellite transponder transmitters are power limited because of the total dc power availability on the spacecraft. In space, power is typically supplied by solar cells, and there is a limit to how many can be built onto the solar power panels used by the craft. Typically, Earth stations tend to have high-power transmitters operating into antennas that have both high transmit gain and a large receive aperture. The problems that must be overcome are mostly losses.

Losses to microwave signals used in Earth satellite communications seriously degrade performance of the system and must be overcome by proper design. We have several sources of loss in the system. First, there is the well-known spreading loss due to inverse square law propagation. This spreading loss can be generalized in decibel form as

$$L = 33 \text{ dB} + [20 \log(d_{km})] + [20 \log(F_{MHz})] \qquad (20\text{-}4)$$

where

L = loss in decibels
d_{km} = distance to the satellite in kilometers
F_{MHz} = operating frequency in megahertz

Working a few examples will demonstrate that the path loss over 36,000 km at frequencies from 1000 to 15,000 MHz can be tremendous. In addition, the loss represented by Eq. (20-4) does not include losses caused by atmospheric absorption (roughly 0.01 to 1 dB/km, depending on frequency) or scattering losses in the atmosphere. The absorption loss only affects the signal over the few dozen kilometers at which there are significant water vapor or oxygen levels present, while the scattering losses depend on localized ionization conditions in the lower atmosphere (see Chapter 2). Designers typically allow a margin for rainy day operation when atmospheric water is much greater.

The system must also account for losses due to system noise. One significant noise source is space itself. During daylight hours, one can measure significant microwave noise from the sun, and at all times there is a galactic noise component. Recall from Chapter 2 that British radar operators noticed that detection range varied according to whether or not the Milky Way was above the horizon. Galactic or cosmic noise is inversely proportional to operating frequency, but contains a minimum offset level that is irreducible. This minimum noise level is a function of where in the sky the antenna is pointed (and thus forms the basis of radio astronomy, by the way).

Because noise power density is a function of resistance, and atmospheric losses can be modeled as a resistance, we can specify atmospheric losses in terms of an equivalent noise temperature (see Chapter 13 for a discussion of noise temperature):

$$T_{eq} = (L - 1)\, T_{atmos} \tag{20-5}$$

where

T_{eq} = equivalent noise temperature
L = atmospheric loss
T_{atmos} = temperature of the atmosphere

We can also calculate an equivalent temperature for the antenna feedpoint. This noise temperature, $T_{eq.\,(ant)}$, includes the combined noise of atmospheric attenuation and galactic or solar noise, but does not include artificial noise and interfering signals from terrestial or satellite sources. The range of the antenna equivalent noise temperature is 2,000 to 10,000 K; obviously, we are *not* talking about the physical temperature of the antenna!

$$T_{eq\,(ant)} = \frac{S_n\,(\text{W/Hz})}{k} \tag{20-6}$$

where

$T_{\text{eq. (ant)}}$ = antenna equivalent noise temperature in kelvins
S_n = spectrum density of the noise signal received by the antenna (excluding artificial terrestrial noise), in watts per hertz (W/Hz)
k = Boltzmann's constant (1.38×10^{-23} J/K)

The total noise losses of the system must also include the receiver noise contribution (see Chapters 13 and 18). In addition to the equivalent antenna noise, $T_{\text{eq. (ant)}}$, we must also include the receiver noise, $T_{\text{eq. (rcvr)}}$. This latter noise must be input-referred by the receiver designer to be used in this context. The total noise loss is

$$T_{\text{total}} = T_{\text{eq. (rcvr)}} + T_{\text{eq. (ant)}} \qquad (20\text{-}7)$$

where

T_{total} = total noise of the system
$T_{\text{eq. (rcvr)}}$ = input-referred receiver noise
$T_{\text{eq. (ant)}}$ is as previously defined.

By combining the antenna gain and the total noise (T_{total}), we can come up with a goodness factor for the Earth station receiver:

$$\text{GF} = \frac{G_{\text{dB}}}{T_{\text{total}}} \qquad (20\text{-}8)$$

or, in decibel notation,

$$\text{GF}_{\text{dB}} = G_{\text{dB}} - 10 \log(T_{\text{total}}) \qquad (20\text{-}9)$$

where

GF = goodness factor in decibels per kelvin
G_{dB} = antenna gain in dB
T_{total} = total system equivalent noise temperature

For commercial Earth stations, the value of GF is on the order of 35 to 55 dB/K, where private TVRO systems have a value of GF on the order of 8 to 15 dB/K.

20-7   SUMMARY

1. Microwave communications can be either terrestrial or through Earth satellites. Services offered include (but are not limited to) point-to-point, one-way radio relay links, one-way studio-to-transmitter links, remote TV pickups, Earth satellite links, ship-to-shore links, two-way voice links, and a variety of other services. The system may be a single-channel design, or it may have dozens of channels multiplexed onto the same microwave carrier.

2. Both time-domain multiplex (TDM) and frequency-domain multiplex (FDM) are used. Communications transmitters may be AM, SSBSC, FM, PM, or pulse modulated.

3. Terrestial communications systems operate over line-of-sight distances of about 50 km; greater distances use relay stations to carry the signal. Relay repeater stations might be either simplex, half-duplex, or (more commonly) full-duplex.

4. Earth satellites for communications must be in geostationary orbits about 36,000 km above the equator. Although great system losses occur, the satellite today is the preferred method of communications, especially internationally or to ships at sea.

20-8   RECAPITULATION

Now return to the objectives and prequiz questions at the beginning of the chapter and see how well you can answer them. If you cannot answer certain questions, place a check mark by each and review the appropriate parts of the text. Next, try to answer the following questions and work the problems using the same procedure.

QUESTIONS AND PROBLEMS

1. List two basic forms of multiplex system: _____ domain and _____ domain.
2. A geostationary orbit is approximately _____ miles above the equator.
3. According to the Nyquist criterion, a 3000-Hz audio channel must be sampled at a frequency rate of _____ Hz in order to be properly reconstituted at the other end of the transmission path.
4. List three forms of pulse modulation.
5. An SSBSC signal consists of a 4-kHz audio channel modulating a 60-kHz subcarrier. If the upper sideband (USB) is used, then what is the frequency spectrum of this signal?
6. Calculate the spectrum in problem 5 if the lower sideband (LSB) is used.
7. A terrestrial microwave system consists of two antennas pointed at each other for maximum power transfer over a 22-km path. Calculate the P_r/P_t ratio at 4.5 GHz if both antennas have a gain of 25 dB.
8. In problem 7, the receive antenna is changed by a technician and is replaced by an antenna with a gain of 40 dB. What is the ratio for the new system?
9. Calculate the equivalent antenna noise temperature for a system when the impinging noise signal (S_n is 5.6×10^{-20} W/Hz).
10. Calculate the total equivalent noise temperature of a system in which the antenna equivalent noise is 5500 K, and the receiver noise temperature is 1325 K.
11. Calculate the loss in decibels of a system in which the satellite is 36,000 km above the equator and the operating frequency is 4.335 GHz.

12. An Earth station transmitter outputs 8 kW on a frequency of 6.2 GHz and feeds it to a parabolic dish antenna that is 16 m in diameter. Calculate the power available at the satellite 36,000 km above the Equator.

KEY EQUATIONS

1. Friis's transmission equation:

$$\frac{P_r}{P_t} = \frac{G_t G_r \lambda^2}{(4\pi d)^2}$$

2. Friis's equation rewritten for P_t:

$$P_t = 8.9 \times 10^9 P_r$$

3. Friis's equation in decibel form:

$$10 \log \frac{P_r}{P_t} = G_{t\,(dB)} + G_{r\,(dB)} + 10 \log \left(\frac{\lambda}{4\pi d}\right)^2$$

4. Spreading loss in satellite communications:

$$L = 33 \text{ dB} + [20 \log (d_{km})] + [20 \log (F_{MHZ})]$$

5. Equivalent noise temperature of atmospheric loss:

$$T_{eq} = (L - 1)T_{atmos}$$

6. Equivalent noise temperature of the antenna:

$$T_{eq\,(ant)} = \frac{S_n \text{ (W/Hz)}}{k}$$

7. Total noise in system:

$$T_{total} = T_{eq\,(rcvr)} + T_{eq\,(ant)}$$

8. Goodness factor for an Earth station:

$$GF = \frac{G_{dB}}{T_{total}}$$

or, in decibel notation,

$$GF_{dB} = G_{dB} - 10 \log (T_{total})$$

INDEX